# The World of Ham Radio, 1901–1950

## A Social History

RICHARD A. BARTLETT

McFarland & Company, Inc., Publishers

*Jefferson, North Carolina, and London*

*The present work is a reprint of the illustrated case bound edition of* The World of Ham Radio, 1901–1950: A Social History, *first published in 2007 by McFarland.*

LIBRARY OF CONGRESS CATALOGUING-IN-PUBLICATION DATA

Bartlett, Richard A.
The world of ham radio, 1901–1950 :
a social history / Richard A. Bartlett.
p.      cm.
Includes bibliographical references and index.

ISBN 978-1-4766-6275-6 (softcover : acid free paper) ∞
ISBN 978-1-4766-1260-7 (ebook)

1. Amateur radio stations.   2. Radio.   I. Title.
TK9956.B34   2015      384.5409'041—dc22      2007019185

BRITISH LIBRARY CATALOGUING DATA ARE AVAILABLE

Cover photograph: Brother and sister Marshall
and Loretta Ensor with their ham radio gear
(Ensor Park & Museum, Olathe, Kansas)

Printed in the United States of America

*McFarland & Company, Inc., Publishers*
*Box 611, Jefferson, North Carolina 28640*
*www.mcfarlandpub.com*

In memory of
Forrest Abbott "Bart" Bartlett, W6OWP,
born March 25, 1914, and a "silent key" as of July 3, 2006

# Acknowledgments

To my brother, Forrest A. Bartlett, W6OWP, I owe the inspiration for this book as well as his professional advice over a period of the seven years in which it was gestating. As a younger brother I watched Forrest build transmitters and receivers, walk rooftops while erecting antennas, noticed the light in his bedroom (where he had his rig) at all hours of the night, and listened to the dots and dashes as he listened, and the tapping of them as he sent out his signals. Forrest, prior to his death on July 3, 2006, read every page of the manuscript and offered corrections, explanations, and clarifications throughout.

To several librarians at the Robert Manning Strozier and Paul A.M. Dirac Science libraries at Florida State University go my heartfelt thanks. They are the ones who pulled out the correct Microfiche of Congressional Hearings—(as a non-librarian, just try to find them!)—helped locate the most unusual sources, provided assistance in locating books, journals, and newspapers and obtaining them via interlibrary loan, and did all this with the most cheerful and helpful attitude.

David Dary, Dean Emeritus of the Journalism School at the University of Oklahoma, offered helpful suggestions. David is a ham. Another ham who contributed information and gracious assistance is Larry Woodworth.

Others deserve credit for their encouragement. Having informed them of my project, they offered suggestions and most of all, they issued a challenge: Was I just piddling with a project or was I dead serious? I could not let them down. Among them, all of whom work (or worked) out with me three times a week at a health spa, are Dr. Jim Cowart, professor emeritus of geology, Dr. Charles Harper, deceased, professor emeritus of English, Chief Master Sergeant John Schmidt (ret.), Sam Mathis, Larry Danalo, Roger Penny, Professor Waldi Anderson, Stan Tate, Luther Lynn, Margarita Grant, Bonnie Mannheimer, Bill Cowart, and Jamie Stafford. Elsewhere Benjamin Morris and John Cosgrove were also aware of, and asked often, of my progress. Miss Anne L. Marsh, a computer expert, stepped in to help when my love-hate relationship with my computer was in hate mode. Thanks also to Doyle Connor who was the Florida Secretary of Agriculture for thirty years and a source of inspiration.

And above all, I appreciate the consistent encouragement of my wife, Marie Cosgrove Bartlett, our daughters, Mary, a medical editor, and Margaret, an English teacher, and my son, Tom, a successful businessman. I have heard this from all of them: "Dad, are you still working on that book?"

All errors and decisions about material are my own, and I take full responsibility for everything appearing in this book.

Tallahassee, Florida
March 13, 2007

# Contents

# Introduction

It is May 25, 1928: The dirigible *Italia*, returning from a trip over the North Pole, runs into a downdraft that forces the great airship toward the jagged ice pack below. Every emergency procedure fails to stop the drift to disaster. The great airship slams onto the ice pack. One man dies. Six members of the crew drift off in a separated part of the dirigible, never to be seen again. Umberto Nobile, the captain, with eight others and his little dog Titina, find themselves marooned on an ice flow in the Arctic Ocean. How to get rescued?

The answer was, by wireless. One of the survivors was Nobile's radio operator, Giuseppi Biago. When he saw the crash coming he wrapped his arms around the small shortwave set, held it to his chest, and closed his eyes. When the crash came he fell to the ice still firmly holding the rig. Finding the radio intact, Biago pieced together steel tubing that had broken loose from the dirigible to build an antenna. Soon he was tapping out *SOS-Italia*. After days of trying, the coded message was heard by a young Soviet farmer-amateur near Archangel, who conveyed the information to the authorities. Nobile and those still on the ice flow were saved.

In February 1925, American media was focused on a Kentucky cavern several miles from the little community of Cave City. A farmer named Floyd Collins had been exploring a cavern when a heavy rock fell on him. He was trapped. Over a period of about two weeks a desperate effort was made to free Collins. It became a media frenzy. The problem was, there were no telephone or telegraph connections at the cave. To the rescue came ham operators, one manning a transmitter at the cave entrance, the other a receiver in Cave City, where the press had access to telegraph and telephone facilities. Thus the latest news of the rescue attempt was conveyed to the world.

When the Ohio River flooded in 1937 it was amateurs, working along with the coast guard, navy, and army engineers, who maintained contact with towns isolated by the flood waters, conveying information about the needs for medical supplies, food, shelter, and rescues.

It was an amateur who helped develop the vacuum tube, the superheterodyne, and FM radio; it was amateurs who developed the single sideband, which greatly aids in ground to air contacts.

It was amateurs who for the first time contacted Europe from America using shortwave.

During the Second World War amateurs were in short supply. They were desperately needed in all branches of the military, plus the merchant marine. Those not in the service were essential at home in case of enemy attacks or to provide communication during disasters when other means of communications were down.

They are terribly important, locally, nationally, and internationally, these amateurs, these hams.

But who are they? How did they originate? Why do they continue to exist in the face of commercial invasions of their meters and kilocycles? Most of us have known hams, or known people who have them as relatives or friends. Some of us have seen their rigs, which fascinate us with so much more equipment than a mere AM-FM radio, or a television, or even a computer. Truth is, a modern ham may have a shortwave rig *plus* a television *plus* a computer in the shack—as the room where the installation is located is always called. Standing there, watching the ham work the rig, either speaking to another ham by voice or by Morse code, our curiosity is whetted even more.

This book unabashedly praises the amateurs, the hams. It traces their first fifty years with emphasis on their social history. Technological changes are kept to a minimum. These were years in which amateurs had to understand radio theory, be able to trace a schematic, and know how to build a transmitter and receiver. They used parts purchased from supply houses and automo-

W9FYK's rig at 2005 Mapleton Avenue, Boulder, Colorado, was in the second story bedroom on the right (photograph by Forrest Bartlett).

bile junk yards, tubes from Woolworth's; they even used the hubs of toilet paper rolls or Quaker Oats boxes. They had to master the use of a soldering iron. They had to erect an antenna, in the process risking their lives on the steep roofs of houses and farm buildings. More: hams had to master Morse code, both sending it at ten words a minute or more, and receiving it and understanding it at an equal or even faster pace. Still more: they had to understand "Q Meanings" such as CQ, QRM, and QRN. And still more, amateurs had to understand the comity of the hobby—how to be decent, law abiding, friendly, and cooperative.

In tracing hamdom's growth we inadvertently note the improvements of automobiles and aircraft. With hams we accompany pilots on their way to Hawaii and Australia and over the North and South poles. We journey with explorers into Borneo and Africa and make our way with pathfinders down the west coast of Mexico. We sail with a seventeen-year-old ham on a round-the-world cruise and visit isolated Pitcairn Island. We trace the concept of relays—the basis for the name of their association, the American Radio Relay League—from contacts twenty-five miles apart to anywhere in the world. We note their valiant services, above any call of duty, in times of disaster. We read of the battles to maintain amateur rights in Congress and in the world at large.

We are present at the creation of the International Amateur Radio Union. We highlight a few of amateurs' defenders, most notably Hiram Percy Maxim. And we attend the radio training facilities of the Second World War.

Why stop at the year 1950? Because by then the avocation, hobby, art—call it what we will—had consolidated into a fixed entity. By then it had its clubs, including the Royal Order of the Wouff Hong; it had its publications, which devote perhaps two-thirds of their space to articles on research and development; it had its lobbyists, who did all they could do to protect the art from incursions by commercial interests; it had its contests and its awards; and it had its provisions for service in the face of disasters. Amateur radio was here to stay.

In the more than half century that has followed, technological changes have continued unabated. The hobby has its own satellites. Amateurs work with their own televisions. Computers have come into wide use. But these are improvements in technology. Hams continue to use what is new and adjust to changes, but the basic establishment, known as amateur radio, continues to exist.[1]

Bless 'em! For their own enjoyment, for their help in times of disaster, for their continued experimenting with new technology which benefits all of us, may the hams continue to prosper!

# 1

## The Beginnings

When the nickel-plated Big Ben alarm clock rang 5:00 a.m., Forrest Bartlett hastened to turn it off. He did not want to awaken his family, all asleep in the upstairs bedrooms of the two story house at 2005 Mapleton Avenue in Boulder, Colorado. Truth is, he had been almost awake anyway. Since September 1930, he had been rising at that early hour to slip on his bathrobe and sit down at his rig. He was working "DX"—amateur lingo for contacting far-off stations. Forrest was indulging in friendly competition to see who among the group of his radio-oriented friends— amateurs, often called hams—could work the most distant station. Between September 1930, and this Saturday, April 4, 1931, his log, which listed all stations worked (and which he still owned as this was written), already listed 907 contacts. Quite an achievement for Forrest, who on April 4, 1931, was just eleven days into his seventeenth year.

As quietly as possible on this Saturday morning in 1931 (it was Easter weekend) W9FYK sat down in the old kitchen chair in front of two dilapidated tables positioned parallel to his bed and hardly a yard from it. On the tables were his transmitter and receiver. To the left of them was a window from which his antenna and ground wires, protected by insulators, entered the room. The boy turned on his rig.

It was crude. The transmitter and receiver were both breadboard style, using end pieces from wooden apple boxes. The front panels with their big dials were of Masonite painted black to make them look like the more expensive Formica or Bakelite. Everything else was open to the air. Forrest's pride lay in the knowledge that he had built both the transmitter and receiver himself. The tubes cost around 89 cents each at Woolworth's. The coil had as its base the hub of a toilet paper roll around which Forrest had painstakingly wrapped copper wire. That wire had come from a Model T spark coil; it had cost 50 cents at a local junkyard. Two coils in the transmitter had been wound with ¼-inch copper tubing. Shield boxes in the receiver had been shaped with the help of a fellow ham named Harvey, who was unusually good at metal working. Other parts had been either traded with local hams for parts they no longer needed, fellows named Chuck and Marvin and Avery and Harvey and Glenn—or had been dug out of Forrest's own big cardboard junk box (all hams have junk boxes). The transformer and filter for the transmitter power supply were purchased with Christmas money and probably cost about $7.50. The spaghetti-like tangle of wires inside the rigs had been carefully soldered where they should make contact. The 50 watt soldering iron had cost a dollar or perhaps a dollar and a half; the spool of solder about 50 cents. Probably the most expensive items in the receiver were the special tuning dial and the Brandes earphones, total about $5.00.

For moments Forrest Bartlett waited for the tubes to glow. He placed the earphones over his head and listened for the rushing sound of the receiver coming to life. Switching to trans-

**Forrest's rig as of 1932 or 1933 (photograph by Forrest Bartlett).**

mit mode, he held the sending key down and checked the transmitter tuning, then tapped a few letter V's (as a test signal) and his call, W9FYK. Then switching to the receiver, he listened for a possible call from a station that might have heard his short test. No answering call was heard but tuning the receiver dial a CQ was coming through signing the call VK7HL. (CQ is the general call universally used by ham operators seeking a contact.) Forrest knew that "VK" calls were Australia; he had made contact with several, but a VK7 had not been heard before. Nervously he switched from receiver to transmitter, moved his hand to the sending key and sent an answering call: VK7HL, VK7HL, DE W9FYK W9FYK W9FYK AR. ("DE" in telegraph usage means "from" and "AR" is the invitation to reply.)

Over the waves came the station answering Forrest's call:

W9FYK DE VK7HL R = TKS For CALL OM = UR SIGS QSA5R7 HR IN WEST HOBART TAS-MANIA? WEST HOBART TASMANIA = HW? AR W9FYK DE VK7HL K ("R" means "received ok" and "K" is the "go ahead" signal once communications have been established. The telegraph symbol "=" is used in similar sense to the period in written text.)

Forrest's response was

VK7HL DE W9FYK R = TKA OM FOR CMG BACK = UR SIGS QSA4R5 XDC HR BOULDER COLORADO? BOULDER COLORADO = VY PSED TO QSO = UR MY FIRST VK7 ES WUD LIKE A QSL = MY QRA OK IN CALL BOOK AR VK7HL DE W9FYK K

The QST English, as it is known, can be confusing for non-amateurs. In plain English their conversation went like this: "Thanks old man for coming back. I read you at a 5 in signal strength [very strong would be 9] with a pure tone quality here in Boulder Colorado [repeated to assure copy]. Very pleased to contact you. You are my first VK7 contact and I would like an acknowledgment card. Address is OK in call signs book." The Tasmanian amateur had thanked Forrest

BOULDER, COLORADO, U. S. A.
2005 Mapleton Avenue

**W9FYK**

Radio.............. ....Ur........QSA........ Sigs Wkd..................................... M.S.T.
Transmitter................................................Receiver................................................
Remarks...................................................................................................................
..................................................................................................................................

TNX FER QSO          73's

THE PRINT SHOP, 962 PLEASANT ST., BOULDER, COLO.

Forrest's original QSL Card, later it was W60WP (photograph by Forrest Bartlett).

for his call, given a report on his signal and promised to send a card (QSL) confirming the contact. (QSL designates the postcard size cards amateurs exchange when they acknowledge contact. The question mark in the midst of the messages was common practice when an unusual word appeared, and a repeat of the word would be given.)

When the coded conversation ended, seventeen-year-old Forrest Bartlett, W9FYK, sat back in the old chair and relaxed. He had achieved one of his goals—to work Tasmania! Later in the day he would telephone his fellow amateurs and give them the specifics of his contact. Within a month he and the Tasmanian amateur would receive each other's QSL cards, proof of their contact.

Forrest has never forgotten that chilly morning in early April, that Saturday of the Easter weekend when he "worked" Tasmania.

If a generation is considered thirty years, then Forrest Bartlett was at the very beginning of the second generation of radio amateurs, or "hams" as they came to be known. No specific date can be established for when the first two amateurs each built a transmitter and receiver and, from across town or possibly from just down the block, contacted one another by wireless. As soon as news items had appeared about wireless, experimenters had gone to work. Could this *really* be conveying messages without direct wire contact? They wanted to know. Throughout the world, and especially in the United States, hundreds of young people, as well as a good many adults, began experimenting with the possibility. The United States, however, was particularly receptive to the wireless enthusiast. Its population at the turn of the last century was remarkably literate. It was an affluent society, which meant that wireless apparatus was within the financial grasp of most of those showing interest. And most important of all, in America the government left them alone. Until 1912 amateurs had the "ether"—as they called the sky above them—free to do with what they wished. They were soon filling the ether with the dots and dashes of Morse code. And so amateurs, especially in the United States, assembled components, built crude transmitters and receivers, and tried making contact with one another. And many succeeded.

Let us date the real beginnings of amateur radio with the most mind-boggling (for the time) achievement of all: Guglielmo Marconi's hearing of the Morse code letter "s" (dit dit dit) at his station on the fog-shrouded, storm-swept Signal Hill at St. Johns, Newfoundland. It was sent from his station at Poldhu, Cornwall, across the Atlantic. It was December 12, 1901.[1]

The achievement made page one of the *New York Times*, which went into considerable detail. It explained how the station at Poldhu had "an electric force more than a hundred times greater than that of the ordinary station." ( Marconi had been experimenting in England for several years.) Via transatlantic cable he had calibrated the time when the Poldhu station was to send signals: from 3:00 p.m. until 6:00 p.m. (11:30 a.m. to 2:30 p.m. at St. Johns), and the signal was to be that simple "s"—three dots (dit dit dit). In preparation for receiving the signal Marconi had to launch an antenna, and this was a problem. First he attached the wire to a balloon. The first balloon was destroyed by wind. The next day he tried attaching the wire to a kite; the kite broke away, but the second attempt succeeded, although it dived and danced in the brisk wind. Listening carefully, with static on the earphones and the wind outside, he heard the three dots: one at 12:30 p.m. and again at 1:10 p.m. Marconi's achievement, his lawyer said, "marks a new era in the history of the world." The tone of the *Times* article left little doubt that the editors agreed.[2]

This was exciting news to the world's gadgeteers, experimenters, and electrical engineers. Increasingly, in the years following the Civil War, electricity and magnetism had emerged as new and intriguing developments in the world of science. Literate, technologically inclined, inquis-

**Marconi in the Operations Room in the old military barracks on Signal Hill, Newfoundland (courtesy of The Rooms Corporation of Newfoundland and Labrador, Provincial Archives).**

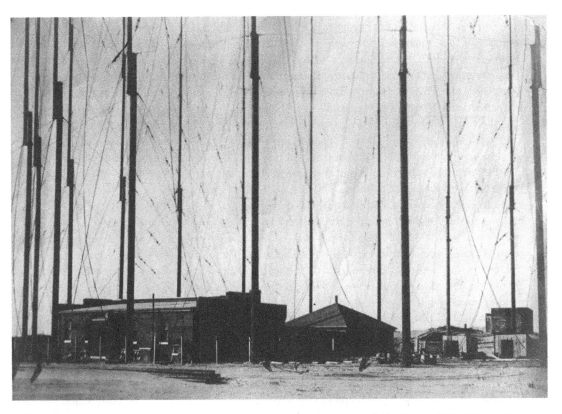

**The Wireless Masts at Marconi Station at Poldhu, Cornwall, England (courtesy of The Rooms Corporation of Newfoundland and Labrador, Provincial Archives).**

itive individuals were fascinated by these new developments. They already possessed some knowledge of the theories of wireless telegraphy. Historians went all the way back to a Greek philosopher named Thales of Miletus, who called frictional electricity "electron," from the Greek word for the sun god. For the experimenters, though, their background knowledge probably began with the speculations of a Scottish physicist, James Clerk Maxwell. He gave the name "ether" to the medium of light and heat which, he speculated in 1867, were electromagnetic undulations that permeated the universe. He said the difference between them was the rapidity of their oscillation (variation between maximum and minimum values)—a difference in frequency. Light waves, for example, were very short, heat waves very long, and between them lay a gap containing an enormous number of wavelengths about which nothing was known.[3]

Next in the experimenters' common knowledge was the achievement of the German physicist Heinrich Rudolph Hertz. Between 1886 and 1889 he confirmed Maxwell's theories. Using a Leyden jar (an electricity storage device) and a coil with a gap, he created a spark—and across the room a ring with a tiny gap responded with a small spark. This was proof of Maxwell's theories, and Hertz was honored by his name being forever applied to radio phenomena: *Hertz*, a unit of frequency equal to one cycle per second; *megahertz*; *Hertzian waves*. Hertz's successful experiment was one of the most notable discoveries of all time. It did not go unnoticed. In Italy a precocious thirteen-year-old boy named Guglielmo Marconi, son of an Italian aristocrat and a Scottish mother (she was a Jameson of the Jameson whiskey family) read of Hertz's successful experiment. It got the boy to thinking: if a spark could be sent through the ether and create another spark many feet away, why couldn't the spark send code and create wireless telegraphy? Thus was Marconi inspired to pursue his career as the pathfinder in the field.[4]

Left out of common knowledge may have been the experiments conducted by Joseph Henry, secretary of the Smithsonian Institution, who sent electromagnetic waves across the Princeton campus in the 1840s; and a dentist, Dr. Mahlon Loomis, who in West Virginia in the late 1860s, using antennas, proved the existence of magnetism from mountain tops eighteen miles apart. Then there was Professor Amos E. Dolbear, who in 1882 conducted successful wireless trials. There were also the exciting experiments of Nikola Tesla, the inventor of alternating current. Tesla may have fabricated the first transmitter and receiver. He demonstrated it in 1893 at St. Louis; at least, the United States Supreme Court has so decided.[5]

Amateur experimenters probably would not have known much about the following men, but their names, or abbreviations of their names, became a part of the wireless operators' vocabulary. Alessandro Volta, an Italian physicist (1757–1827) whose name is applied to the *volt*, a unit of electrical measurement; Andre Marie Ampere, a French physicist (1775–1836) whose name is applied to the practical unit of electrical current, *amperes*; George Simon Ohm, a German physicist (1787–1854) who formulated the law that governs the flow of electrical current, *ohms*; Michael Faraday, an English scientist (1791–1867) who discovered that electricity could be induced by magnetism and who gave us the words *electrode*, *anode*, and *cathode*; and Edouard Branly, the French physicist (1844–1940) who invented the Branly Coherer, the first detector of wireless waves.

Within two decades into the new century other men's names would become commonly known. The electric light bulb, developed years prior to radio, was actually a "vacuum tube" in the sense that it was a filament within a glass bulb that had been exhausted of air. Experiments had been made in which currents could be detected when a second element was inserted in the lamp bulb. Alexander Fleming invented the vacuum tube in 1904; Irving Langmuir developed the high vacuum technology which led to improved versions of Fleming's invention; and Lee de Forest added a grid that made practical the modern vacuum tube. (De Forest's claim is flawed: many a radio engineer believes that Edwin Howard Armstrong really deserves the credit.) Reginald Fessenden developed the alternator and sent the first music by radio. Armstrong again is known for his many inventions including the superheterodyne and, in the late 1930s and early '40s, frequency modulation (FM). These are but a few of the many outstanding men who discovered or developed improvements for wireless communication. Literally hundreds of inventions for radio were patented and, sadly, many resulted in infringement lawsuits, some of them bitterly contested well into mid-century.

A good measure of the interest in wireless telegraphy—the word "radio" does not come into common usage until the years 1910–1915—is the number of pertinent articles that appeared in the nation's periodicals. According to Clinton B. DeSoto, a historian of the early years of amateur radio, 115 articles appeared on the subject of wireless telegraphy during the years 1900–1904 and 150 more articles to 1909. Occasionally one of them included instructions for building transmitters and receivers, with the schematic of the set included. *Delineator*, a popular magazine of the time, ran an article in its August 1911 issue titled "How Boys Can Make Their Own Wireless," and *Woman's Home Companion* in its February 1911 issue ran instructions for "Wireless Telegraphy: A Five to Ten Mile Sending Outfit."[6]

The common image of the experimenter in the early 20th century is of a boy standing on the edge of the barn roof with an opened umbrella, prepared to float back to earth. Providing the lad survived the fall, he might well have turned his attention to other things—technological things. He may have tinkered with labor-saving devices for the farm, or possibly he experimented with the new internal combustion engine, or he could have become interested in wireless telegraphy. The boy was literate, gadgetry was all around him, and the possibilities of contacting a friend with similar interests, without telephone or telegraph wires, intrigued him.

Marconi's transatlantic wireless achievement was exciting news to the world's boys and men who were gadgeteers, experimenters, and electrical engineers. Many of them already possessed some knowledge of magnetism, electricity, and the potential of wireless communication. They were inspired by what they knew of Marconi, who had become a celebrity. First off, they read that he had begun his experiments in the attic of his home in Italy. They knew that the parts for his wireless experiments had not been expensive. They read that he was just twenty-three when he convinced English capitalists to invest in his proposals, and that by age twenty-seven, when he made his transatlantic experiment, he was becoming wealthy in his own right. If Marconi could achieve such success, why couldn't a man or teenage boy accomplish moderate success with wireless in his own home? So, while "older heads murmured in awe and consulted their Bibles" (as related in the *Radio Amateur's Handbook*), "our youthful electrical experimenters ... perceived immediately that here was something a hundred-fold more engrossing than 'electricity.' With one voice they asked, 'How does he [Marconi] do it?' and with one purpose of mind they proceeded to find out for themselves."[7]

The experimenter in small-town Ohio or Colorado, or even in New York City, had learned this much: that a radio wave could be picked up only by a receiver tuned to the same wavelength. The analogy often used to explain this phenomenon was that if someone pressed the key of C on a piano, the C string of a violin in the same room would vibrate—but the other strings would remain still. Thus the waves sent out by the transmitter had to be consistent, while the receiving set had to zero in on that particular wave. With just this primitive floor of knowledge the young experimenter knew that with relatively few parts soldered together, attached to a battery, an antenna and a ground, and with dials and headphones, it was possible to transmit code to someone with a receiver, and in turn to hear that person's message, sent by his or her transmitter, on one's own receiver.[8]

If this seems fairly simple, it must be added that in those early years the task was much more complicated. Essential to the receiving set was the detector. As the name implies, this is the circuit device that detects the radio wave. Of several designs, the most likely choice would be a complex, highly inefficient unit known as the coherer, often called the Branly Coherer after its inventor, Edouard Branly. If the young technician could build a spark coil for his transmitter, which was not too difficult, and a much more cantankerous coherer for his receiver, then possibly he could contact a friend with similar appliances in the next block or neighboring farm.

Remember that Hertz had used a spark to prove that ether contained waves? And that young Marconi conceived of the idea of sending code by way of sparks and made it happen? So did the young technician take the idea and set about building his own spark transmitter. He accomplished this as Forrest Bartlett, W9FYK, had done it, beginning with the wooden end of an apple box as the base. He assembled the other components in much the same way as Forrest had, soldered the parts appropriately, added the batteries, antenna, and ground, adjusted the headphones over his head and began tapping CQ on his key. If the potential amateur was somewhat affluent, all the components could be purchased, for entrepreneurs were incredibly fast at producing wireless parts. Even so, the costs were not excessive. Headphones may have cost as little a $1.50.

The same antenna, ground, battery and headphones served the receiver. Necessary for the tuning mechanism was wire for a homemade coil with a sliding contact to adjust the coil's inductance. Next came the real problem: the ham needed that darned coherer! In this device electromagnetic waves fused metal filings, which gave the sound, and in the interval between the waves a tapper was used to defuse, or shatter, the filings. The coherer was difficult to build and was sure to cause ceaseless trouble. Everyone interested in wireless fussed and fumed over this problem. A variety of methods were used to tap the filings so that they fell, ready for the next dot or dash. None worked well. Even so, as instructions in the British magazine *The Boy's Own Paper* prove, a coherer accomplishing this could be built at home. So, with a drill, solder and soldering iron, some screws and clamps and a wooden baseboard and a board for the dials, the amateur actually could create a receiver with a coherer. Having accomplished all this, the ham was ready to test his rig.[9]

It took the minds of bright people to understand the schematics and accompanying plans in instruction books, journals, or even in the kits almost immediately sold by wireless entrepreneurs. Logic and some understanding of the theory of wireless were demanded, and beyond that, the ham had to transfer these abilities into the mechanical tasks of grouping and soldering the wireless components into workable units. Just to understand the dots and dashes heard in a receiver demanded hours of concentration, and to send the code was a further challenge. Defying these obstacles, a few amateurs were on the air by 1901. Their technology improved and their numbers increased steadily. Yet it is safe to say that there were not many hams around. It really was a bit too complicated.[10]

Then, in 1906, the wireless revolution received a great boost. It was discovered that certain minerals, known in the trade as crystals, had the ability to detect radio signals loud enough to be heard. The principal crystal used was galena. Suddenly a receiving set was both cheap and simple to build. When the coherer was replaced by the simple crystal detector the world opened up to amateurs. The galena crystal, about the size of a dime and with a tiny wire called a cat's whisker to play across it until sound could be heard, immediately made the wireless receiver a device of utmost simplicity.

We have no way of knowing how many amateurs there were in the first years after 1901, but we do know that after 1906 their numbers increased rapidly. Their activities were just one manifestation of a frenzy of activity in the wireless field, for the commercial possibilities of wireless were quickly grasped. One authority compares the enthusiasm of wireless oriented engineers in the years 1900–1910 to the furor of the '49ers to get to California. The number of patents involving wireless greatly increased as did the lawsuits claiming patent infringements. Wireless suppliers sprang up. Businessmen, especially those involved in transoceanic commerce, grasped the potential of this new means of communication as a quantum leap toward improving the safety of shipping and traveling on the high seas.

Then as now, thousands of ships—freighters and passenger liners, naval vessels and yachts, tug boats and revenue cutters, army transports and lightships (vessels stationed at harbor entrances or near shoals or other dangers, which cast a beam or beams of light to warn ships)—plied a world already interconnected by trade. Before the advent of wireless, bell buoys, searchlights, Morse signal lamps, semaphore flags, foghorns, the ships' own whistles, lighthouses, lightships, plus mathematical reckoning of latitude and longitude constituted the measures existing to ensure that a vessel did not run aground. But what of collisions with other ships? In dense fog and on dark nights ships plying the great sea lanes reduced their speed, used all the facilities available—and their officers prayed. They prayed that out of the dark mist another boat's ghostly outline did not appear, headed straight for their ship.

To Marconi, to the executives of the great maritime companies, to insurers such as Lloyd's of London, and most of all to ships' officers, the possibilities held out by "wireless telegraphy" was as electrifying as the concept itself. A wireless aboard ship, in contact with other ships and with land stations, could determine the vessel's exact location, be in close touch with other ships in the vicinity, and plow ahead in the fog and darkness as if it had eyes in its prow. If tragedy did occur—a collision or foundering on a reef—wireless could contact help.

Moreover, a ship could inform a shore station or a lightship to convey its estimated time of arrival entering a harbor, or have a doctor send instructions by wireless to aid a sick passenger. The Marconiman (as the wireless operator was called) also exchanged messages for passengers, ship to ship, as well as ship to shore. These Marconigrams were often sent by the ship's wireless operators to seaboard telegraphers who then relayed the messages via telegraph to the specified addresses. As early into the new century as October 1903, it was found possible to supply news to the Cunard liner *Lucania* during her entire crossing from New York to Liverpool. News on the eastern side was sent from Marconi's station at Poldhu, Cornwall, and on the western side from his station at Cape Breton Island in Canada.[11]

On Cunard White Star Lines, stock quotes and the latest world news were received by the wireless operator and were printed every morning in the ship's newspaper, or posted for passenger's information. So new was the technology that any out-of-the-ordinary incidents involving the use of wireless made headlines. When a man named Crippen murdered his wife in England and took passage to Canada with his mistress, he was surprised to be arrested before the ship docked at Halifax, all because of contacts made by wireless. Crippen was returned to England, tried, found guilty, and hanged.[12]

Land based amateurs were fascinated by the descriptions of commercially used spark transmitters and primitive receivers. Although highly inefficient, they were state-of-the-art until the First World War. In 1903 a writer named Lawrence Perry pronounced unequivocally in the *World's Work* that "the experimental state of wireless telegraphy has passed." Although Mr. Perry could not have been more incorrect, it is true that at the time he wrote, many ships, especially the great passenger liners, were making radio transmitting and receiving apparatus a permanent part of their marine equipment. Wireless had become a part of the routine of shipboard activity.

The wireless operator was a hero and an inspiration to amateurs. (The operator was often an amateur wireless hobbyist himself.) They read descriptions of passengers listening to the staccato sounds of dots and dashes from the wireless room, of voyagers allowed into the shack, watching, listening, fascinated by the "blue-white lightning flashes as the current leaped from the 'sparker' at each bend of the [operator's] wrists, causing the blue flames to play about the six Leyden jars." The physical spark was impressive, about two inches long. "With the blinding flash accompanying each movement of the key occurs a report to be compared accurately with the noise attending the charge of a Krag-Jorgensen rifle," wrote one observer. Another described the sound as "like sputtering gun shots."[13]

Merchant vessels were slower to use wireless. One company stands out as a pacesetter. This was the United Fruit Company, whose primary cargo, bananas, was extremely perishable. Its executives were quick to grasp the potential of wireless communication. By 1905 the firm had established shore stations in the United States and Central American countries which were in contact with wireless units on board its ships.[14]

As early as 1900 land wireless stations were established to make contact with incoming and outgoing vessels. By 1912 the United States Navy maintained forty-five of them along U.S. coasts, while commercial wireless companies—and there were a half dozen or more competing with each other—maintained still other coastal stations contacting ships. A few worked in conjunction with lightships, which were among the first vessels to be equipped with wireless by which they could contact and convey information to ships (providing the ships had wireless) and with land stations. It was hardly a romantic assignment. The Nantucket Shoals lightship, for example, was anchored from November to May in continually rough waters fifty-two miles from land. It communicated with ships coming and going and with a land station at Siasconset on the south shore of Nantucket Island.[15]

In step with the acceptance of wireless by the world's passenger liners, and many merchant vessels, were the world's navies. By 1905 eighty British and thirty Italian warships were equipped with wireless and other countries, particularly Germany, France, and Japan, were busy installing wireless equipment on their vessels. The Unites States Navy's interest dated from 1899; in 1903 a Slaby-Arco rig, so named for the company building it, was placed on the *Baltimore*. A photograph survives: it "shows a conglomeration of machines festooned with wires, switches, and dials that resemble something assembled in Hollywood for a Frankenstein movie of the 1930s." By 1910 the navy was emerging as a prime user of wireless and also as the chief critic of amateurs.[16]

This is the first lightship, Vessel Designation LV 66, equipped with a radio. Note the antenna (courtesy United States Coast Guard Historian's Office Photo Collection).

Maritime interests were not alone in embracing the new art. On land the Weather Bureau and the Department of Agriculture were early on experimenting with the potentials of wireless as an aid to their tasks. With breathtaking rapidity in those early years factories proliferated, producing condensers, rheostats, headphones, insulators, wire specifically made for antennas and grounds, batteries, dials, and other gadgets, some necessary and some not, some practical and some practically worthless or quickly made obsolete by still newer contraptions. Supply houses multiplied, catering to both professional and amateur operators. (Most professionals had started as hams and many still worked personal ham radios.) The prospectuses of some companies described in glowing terms the profits anticipated by sales of their products. Others set up as service companies, with land stations contacting ships with rigs likewise installed and serviced by the company. There were many failures, and more than a few cases of stock jobbing. Lee de Forest with a man named White were deeply involved in one of these "fast buck" schemes.

An important component of this early milieu were the amateurs. Although men and a few girls and women participated in the new pastime, it was known generally that the hobby of wireless transmitting and receiving was the special bailiwick of boys—teenage boys. The crystal set used no batteries. The expense was minimal. Many thousands built these simple receivers, and went no further. Their great interest lay in pulling in the farthest station. Prior to the advent of commercial radio, they listened to Morse code originating with maritime shipping or the navy and, in the very few instances when telephony was in use, heard the spoken words of the person transmitting. Even into the 1930s many a boy spent long hours at his receiver (now with tubes and electrical hook-up) pursuing this quiet hobby, and attempting no more than that. But once Morse signals were heard, a substantial number advanced their obsession further. Intrigued by the dots and dashes coming over his earphones, many a youngster was inspired to learn the Morse code, build a transmitter and realize his dream of "talking" to a fellow ham miles away.

These young "nerds" and "dorks" of the past century had found something to arouse envy and awe amongst their peers and elders. Athletes had to vie for the adulation of pretty girls with boys who couldn't throw a football five yards or catch a baseball or shoot a basket. The young ladies were fascinated by the dials, the blue spark and staccato sounds made as the ham tapped away on his key, the squeals and buzzes and pops as the boy tuned his receiver. They heard the code and watched as the operator wrote down what was being received. Most boys became acquainted with other local boys who shared their interests. Soon they formed amateur radio clubs, helped each other with their rigs, traded parts from each other's junk box, risked their lives on steep roofs putting up antennas, experimented, and competed for the credit of working the most distant station. Thousands of parents encouraged their sons. The hobby kept them off the streets and out of trouble, and besides, Dad and Mom were themselves intrigued by their son's hobby.

Working transmitters and receivers automatically linked the boys with a new breed of hero—ships' wireless operators. Again and again down to the First World War these brave souls were honored for the part they played in saving lives during disasters at sea. Two examples of many will suffice.[17]

The first began at 6:15 a.m., January 23, 1909. A dense fog encompasses the Marconi Company radio shack at Siasconset, on a spit of sand extending into the Atlantic from Nantucket Island, off the coast of southern New England. The wireless operator, A.H. Ginman, awaits dawn; all he can hear is "the distant pounding of the breakers, the wind singing through the wires overhead, and the steady, uninteresting click, click of the [radio transmission from] chattering ships." Then, suddenly, he hears the letters CQD. CQ stood for "all ships" and D for "distress" (although as a colloquialism it meant "Come Quick, Danger"). Then followed in code the message: "We were struck by an unknown boat: engine room filled; passengers all safe; can stay afloat; latitude 40.17 longitude 70. *Republic*."[18]

The message was conveyed by wireless operator Jack Binns on the White Star Line's four-year-old *Republic*. It was bound from New York for Mediterranean ports with a full contingent of passengers. Proceeding slowly in an "almost impenetrable fog" with Captain William L. Sealby at the bridge, there were suddenly "a dozen quickly repeated blasts on a fog siren.... Almost at the same instant a hazy shape loomed up in the mist bearing down on the *Republic*." It was 5:30 a.m. when the Italian Lloyd passenger liner *Florida* crashed into it, "lurching her over to one side as the sharp prow of the colliding vessel gouged through the iron plates of the engine room. Then the vessel pulled away, righted herself, and staggered off into the fog." But the *Florida*, although badly damaged and filled with immigrants, did not stagger far; it halted nearby and made contact with the stricken *Republic*. The location was about twenty-six miles southeast of the Nantucket Lightship, two hundred fifty miles from New York City.[19]

Captain Sealby called the crew to quarters and had the collision bulkheads closed, shutting off the engine room. Then he contacted John M. (Jack) Binns, the Marconi operator. Binns made his way to what was left of the radio room, for three of its four walls were shattered. Fortunately the radio still worked, although in the pitch darkness Binns broke the sending key and had to hold part of it with one hand while tapping out code with the other. When the dynamos on the *Republic* went out the Marconiman went to a storeroom and groped for extra batteries, found them and within a half hour continued sending out CQD. Although he contacted other

"SOS ... 'Saving of Souls'"; "Wireless, the Alarm-Giver": A typical radio shack in a steamship (this is not the *Titanic*) (courtesy *The Illustrated London News*, April 20, 1912, p. 589).

ships, his most important connection was with the Siasconsett land station. A.H. Ginman, the operator, in turn contacted several ships in proximity to the *Republic*: the *La Lorraine*, the *Lucania*, several revenue cutters, a torpedo boat, and, most significantly, the White Star Line's *Baltic*. By 7:30 a.m. the *Baltic* had reached the *Republic* and the *Florida*, which, though badly damaged, had taken on board the passengers and much of the crew. Subsequently the *Baltic* took aboard the passengers from both the *Florida* and the *Republic*, seventeen hundred in all, and made its way to New York harbor. Just four (some say five and some say six) lives had been lost.[20]

To a world accustomed to terrible marine disasters, the rescue by wireless of the *Republic's* passengers and crew (although the ship was lost) was miraculous. "What hath wonders wrought?" The hero of the tragedy was Jack Binns. Congress gave him an ovation after Representative Henry Sherman Boutell of Illinois eulogized this young man "who had the cool head and steady hand to send forth on the willing winds of air the message of disaster that saved hundreds of lives." A photograph of Binns reveals a quiet, rather good-looking young man of average height. He later turned to writing, became a champion of amateur operators, and even wrote an introduction to the first of the *Radio Boys* books, a series popular in the 1920s.[21]

The adulation by boy amateurs of maritime wireless heroes was even more pronounced with the sinking of the *Titanic* on April 14, 1912. This most tragic marine disaster of modern history has been the subject of more books, articles, and motion pictures, and has engendered more speculation and controversy, than any other marine tragedy. As is already known, on its maiden voyage this state-of-the-art passenger liner was brushed by an iceberg and subsequently sank. Perhaps the continuing fascination can be attributed to the macabre human tendency to contemplate the terrible—what it must have been like to wait for more than two and a half hours (from 11:40 p.m. Sunday night, April 14, until 2:30 a.m. Monday morning) for help that failed to come, all the while watching in horror as the ship was listing more and more, the water level rising—until the magnificent *Titanic* disappeared prow first beneath the waves and passengers were spilled into the frigid waters of the North Atlantic. They faced certain death within twenty minutes in the cold water even if they could swim.[22]

Three wireless operators emerged from the *Titanic* disaster as heroes; another had gone to bed, his tour of duty at the wireless completed. Two Marconi operators aboard the *Titanic* were young men from the British telegraphy school: Jack Phillips, senior Marconiman, was twenty-five, and Harold "Judy" Bride was twenty-two. Phillips died of exposure or drowning during the disaster, but Bride survived to tell his tale. It was one of courage, tenacity, danger, and dedication.

Phillips and Bride were both Marconimen, which meant that they were employees of the Marconi company, not of the White Star Line. Their rig was a Marconi make with a range of 350 miles by day, more at night. It was housed in a "Marconi house" on the boat deck, had a double antenna strung between both masts, 205 feet above the sea, and had two separate circuits plus reserve storage batteries in case of a power failure. Prior to the ship's sailing the two had worked with the rig and found it in excellent condition. During the voyage they worked six hours on and six hours off, and were kept very busy. They sent and received messages from other ships, including information telling of ice fields at various latitudes and longitudes along the route; these were conveyed to the proper ships' officials. First class passengers kept Phillips and Bride busy sending and receiving messages in mid–Atlantic, a form of snobbery that was costly but apparently most gratifying.[23]

Possibly they used the wireless too much. According to Bride, the radio broke down on Sunday and he and Phillips spent seven hours repairing it. At one point Phillips ordered Bride to bed for a few hours, but at some stage he woke up. "I remembered how tired he [Phillips] was and I got out of bed without my clothes on to relieve him," Bride recalled. As for the collision with the iceberg, Bride "didn't even feel the shock.... There was no jolt whatsoever." Then the captain "put his head in the cabin." "We've struck an iceberg," he said, instructing them to prepare to send out a call for assistance. Ten minutes later the captain returned: "Send call for assistance," he ordered. Phillips began sending CQD and then, at Bride's suggestion, sent the more recently accepted SOS. Meanwhile Bride dressed and obtained an overcoat for Phillips, who by now had contacted the *Carpathia*. The captain again appeared, informing them that the engine rooms were taking water and the dynamos would not last much longer. They fetched batteries and kept sending SOS, although the sending was growing weaker. Bride went out on deck and observed the chaos. The water, he related, "was pretty close up to the boat deck. There was a great scramble aft, and how poor Phillips worked [the transmitter] through it I don't know.... He was a brave man.... I will never live to forget the work of Phillips for the last awful fifteen minutes." Even after the captain had released the two men to fend for themselves, when they had done all they could do, Phillips kept on until the radio stopped working. Then both men searched for a way to safety. Only Bride survived, first by swimming to a collapsible boat, then boarding one of the crowded lifeboats. Survivors sat on his legs, which were already painfully wedged between the slats. When rescued by the *Carpathia*, the first ship to come to the *Titanic*'s aid, he spent his first hours in the ship's hospital. When he was told that the *Carpathia*'s wireless operator, Harold Thomas Cottam, the third operator considered a hero, had been at his rig so long that he was getting "queer," Bride rose from his bed, painfully made his way to the radio room, and took over. He remained there, tapping out messages of passengers safe or missing, even after the ship had anchored in New York harbor. Marconi, in New York City at the time, decided to go to the *Carpathia* and see the young Marconiman. He made his way up a long stairway and fairly ran to the radio room. As reported in the *New York Times*, Marconi found this scene:

> One lamp burning and a young man's back was turned to him and between two points of brass a blue flame leaped incessantly. Slowly the youth turned his head round, still working the key. The

*Left:* Jack Phillips, wireless operator on the *Titanic*, who died aboard the ship (courtesy *Illustrated London News*, April 20, 1912, p. 594). *Titanic* wireless operator Harold Bride (courtesy *The Illustrated London News*, April 27, 1912, p. 623).

hair was long and black and the eyes in the semi-darkness were large, strangely large. The face was a small and a rather spiritual one which might be expected in a painting. It was clear that from the first tragic moment the boy had known no relief.

"Hardly worth sending now, boy," said Marconi, hoping to cause the youth to stop.

"But these poor people, they expect their messages to go," and the boy saw Mr. Marconi's face and his hand extended."[24]

"Carried ashore with feet crushed and frostbitten, Mr. Harold Bride ... leaving the *Carpathia*" (courtesy *Illustrated London News*, May 4, 1912, p. 686).

Twenty-year-old Cecil Evans, the operator on the *Californian*, which was closer to the *Titanic* than the *Carpathian*, had served his tour of duty for the day and gone to bed. However, there is controversy over his last few minutes at the wireless, as well as a question about the *Californian's* skipper ignoring distress rockets sent up by the *Titanic*. Had messages been received or had the captain headed his ship where the rockets had been observed, the *Titanic* tragedy could have been less severe.[25]

Such courageous actions on the part of non-athletes and non-warlike young men employed on ships served to link the boy amateurs on land with their fame. It increased the incentive to build or buy a transmitter and receiver, master Morse code, and join the growing amateur family. Remember that until at least 1920 amateurs heard code almost exclusively, so the incentive was strong to be able to understand it first, as it came through the headphones, and then to send it. Thousands of boys, a few girls, and a few men and women entered the ranks of amateurs in the years between 1906, when the crystal detector came into use, and the outbreak of the First World War. Because there were no regulations nor amateur licenses prior to 1912, there is no way of knowing how many amateurs there were in the United States. At Congressional hearings numbers from ten thousand to sixty thousand were bandied about.

Writers of boys' and girls' books were quick to capitalize on the boys' new interests. (In the pre-television and pre–Pokemon age, teenagers' books were a huge market.) Some of the titles were *Tom Swift and His Wireless Message*, by Victor Appleton (1911); *The Radio Boys*, a series of thirteen books by Allen Chapman, published in the 1920s; and *The Radio Detectives in the Jungle* (1922), just one of a series written by A. Hyatt Verrill for his *Radio Detectives* series. *The Short-Wave Mystery* was a Hardy Boys title by Franklin W. Dixon (1945). For girls there was *Kay Everett Calls CQ*, by Amelia Lobsenz, a Junior Literary Guild selection (1951). Indeed, girls and women, although few in number, were into wireless. As early as 1912 the *New York Times* took note of twenty-two-year-old Miss Anna A. Nevins, an employee of one of the commercial wireless companies, who operated from the roof of the Waldorf-Astoria Hotel "sending and receiving messages through the air to ships at sea."[26]

Most parents were proud of their children's showing interest in amateur radio. It kept them off the streets, challenged their minds and tested their abilities at reading schematics and building the transmitters and receivers. It would seem that teenagers finally had an innocuous hobby, one that could never get them into trouble. But they were wrong. A proliferation of accusations against amateurs brought on demands for control over their activity, if not its complete elimination. The result was proposed legislation seriously discussed in Congress between 1907 and 1912. The many hearings and the public reaction, as revealed in editorials, tell us much about the public's view of amateurs in the early days of wireless communication. Finally Congress in 1912, spurred on by reports of amateur chicanery during the *Titanic* disaster, passed an act that was believed to end the amateur problem.

In 1907 the navy was exasperated about events taking place in Alameda, California. The trouble involved naval vessels anchored nearby, and the activities of two teenage boys. It seems that Henry C. Heim, fifteen years old, and Alfred Wolf, fourteen, were listening in on orders to or from the fleet, "more or less confidential," and giving the information to the press. An article about the young hams in the *San Francisco Examiner* made clear that the boys were listening to far more than just "orders to the fleet." They were said to be "compiling a book of their choicest confidences" as well as giving information to an Alameda newspaper. Certainly some of their intercepted messages were embarrassing.

> Miss Brown, Oakland—Can't meet you tonight. No shore leave. Be good in the meantime.
> Mrs. Blank, Alameda [real name not given]—Will see you sure tomorrow night.
> Didn't like to take too many chances yesterday. We must be discreet.
> A married woman on Mare Island addressed another woman's officer-husband in this mysterious way: "All lovely. I'm sure you are mistaken. Call again. Your P.O."

Another of the officers addressed his inamorata at Mare Island:

"Honestly, could not show last night. Am arranging so I can see you oftener. Will take you to dinner Wednesday afternoon."

Navy tempers were also raised because the boys had also been up to tricks with their transmitters. If they were not abusing their freedom to hear, they certainly abused it with regard to what they sent. They used deception. Young Heims imitated the name of Admiral "Fighting Bob" Evans when contacting a warship for information. "Once, just for fun," as he expressed it, "he ordered a cruiser about to sail to delay its departure," the *Examiner* reporter wrote. "Something was evidently suspected, for the cruiser sailed and Heims was deprived of experiencing how it felt to be an admiral."[27]

Certainly these two precocious youngsters represented the very epitome of boy experimenters. "The technical proficiency of both boys at sending and intercepting wireless messages is second only to the ingenuity they have displayed in manufacturing the apparatus for this purpose," reported the *Examiner*, for they had built their own sets from instructions derived from library books. They were able to "send and receive messages from as far away as Acapulco [Mexico], Bremerton [Washington], and the Farallones [islands about twenty-six miles west of the Golden Gate where the Navy maintained a wireless station]." While complimenting the boys on their ingenuity, the *Examiner* also suggested that their achievement was dangerous "when there is disquieting talk of war with Japan in the air."

Two years later the navy and commercial wireless companies had reason to be infuriated. In 1909 the Great White Fleet of the United States Navy was returning to Hampton Roads, Virginia, after circumnavigating the globe. Newspapers speculated on the date of return. Would America's pride arrive prior to President Theodore Roosevelt's departure from office, March 4, 1909? And what happened? A boy amateur in Portland, Maine, received the first message from the *Connecticut* announcing the estimated time of arrival, and he gave out the news. In testimony before a Congressional committee, Mr. W.E.D. Stokes, Jr., representing the Junior Wireless Club of America, said the boy's achievement "put to shame and rage all the professional operators and stations [and] these stock-jobbing, so-called 'wireless companies' have not ceased to make war on us boy operators, and unjustly accuse us."[28]

Both of these examples are admittedly frivolous, but they demonstrate how easily the "innocent" hobby of amateur radio could anger those involved in the serious business of radio communication. The complaints coincided with problems arising from the proliferation of radio communications carried on by naval and shipping interests, not just by the United States but by all countries with maritime interests.

The new century was hardly under way before the necessity was apparent for some international agreement to control wireless. The first two international radio conferences were held in Berlin in 1903 and 1906. The proposals involved directing wireless stations, on ships or on shore, to accept all messages, regardless of the wireless companies involved. The United States, concerned about constitutional questions, did not accept the suggestions. Other nations had monopolies over radio which the United States did not have, and under the Constitution could not have. Just how far could the federal government go in regulating radio? The 1903 proposals were seriously considered by American boards appointed by the president in 1904 and 1905 as well as by committees of Congress. Clearly, something had to be done.[29]

More elaborate proposals came out of the Berlin Conference of 1906, including the recommendation for the distress signal SOS (dit dit dit dah dah dáh dit dit dit: three dots, three dashes, and three dots) which was eventually adopted universally. The conference proposals were signed by twenty-seven countries including the United States. On December 7, 1907, President Roosevelt submitted to the Senate for its advice and consent the ratification of this Berlin Convention of 1906, but the Senate failed to ratify it at that time.[30]

The public, through newspapers and journals, as well as through statements by politicians, was made increasingly aware of the need for legislation. In the years down to 1912 several subjects dominated media articles on radio. The first were the stories of lives saved at sea through the dedication of valiant wireless operators who stuck to their rigs until help was on the way. The rest were critical. Stories related false reports given out by rogue amateurs who caused, or at least could have caused, disasters at sea. There was criticism of Marconi's and other wireless companies, in stiff competition with one another, refusing to honor each other's signals, even if they involved catastrophes at sea. Another common complaint concerned the jamming of the air waves, thus impairing legitimate transmitting and receiving. There were wide complaints of the stock jobbing (the selling of worthless securities) by wireless companies that had cost investors thousands of dollars. During the several Congressional hearings on these problems, both the navy and the commercial companies viciously attacked the amateurs. Only occasionally did one of their number testify in their defense. But, surprisingly, the unorganized amateurs brought *their* defenders to the hearings, and they were heard.

Reading the Congressional reports reveals much repetition and a consistent recognition of the problems involved. Reports multiplied of amateur interference with communications involved with the safety and location of ocean-going vessels. A January 1909 *New York Times* editorial entitled "Mischief in 'Wireless'" began by stressing the increasing numbers of amateurs. It said that sixty thousand individuals had discovered that for about $40 they could build a rig to transmit and receive messages. Then it quoted a *Boston Transcript* article describing a typical incident. The steamship *Bremen* was approaching the New England coast while the Nantucket Shoals Lightship was out of commission. Unaware of this, the *Bremen* operator contacted what he thought was the Lightship as to weather and location. The operator received a surprising reply which, if correct, placed the ship's navigator in error; moreover, the weather as given to the *Bremen* operator was different from that being experienced by the steamship. "On reaching New York the matter was looked into," the editorial noted, "and it was discovered that the Lightship had been sunk a week prior to the sending of the massages and the work was that of amateurs." Had the captain of the *Bremen* not been certain of his location he might well have changed course and lost his ship and passengers.[31]

The *Outlook*, one of the popular journals of the time, ran a scathing article against amateurs in its January, 1910 issue. "By ... constructing apparatus out of all kinds of electrical junk, [the amateurs] have built wireless equipments that in some cases approach the naval stations in efficiency," the author noted. The amateurs had also mastered the code. The result was the problem the navy called "amateur interference."[32]

It is interesting to note that the writer for the *Outlook* did not advocate eliminating the amateurs. He did suggest federal regulation, but recommended also that the navy and commercial companies come up with their own scientific solutions.[33]

The navy in its testimony was especially rough on amateurs. Some of its testimony was damning. "Sir, I am impelled to mention to you the following: At 12.20 p.m. on December 6, 1909, a wireless operator of the *Connecticut* intercepted the following message: 'One of those god dam battleships was in there too. I'd like to choke the son-of-a-bitch.'"[34]

To the complaints of the navy and the commercial companies the amateurs mounted a strong rebuttal. The statement of a young man named Stokes in behalf of the amateurs included an acknowledgment of the need for some regulatory legislation, including licensing costing just fifty cents or a dollar; under some circumstances the licenses could be revoked. Licensed operators should promise to forward government messages when requested to do so, be citizens of the United States, know Morse code, and state annually the kind of apparatus used and the wavelength employed. He said that the boy operators stood ready to extend their clubs nationwide and help the government in case of emergencies. This was the first hint of the concept of running relays from amateur to amateur across the United States.

Finally, Stokes defended the American boy:

> We feel that the greatest objection against this bill is that if it is passed it will stifle the ambition and great inventive genius of American boys. We boys of today are the citizens of tomorrow. We have, many of us, already chosen wireless as our line of work. There are vast possibilities, great discoveries, and marvelous inventions yet to be revealed in the study of radio communications. We boys want a try at the great rewards that are to come to the successful experimenter and inventors in these lines. Wireless is not mere play for us boys, as some seem to think. We love the work, hence the name amateur; but it is always the amateur or lover of a line of work who produces results.[35]

And solutions began to be made. Amateurs agreed to licensing, with restrictions that included revocation in cases of malfeasance. Hams suggested that certain wavelengths be restricted to commerce and naval affairs, while other wavelengths should be allotted to amateurs. They suggested that the linkage of amateur clubs would make them useful to the government. Finally, in spite of the many harsh complaints against amateurs, it was clear that American boys who were involved in wireless had their defenders, and may have surprised the opposition by the vigor with which they defended their rights—or rights they insisted they should possess.

When the 62d Congress, 2d session (December 1911–August 1912) convened, committees once more scheduled hearings. Those on the International Wireless Treaty of 1906 (also known as the Berlin Convention) resulted in ratification by the Senate on April 6, 1912. According to historian Susan Douglas, thirteen bills were introduced in the 62d Congress dealing with regulation of radio.[36]

Senate Bill 6412 embodied the gist of several other bills that had been introduced. In a separate section on "Amateurs" the committee "regretted" the legislation but said it had no alternative: it was an indisputable fact that amateurs had on occasion interfered, "doubtless unwittingly," with the transmission of distress messages. The bill restricted them in their use of transmitters, especially of those hams near seaports, to just one kilowatt of power and limited their wavelengths to 200 meters and less.[37]

The bill was passed by both houses of Congress and signed into law by President Taft on August 17, 1912. By its provisions the secretary of Commerce and Labor was to issue licenses to commercial stations and amateur operators. Licensees were to state their location and wavelength, and hours during which the station could operate. In time of war all radio apparatus was to be removed. Special provisions could be made to stations involved in experimentation. And most important of all to amateurs, they could operate at 200 meters and below, using just one kilowatt of power.[38]

Two hundred meters and below, just one kilowatt of power—1,000 watts maximum. To the opponents of amateurs, these two restrictions dealt a death blow to that branch of radio. In their minds it meant that the amateur could do little more than exchange messages with a friend five or ten miles away. It would end long distance operations. Amateurs, they thought, were now on their way out.

They misjudged the ingenuity of the amateurs. Those boys were going to keep right on experimenting, and behold! What a world of wireless they discovered!

# 2

## Hiram Percy Maxim and the Relay Concept

July 17, 1930, was the date set for the Amateur Radio Operator's examination at the Department of Commerce Ninth District Radio Office in Room 538 of the Denver Federal Building. Forrest Bartlett and his fellow amateur Harvey Wilcox drove to Denver in the Bartletts' 1928 Essex Super Six. On their way they picked up another ham, Marvin Juza, who had moved to Denver the year before. Forrest parked the car and the three walked to the nearby building and took the elevator to the fifth floor. They were quiet, apprehensive, and hopeful.

Looking back seventy-two years demands quite a memory. Forrest remembers that the test was given in the morning but he does not know if he was wearing a suit. He is sure his mother implored him to wear one, but he bets that if he did, he removed the coat and vest prior to taking the exam. In addition to Forrest, Harvey, and Marvin, a boy of similar age—name forgotten—from northern Colorado took the test. Forrest does remember the radio inspector's name: Glen Earnhart.

Mr. Earnhart was a kindly man. He noticed the boys' nervousness and told them to relax a bit before he administered to them the ten-words-per-minutes code test which they had to pass prior to taking the written part of the examination. One minute copied without an error was the passing requirement. Forrest and Harvey passed, but Marv didn't copy the required number of words for a passing grade. "Don't worry," said Inspector Earnart, "go for a walk and come back and take the test again." On his second try, Marv passed.

The second phase consisted of writing essays on the theory of radio, drawing diagrams, and explaining how receivers and transmitters worked. All four boys passed.

Then it was home again to await receipt of their licenses and call letters. Harvey and Marv received theirs within a month and a half, but there was a hitch with Forrest's. He had been born in British Columbia, so he had to furnish the birth certificates of both his parents along with a notarized statement that they were temporarily residing in Canada at the time of his birth. This red tape proved satisfactory and in due time Forrest also received his license and call letters: W9FYK.

Jump from the summer of 1930 to the weekend of August 26, 1933. Forrest was twenty-one years old. Among his many interests were radio, cars, fossils (a nearby bluff exposed fossilized sea creatures) and he had a girl friend. He had been editor of the high school newspaper, the *Prep Owl*, and had known a fellow student named Jean Stafford who would become a well-known fiction writer. In college he had majored in electrical engineering. He had pledged Theta XI fraternity, where he had met Willard Fraser, from a prominent Montana family. Willard had urged Forrest to introduce him to a patient named Marjorie Frost at the nearby tuberculosis sanitar-

**Hiram Percy Maxim (courtesy ARRL).**

ium, Mesa Vista. Forrest's parents had urged their high school English teacher, the poet Robert Frost, and his wife Eleanor to place Marjorie there. She not only recovered, but married Willard and went to Montana with him. She was ecstatically happy with her young husband, but died shortly after giving birth to their daughter Robin.

Now Forrest and Harvey were headed for the regional American Radio Relay League (ARRL) convention at Colorado Springs. Driving along in Harvey's family car—a 1932 Chevrolet so quiet and modern that Forrest wondered why his parents put up with that horrible old Essex—Forrest remembered events involving radio that had already occurred in his life. There was the fellow ham in Boulder named Max Richmond whose theory was that the greater the circumference of antenna wire, the better the signal would be. He asked Forrest to help him test his theory. In one of the hardest day's work of his life, Forrest helped Max raise a sixty-six-foot-long antenna consisting of one-fourth circumference copper wire. Somehow they got the aerial installed, but as far as Forrest and the rest of his radio friends could determine, Max's signal was no better with that thick copper wire than anyone else's.

The boys probably registered at the Acacia Hotel in Colorado Springs. Amateurs from New Mexico, Texas, Utah, Colorado, and Wyoming were there. This was an official ARRL convention with a headquarters representative, Mr. A.A. Hebert, on hand. Forrest's mental image of the banquet, and the way the tables were arranged, warrants a guess at about a hundred participants. Time was filled with technical sessions that covered subjects of interest to the amateurs. Moderators were professional engineers who were also hams. There was a picnic on Mount Manitou, scenic trips that included the Garden of the Gods and the Narrows, and visits to stations owned by local hams.

The event most remembered by Forrest and Harvey was their initiation into the Royal Order of the Wouff Hong. This honorable group is made up of any licensed amateurs in good standing. Forrest does not remember the mumbo-jumbo of the initiation (and if he did, he was sworn to secrecy) but he well remembers the site. At midnight on August 25, 1933, in the Cave of the Winds outside of Colorado Springs, at a depth of 300 feet below ground in a grotto dubbed the "Bandit's Hall," Forrest and a number of other young amateurs became members in good standing of the amateurs' fraternity. Forrest and Harvey returned from Colorado Springs with greater interest than ever in amateur radio.[1]

One of the signatures at the foot of the Wouff Hong certificate was that of "The Old Man." Who was "The Old Man?" His name was Hiram Percy Maxim, and if amateur radio has a patron saint, he is the man. Without his dedication and active leadership, amateur radio could well have been so stifled, so heavily regulated, as to have been crushed. The part he played in defending amateur radio demands a knowledge of its enemies.

In retrospect it is passing strange that the boys' hobby of amateur radio should have survived the onslaught of well-organized and well-financed opponents. Private wireless companies such as Westinghouse, Western Electric and General Electric opposed ham radio. In the 1920s commercial radio station owners as well as private citizens protested amateur interference to AM receivers. Maritime companies lobbied against amateurs because they were accused of interfering with safe, consistent communication. Even the United States Navy and other federal agencies campaigned at one time or another, and often all together, against amateur radio. They introduced bills in Congress. They demanded everything from total destruction of the hobby to controls so strict as to suck the life blood from the art. True, amateurs did have their defenders in Congress. Lay people wrote letters to the editor and some newspapers defended them in their editorials. Even the occupant of the White House occasionally spoke out in their behalf. Radio supply companies manifested a strong interest for the hams. Yet all this help was not much of a defense in the face of such formidable opponents. To save amateur radio a strong-willed competent person and a national organization were needed. Without an intelligent, dedicated, courageous champion, someone serving as a catalyst for the movement's defense, it is hard to envision amateur radio surviving the attacks of its many opponents. This is where Hiram Percy Maxim enters the story.

By any measure Hiram Percy was a brilliant man, and by some he would be considered a genius. His father, Sir Hiram Stevens Maxim, had invented the Maxim machine gun and had moved to England, leaving his family in America; Hiram Percy was just twelve years old at the time.[2]

Yet his father in the early years of the lad's life had a life-long influence upon him. Certainly the Maxim curiosity, imaginativeness, ingenuity, and determination that earned for Hiram Stevens more than 250 patents in England and the United States, and several patents for his brother, Hudson Maxim, were inherited by son Hiram Percy. At age sixteen, in 1886, he graduated from the Massachusetts Institute of Technology, the youngest student in his class. Thereafter he was employed as an electrical engineer by several firms. By 1895 he was chief engineer for the Electric Vehicle Company of Hartford, Connecticut.[3]

The young technician became absorbed in the possibilities of the internal combustion powered horseless carriage. The story is well told in his book, *Horseless Carriage Days*. From the most basic of beginnings, within six years Hiram Maxim had developed an automobile on a chassis built just for a motorized vehicle, with a radiator and a motor in front, the motor connected to a universal joint ahead of the rear axle that powered the wheels, with a steering wheel (rather than a tiller) on the left side of the vehicle, a carburetor basically the same as those used on cars until after the Second World War, a clutch and transmission with three gears forward and one back, and a brake. In other words, in about six years, starting from an experiment with a drop of gasoline, Hiram Percy Maxim had developed the basic concept of the modern automobile. More than forty patents relating to automobiles are attributed to him.[4]

In 1908 he patented a silencer suppressing the sound of a discharging firearm and in later years devoted much of his work to developing apparatus to overcome noises from pipes, engines, and ventilators. He was a successful businessman, married the daughter of a Maryland gover-

Kenneth Bryant Warner was secretary and general manager of the ARRL from 1919 until his death in 1948. His loyalty to amateur radio, and his dedication to its continuity in the face of strong opposition, contributed to amateur radio's survival (courtesy ARRL).

nor, and had two children. On February 17, 1936, hospitalized in La Junta, Colorado, this father of the ARRL, founder of the amateurs' journal QST, vigorous defender of amateurs' rights, and polished amateur in his own right, died.[5]

It is no surprise that such an imaginative, tenacious, technically-oriented individual should have manifested an interest in amateur radio. According to Alice Clink Schumacher, his biographer, Maxim became a ham in 1911 along with his eleven-year-old son, Hamilton. "His first station," she writes, "was SNY, then 1WH, then 1ZM [and] after World War I, 1AW." His first station had a range of one city block, but, challenged and experimental by nature, he soon had what may have been the finest, most up-to-date rig in Hartford, Connecticut; his range soon spread to a hundred miles, and he was determined to extend it.[6]

He soon knew of the hams working in the vicinity, and in January 1914 chaired the first meeting of the Radio Club of Hartford. Elected secretary was eighteen-year-old amateur Clarence Tuska. The club was an immediate success, by March having thirty-nine members, a constitution, and by-laws. This was less than two years after the Radio Act of 1912 had been signed into law by President Taft. By restricting amateurs to two hundred meters and below, with just one kilowatt of power, their opponents thought they had shelved the problem of amateurs for good. Instead, they had challenged them. What, the amateurs asked themselves and each other, can be done with such "super-short" wavelengths? The answer, they were soon able to give, was *plenty*.[7]

The restrictions did not stop them. According to Clinton B. DeSoto, whose book *Two Hun-*

*dred Meters and Down* is the one reliable source of such information, in the four months after passage of the 1912 act, 1,185 licenses were issued and by the end of 1913, more than 2000. By June 1914, more than 5,000 licensed amateurs were dot-dashing through the ether and another 5,000 were thought to be operating illegally. Enforcement of regulations was restricted by the small number of radio inspectors. Violations of the allowed wavelengths by amateurs using 250, 300 and more meters rather than the 200 and below was widespread. Enthusiasm was rampant as evidenced by the proliferation of amateur radio clubs. They existed in all the big cities and one can safely assume they were prevalent elsewhere. Pennsylvania even claimed a statewide organization. The art was such that those interested just naturally got together to "chew the rag."[8]

It was in the early spring of 1914 that Maxim, besides being a leading ham in Hartford, made a major, lasting contribution to the hobby. Until then ham radio had been pretty much a pastime without a social purpose. Hams experimented, tried to extend their range, worked at improving their sending and receiving and the clarity of their signals, and did "rag chewing" with ham friends. Without contemplating the possibilities, some hams did "message handling." It consisted mostly of doing favors for neighbors and friends. A ham would convey best wishes to an amateur in another town, who was to deliver the message to the friend's friend. That such message handling could take on greater importance had been mentioned in testimony before a Congressional committee in 1912, but its significance had been lost amidst the mass of evidence presented.[9]

A problem arose that caused Maxim to concentrate on this matter of message handling. He wanted one of the new audion tubes. These were primitive vacuum tubes with enormous potential. When Maxim heard that one of them was for sale by an amateur in Springfield, Massachusetts, he tried to contact that ham with his transmitter, but failed. (Unusual conditions down to the present day make contacting Springfield, thirty miles north of Hartford, difficult.) Suddenly Maxim remembered that a young amateur from Windsor Locks, a small town about half way between the cities, had attended a Radio Club of Hartford meeting. His call letters were known. Maxim was able to "raise" him, tell the young man in Morse code what he wanted, and—could he contact the ham in Springfield? The favor was done, and we presume that Maxim received his vacuum tube.[10]

Hmmm. Just as peddling a bicycle home from a date on a dark night had got him thinking about a motor driven vehicle, Hiram Percy now became absorbed with the glaring need for cohesion among amateurs. Some kind of mastic was needed to link them together coast-to-coast! No national organization existed, however. Such an organization, Maxim reasoned, would set up standards, register amateurs committed to operating their rigs at specified times and agreeing to carry out a fixed pattern of relays. Such an organization would make a positive contribution to society. It would elevate amateur radio from its status as a mere hobby to acceptance as a major contributor to modern communications. Maxim's later activities on behalf of amateurs lead us to believe that he knew that a relay league could accomplish something else: by its contributions it would increase the security of amateur radio from attacks by its enemies. He knew that it would also create a national force of amateurs better prepared to defend their rights.

He had already thought of a name for the organization: The American Radio Relay League. In the spring of 1914 he presented the idea to the Hartford Radio Club, which accepted it with great enthusiasm. In typical American style, a committee was formed. By May members were sending application blanks to every ham operator whose address was known. There was quite an incentive to join, for there were no dues; requirements, however, were fairly stiff. A ham had to know his Morse code and the rules and regulations established by the government. Even so, the response was so strong that the club appropriated $50.00 for further development (Maxim and Tuska had footed the bills up to then).[11]

When a qualified member sent in the application, that person was automatically considered an operating member of the league. Maxim saw to it that the relay concept was quickly made operational. De Soto writes that by August 1914, more than two hundred relay stations had been established, "Maine to Minneapolis and Seattle to Idaho." A map published by the league in September showed 237 stations in thirty-two states and Canada. In October, the League published a *List of Amateur Stations* giving "the names, addresses, calls, power, range, receiving speed and operating hours for four hundred stations." The distances claimed by stations depended primarily upon their power: one kilowatt rigs worked up to 350 miles while smaller stations had a range of as little as ten miles.[12]

The success of Maxim's relay idea apparently caused trouble with the Hartford Radio Club. In February 1915, Maxim, Tuska, and the club's president, David L. Moore, resigned, taking the ARRL with them. Maxim is said to have reimbursed the club for all expenses involved in creating the national organization. Subsequently, H.P.M. (as he was often referred to) had the ARRL incorporated under the laws of Connecticut. All the while membership increased to over six hundred by the end of 1915, while the range of some stations, especially those boasting one kilowatt of power, was extending to as much as a thousand miles. This was an extraordinary achievement considering the two hundred meter and one kilowatt restrictions. It was even more amazing when it is considered that most of the rigs were home made, and probably cost the owners less than one hundred dollars.[13]

Meanwhile Maxim had been busy acquainting key government agencies with the existence of the new ARRL, its purpose and its achievements. He was already a well-known citizen, son of Hiram Stevens Maxim and in his own right a respected inventor. In promoting his gun silencer he had made important contacts in Washington, D.C., and through his wife he had made still other connections. In 1914 H.P.M. traveled to Washington and conferred with the commissioner of navigation of the Department of Commerce, who was assigned the task of regulating amateurs. Maxim's aim, besides acquainting the commissioner with the existence of the ARRL, was to persuade him to allow certain inland ham stations to operate up to 475 meters. This was necessary, he explained, because some of the relays were remote and needed the additional authorization to contact the next relay member and thus retain the necessary unbroken chain. The request was granted with the provision that only bona fide messages be allowed.[14]

As would be expected, the process of relaying messages did not function smoothly. Efforts to solve the problems resulted in a package of ideas that, when activated by the ARRL, increased the league's importance. Services by the league grew due to need, not by planning. A booklet listing hams, their call letters and addresses, was in the works. Certificates were being designed for outstanding relayers. As time went on a bureaucracy arose, still existing, serving the many needs of amateurs, their relays, their clubs, their concerns.

Many of the problems, Maxim realized, could be solved or at least controlled through a periodical initially distributed to relay members but devoted to amateur radio in general. In December 1915, at his and Tuska's expense, all six hundred members of the ARRL received a copy of QST (the amateur's abbreviation for "General Call to All Stations"). Its object, the announcement on page two stated, "is to help maintain the organization of the ARRL and to keep the Amateur Wireless Operators of the country in constant touch with each other." In its final paragraph the notification requested financial help by each member sending in twenty-five cents for a three month trial subscription.

Maxim's awareness of the precarious situation of amateur radio vis-à-vis government, especially military and naval interests, was further reflected in QST's first lead article. It reminded readers that the country was faced with "extremely serious national questions," by which he was referring to problems of national defense. It was hardly necessary to remind his readers of the terrible World War then in progress. "One of the most important factors [in preparing our own defense] is Radio Communication," he wrote. He then inserted copies of letters sent to the secretary of the navy and the secretary of war. "In connection with your plans for national defense, it may be that the ARRL may be of service," his messages began. Maxim then described the ARRL, how it had existed for over a year and had more than six hundred members in thirty-eight states who were capable of communicating with "all important points" except for a few gaps in the South. Included with the letters were the *List of Stations* and a copy of the official ARRL message blanks. The secretaries of war and navy were further reminded that members were middle-aged, young men, and boys, and that many of the men were wealthy. A businessman, it was stated, was in charge of management. (That was, of course, Maxim.)

The letters emphasized contributions made by amateurs in times of natural disaster, such as the terrible floods of March 1913. When all other means of communication were cut off, he reminded the secretaries, amateurs were still able to contact the outside world. "Most of our membership is along the Atlantic and Pacific coasts [and] it is not impossible that we might be of value to our fleet standing off our coasts in time of war," he wrote. "We respectfully offer the services of our organization, and its facilities."

Replies had been received and were printed following Maxim's letters. Secretary of the Navy Josephus Daniels suggested that Maxim confer with the navy's superintendent of the radio service, especially informing him "of the method employed for the interior control of the amateur stations constituting the League." This rather vague statement, apparently expressing Daniels' mistrust of everything amateur, was a portent of navy attitudes toward amateur radio for years to come. That service, under Daniels's leadership, envisioned total control of American radio. After the Armistice and into the early 1920s the navy did everything in its power to prevent the reintroduction of amateur radio into American life.[15]

Clearly the part to be played by amateurs in time of war was a matter of much discussion. Maxim's concern with government-amateur relations was based upon his knowledge of the opposition that had fought amateurs prior to passage of the Radio Act of 1912. He was trying to improve amateur radio's reputation. Possibly opposition could be reduced by the existence of a smooth working system of nation-wide relays. This would also elevate amateur standings with the public and make it more acceptable at the federal level.

In the space of just two months—December 1915 and January 1916—responses to the trial issues of QST had been so strong as to warrant the launching of QST as a monthly magazine. Its format had already taken shape. It ran a technical article or two, editorials, and amateur news; descriptions of ham radio stations, often with illustrations; a listing of new ARRL members, their call letters, names, and addresses; new radio clubs; reports of amateurs brought to justice for violation of Federal regulations; letters to the editor; swap want-ads; and advertisements. It was clear that a periodical exclusively for amateurs filled a need, and QST was filling it. The subscription price had been set at one dollar; for two dollars the subscriber received in addition a Certificate of Appointment, the 1915 *List of Stations* book, and a pad of message blanks. The future of QST as the radio amateur's magazine seemed assured.

All of these developments took second place in H.P.M.'s mind to the relay idea. "Practical Relaying" was his lead article in the February issue. It included a map of proposed trunk lines across the country. (It also included Maxim's portrait: a slim man of average height with a shock of graying hair that tended to grow upward.) He envisioned six relay lines labeled A through F (see map), each with a point of origin and destination. Trunk line A ran from Portland, Maine, to Seattle, Washington, with relays established at eleven cities en route. B ran from Chicago to San Francisco with seven relays en route; C from Boston down the eastern seaboard to Jacksonville, with nine relays; D ran from Philadelphia to New Orleans with nine relays; E from St. Louis to Los Angeles with nine relays; and F from Vancouver south to San Diego, with five relays.

Some cities—Chicago, Philadelphia, and San Francisco to begin with—would be designated as junction points for two or three of the lines. Their operators would be known as Star

**The beginnings of the relay idea. Courtesy *QST*, February 1916, p. 21.**

Relays and be given special authority for operating up to 450 meters. Maxim perceived a smooth-running system:

> If it were possible to relay from point to point along these trunk lines, there would be a practical traffic established to most of the principal points in the country. Msgs. [messages] could be routed via AF to go from Boston to San Diego, or via ABF or via ABEF.... If eachtrunk line were kept open by daily test, any combination of them is naturally also open.[16]

Maxim, a good businessman, had to know that the problems of creating a workable, reliable relay system were formidable. Possibly it was an impossible dream. The first and foremost of the problems he never mentioned. This was the necessity of relying entirely upon volunteers, amateurs who were willing to subject themselves to the discipline of fixed schedules two or three times a week. There was no reward beyond credit, perhaps, in *QST*, or possibly a nice-looking certificate to tack on the wall. At the very least, volunteers for relays had to be highly reliable individuals with good rigs. And there was the problem of timidity. Many a ham was by nature an insecure person, afraid to go on the air with a QRU (have you anything for me?). "Some people," Maxim commented, "would rather take a threshing than touch their keys for this purpose."

Assuming that relay stations were established, how could their reliability be secured? This could be done, it was suggested, by running proof tests, much as a fire station called every day at six and twelve o'clock just to make sure it was on line and reliable. At a specified time every relay operator should send out a QRU. "If a fixed set of trunk lines were laid down and those on these lines kept at it until they could run a test message out to the end of the line and back at certain stated times, it would mean that delayed messages would be enormously reduced," ran an editorial in the February 1916 *QST*.

Maxim's follow-up article in the March issue was highly optimistic. He had received letters on the subject sufficient to sense that the six trunk lines met with approval. The next step was to establish a district headquarters for each trunk line and commit it to organize that unit. He suggested cities that were junction points because they could run tests of several trunks from a single base. He gave Chicago as an example. It was approximately the central point for Trunk Line A and a starting point for Trunk Line E. A test message starting from Chicago could go west to Seattle, the end of Trunk Line A, east to Portland, Maine, the other end, and southwest to Los Angeles, the end of Trunk Line E. Chicago would also cover a line down the Mississippi to the Gulf and west to the Pacific Coast. Similar links would be established at other trunk district headquarters.

He emphasized the need for reliable people to man the district headquarters. They must have letter-writing ability and be good correspondents, for much contact would be required with headquarters and other district offices. Above all, such people must be on the job at the times specified. And, recognizing that most amateurs were teenage boys, he pontificated:

> The failure to keep regular hours is the greatest fault we amateurs have, and it is the great big factor which limits our radio relaying work. Tom is never on when Dick and Harry are on, and Dick seems to make a point of going to the movies when Tom and Harry most need him. Harry invariably reserves his off night for the time when Tom and Dick depend upon him being at his instruments.

Along with a specific day of the week for relaying, an hour had to be chosen. Midnight in the east was a rather late hour for boys, but it was just nine o'clock on the West Coast. H.P.M. discussed the matter at length but left the final decision to the district managers. Then there was the matter of the test message to be sent out and checked to know who received it. It should include a secret password, he suggested, while the return receipt (or QSL) should bear the call letters of every station that had received and relayed it. QST should offer a prize each month to the trunk line with the most credible showing. Finally, H.P.M. implored Chicago, Philadelphia, and San Francisco hams to write him with suggestions for the district set-up, and the choice of directors.

By March 1916, nearly a thousand amateurs were members of the ARRL. Competing organizations were weakening. These included the National Amateur Wireless Association, which had as a journal the successor to the *Marconigram*, the *Wireless Age*, and Hugo Gernsback's Radio League of America which had as its house organ *The Radio Experimenter*. QST warned amateurs about the proliferation of amateur wireless associations, suggesting that the ARRL, like the American Automobile Association, had come along to help the uninformed choose wisely. "Amateurs should consider carefully what these different associations offer," it warned, suggesting that membership including a subscription in another organization's amateur journal was usually overpriced.[17]

The package of ideas that became the ARRL bureaucracy was dominated by the relay concept. It had engendered tremendous interest. The concept, in fact, was being tried by other hams even as the ARRL's ideas were still forming. As early as March 1916, just as QST was going to press, it was able to report an accomplishment by William H. Kirwan, 9XE, of Davenport, Iowa. On New Year's Eve, 1915, he had tried a nation-wide emergency relay to see how rapidly, and how thoroughly, such a relay could cover the United States. The extent to which this was accomplished slipped by QST, but in the April issue QST described Kirwan's effort as a success.[18]

The real significance of Kirwan's test was, first, that it proved that just three years after passage of the Radio Act of 1912 amateurs restricted to 200 meters and below, and to power of just one kilowatt, with rigs usually built by themselves, were contacting stations hundreds, even a thousand miles and more away. Second, Kirwan's success led to a coordinated attempt to convey nationwide a "Washington's Birthday Amateur Relay Message." Again, this was not inspired by the ARRL although Maxim supported the endeavor and the ARRL, as well as Gernsback's association, participated.

Details of this national amateur operators' exercise are skimpy. It appears to have been the brainchild of Colonel W.P. Nicholson at the Rock Island Arsenal along with Kirwan, who lived across the Mississippi in Davenport, Iowa. The army was involved as was the navy, which agreed to send out, from its Arlington station NAA, QRT (stop sending) and QRM (I am being interfered with) signals, in amateur lingo requesting that all hams not involved in the relay cease operations for a specified period on February 22. The plan was ambitious: hams were to deliver in as short a time as possible a wireless message—a rather dull message—to the president of the United States, the governors of the forty-eight states, and mayors of cities. "Never before," wrote Kirwan, "had the amateurs been flattered by such cooperation." According to *Scientific American*, the purpose of the experiment was to determine how quickly an army of three million soldiers could be mobilized; "it is safe to predict that the efficiency and value of the amateur wireless stations and their personnel will be proved to the Federal authorities."[19]

Maxim's concept of trunk lines was not followed in this first Washington's Birthday Relay. Stations had been approached and divided into districts. By means of geographic circles the range of Special, First Class, and small amateur stations ("the little fellow with the gas engine coil and a few dry batteries") was determined. Jealousies had to be smoothed out. Some amateurs protested the day chosen, "some wanted friends of theirs, second cousins, appointed in Squedunk," and some wanted a copy of the message before it was transmitted—a request rejected because the relay was to be considered an emergency call. A printed circular sent to every participant gave the final instructions. Test operations were made on February 18 and were a total failure. Undeterred, the experiment was again conducted, beginning at 11:00 p.m. February 21. 9XE (Kirwan) wrote that "for the first time in the history of wireless telegraphy, the air was clear with only an occasional QRN [atmospherics].... Every amateur in the country was quiet—truly a remarkable thing ... and out the message came."

> QST, QST, QST, de 9XE
> QST relay MSG [message]
> A Democracy requires that a people who govern and educate themselves should be so armed and disciplined that they can protect themselves.
> (Signed) Colonel Nicholson, U.S.A.

The two aims of the relay were speed and a nationwide spread. As for speed, it took just fifty-five minutes for the message to reach the Pacific coast and just one hour for it to reach the Atlantic. New Orleans heard it in twenty minutes. "Every amateur may feel with pride that he is one of the efficient system which bounded [sic] the United States in one hour," boasted 9XE. Only a handful of the amateurs—those closest to executive mansions or to the homes of mayors—actually delivered messages to those notables. In several communities Boy Scouts carried out the task.

Kirwan 9XE had to acknowledge that errors were made. "Colonel Nicholson became Colonel Nichalson, Colonel Michalson, Colonel Micholson, Colonel Nichols, Coloner Nick, Colonel Richardson, and several other colonels amplified to the nth power," he reported. After all, he added, "it was a common cause, rich men, poor men, young men, old men, two ladies, a host of boys, and several ministers [taking part]. All the talk about preparedness shows that young America is on the job when aroused."

Some amateurs had amusing tales to tell of the circumstances under which they received the messages, and a few had interesting stories of how the note was delivered to executives. H.E. Rawson of Kuna, Idaho, received the message at his ranch, jumped in his Model T Ford and sped the eighteen miles to the governor's residence at Boise. "The governor," it was reported, "was so surprised at the strange maneuver that he was unable to express himself for several days." William C. Colburn of the Colburn Ajax Mines at Victor, Colorado, received the message at 10,800 feet. John C. Stroebel of West Virginia had to fall back upon Western Union Telegraph to convey the brief statement to the governor. At 1:30 a.m. in Williamsburg (a part of Brooklyn) a ham named A.R. Boeder handed the message to the mayor of New York City. Two operators recorded it in Philadelphia, where Boy Scouts delivered the message to the mayor. The Harvard Radio Club heard the message at 1:45 a.m. and again, Scouts delivered the message to the chairman of the board of Selectmen at Lexington.[20]

Possibly most important of all, as reported by the *Washington Post*:

> Preparedness of amateur radio operators of this country was demonstrated yesterday by the successful transmission from Davenport, Iowa of a message....
> At two o'clock yesterday morning Walter A. Parks, 1220 Jackson Street, Brookland, picked up the message which came faintly but steadily. Eight minutes later it had been transcribed and Mr. Parks personally delivered it to the executive offices of the White House at 4 o'clock yesterday morning.

(There is a discrepancy here: Someone committed literary license. According to *QST*, Parks "copied the message, jumped on a waiting motorcycle, opened up the machine, and after nearly frightening several policemen, especially the one at the Executive Mansion, dashed up the steps of the White House with the msg. written on an ARRL blank, for the President of the United States, whom we hope was peacefully sleeping as it was about 2:00 a.m.")

*QST*'s report of the Washington's Birthday Amateur Relay Message, which, by the way, was considered a major success, ends with a slate of the relay points. At least 180 names are listed covering thirty-four states and the District of Columbia. The periodical also noted one rather embarrassing fact: many of the stations involved were working up to 500 meters. If these stations did not have a Special License, they were in violation of federal regulations. *QST* ventured that the relay would not have succeeded without stations using more than 200 meters and then added some hopeful news. From the relay's success the regulators realized the necessity of stations working up to 475 meters, and now appeared to be making more Special Licenses available. Do not apply if you are not a consistent amateur, *QST* emphasized, and when applying use a typewriter and fill in the blanks with great care.

Unfortunately this first Washington's Birthday Relay received little press coverage. The bloody Battle of Verdun had just begun, there was much foreign news, and controversy raged over the appointment of Louis Brandeis to the United States Supreme Court. With such events, the simple experiment of relaying messages within two hours across the country by wireless was not considered newsworthy. Apparently local papers made some mention of the relay from the local angle, but besides the *New York Times* only the *Washington Post* among major newspapers carried the news, and the *Post*'s article was on the last page: "Local News." *QST* gave the achievement the credit it deserved. Along with Kirwan's lead article, a full page cartoon in the April issue showed Uncle Sam looking at an amateur station and commenting: "I Reckon My Amateurs Lead the World."

Although it did not receive deserved publicity, the first Washington's Birthday Amateur Relay was significant in developing amateur awareness that theirs was a national pastime. It showed what could be done with cooperation by hams on a nationwide basis. It also highlighted the rapid advances being made in technology. Whereas just a few years past, contact with another ham twenty-five miles away was considered an accomplishment, now there were hams regularly contacting other hams a hundred and more miles away, and on a regular basis. The relay also proclaimed that, with all its weaknesses, the idea was valid; it could work.

Although the transcontinental communications experiment had not made use of Maxim's concept of trunk lines, the exercise was helpful in promoting the idea. The April issue of *QST* noted the appointments of a director of Trunk Lines G and D and for A and E. By May 1916, the line between Boston and Philadelphia was operational. The goal had been to have a relay every twenty-five miles but with an occasional "Star Station" (as it was designated) that was

QST's reaction to successful relaying (courtesy *QST*, April 1916, p. 2).

known to work up to one hundred miles. A listing of stations from Baltimore to Jacksonville revealed a dozen serious gaps as the relay system worked down the Atlantic coast. QST urged amateurs in the region to write in and volunteer their services. As for California, no one had come forward to be the director; again the journal pleaded for amateurs to offer their services. Some cities, with relays established, were bracketed with nearby municipalities indicating that two amateurs were involved and if one was absent, the other could fill in. Two contact days for relays had been designated: Mondays and Thursdays with the usual time 11:00 p.m. Eastern Standard Time. All this activity reflects the enthusiasm of ARRL officials. They were doing all in their power to push their relay concept.

The obstacles were enormous. As the June 1916 issue stated, in those days the problems of static were so bad in the summer that many a ham shut down his rig, turned to seasonal amusements such as fishing or baseball, and awaited the coming of fall. QST even considered ceasing publication for the summer months while one relay director simply announced the closure of his trunk line for the summer. The Literary Digest even ran an article on "Summer's Blight on Wireless." A few of the trunk line directors resolutely faced the static and attempted to keep their lines open. Maxim and Tuska refused to give up, summer and static and fewer active hams notwithstanding, and continued to promote their concept of coast-to-coast relays.[21]

They were aware that many amateurs manifested little interest in the relay idea and made no attempt to participate in that aspect of ham radio. Most of the departments in QST had nothing to do with relaying, nor did many of its articles. Yet the relay concept was the mastic that instilled in amateurs a sense of being part of a nation-wide organization.[22]

From passage of the Radio Act of 1912 until America's entrance into the First World War, relations of amateurs with government agencies were generally amicable. In 1916 radio inspectors employed by the Bureau of Navigation in the Department of Commerce arrested and brought to trial a few violators coast to coast. With their small staff one suspects the inspectors did this to frighten amateurs into obeying the regulations.

The army manifested dissatisfaction at the time of General Pershing's pursuit of Pancho Villa in northern Mexico. Maxim was informed that no amateurs south of Austin would be issued licenses nor would existing licenses be renewed. The official wrote Maxim that he "was very much disappointed in the behavior of the amateurs in that vicinity and sincerely hope[ed] that this will be a lesson to the amateurs, in general, for the future.[23]

In an editorial, amateurs were warned that "this is the handwriting on the wall.... There will not be much to Amateur Wireless in these U.S.A. if we once arouse the Government."[24]

Change takes place so rapidly today that it is difficult to envision "future shock" happening three or four generations ago. Yet improvements in the automobile, the airplane, electrical appliances, and above all in radio were advancing back then with almost breathtaking rapidity. With their upgrading came the advance of auxiliary services. An example of the growth was QST, which served amateur radio. Its thirteenth issue (December 1916), considered its first anniversary issue, was indicative of the growth of radio and the success of the ARRL. What had begun as a mere bulletin was now a journal numbering seventy-four pages with a colored front cover and numerous advertisements. Its lead editorial placed credit with the subscribers. Then it announced that the ARRL would be asking for dues of a dollar, apart from subscription to QST, to help make expenses. Other journals devoted to amateurs were falling by the wayside. Even Gernsbeck's The Electrical Experimenter was a poor second with subscribers, and his American Radio League likewise lagged far behind membership in the ARRL.

Of greatest interest in the anniversary issue was H.P.M.'s suggestion for the "First Trans-Continental Relay." (The Washington's Birthday Relay had simply sent messages to be delivered to government officials; Maxim's proposal was for a scheduled relay of a single message from coast to coast and back, in the shortest possible time, using the trunk lines.) "Suppose six or eight of us were to get our heads together and work up a plan to handle a message from some point on the Atlantic Coast to some point on the Pacific Coast, and get an answer back the same night," he wrote. "Would it not be something to be proud of in the years to come? Why not try it?" He reminded his readers of the first transcontinental automobile trip, which was "hailed ... as a wonderful exploit," as was the American Telephone and Telegraph Company's first transcontinental telephone call. An equally great honor, Maxim suggested, "awaits those of us in the ARRL who successfully handle the first bona-fide relay message across the continent and receive an answer back the same night." Although such relays were rumored to have taken place, none were confirmed. "There are not as many of these big things to do as there were a few years back," he added, hoping to arouse enthusiasm for such a trial.

He saw success as a definite probability. "Every day," he wrote, "somebody adds something better in the way of transmitting and receiving equipment," and new stations were cropping up even in the sparsely settled West. "Things are being done right now which were impossible at this time last year ... in no art being practiced today is advance so rapid as in amateur wireless telegraphy." Moreover, relay plans had progressed to the stage where success was a probability. He had impressed on trunk line managers the necessity of carrying out directions and adhering to schedules. A new, helpful, but unexpected aid was the increasing policy of amateurs exchanging cards with those hams they had contacted. This was rapidly blossoming into the international use of QSL cards. Relay participants needed to keep in touch with each other, and cards and letters sent by the reliable United States Post Office answered the problem. In the same issue a brief editorial urged subscribers to read Maxim's article and then participate in the challenge of being "PIONEERS in the FIRST AMERICAN TRANS-CONTINENTAL AMATEUR RADIO RELAY."[25]

Experience had already proved the obvious—that it was asking a lot of a volunteer amateur to be at his or her rig at a specified time on two designated days of the week. On Thanksgiving day, a Thursday, of course, one of the two days specified for relaying, one manager reported that "everyone seemed to be eating." Other problems were listed under the general designation of interference, or QRM. Five types were noted: conversation, improper calling, unnecessary repetition, thoughtless testing, and infantile efforts of the beginner with the small spark coil. Static also caused difficulty. In those days, bad weather could make amateur efforts a total failure.[26]

Even so, manager's reports appearing in the January 1917 issue were optimistic. Some of the trunk lines had advanced far beyond the testing stage. One trunk registered forty-six messages sent in ten days; another listed a hundred and ten messages sent, and a third, seventy-six. It was noted that plans were progressing for the coast-to-coast relay, planned to take place in the near future. One of the outstanding relayers was a woman pioneer in amateur radio: Mrs. Emma Candler, 8NH, of St. Mary's, Ohio, operating with a three inch spark.[27]

It was the near future indeed. The first ARRL transcontinental relay was attempted on the night of January 4, 1917, with a repeat the next night. ARRL president Maxim, station 1ZM at Hartford, was the Atlantic terminal. The manager designated to run the test was Mr. Mathews of Chicago; the Pacific terminal was the station of Henry W. Blagen, 7DJ, at Hoquiam, Washington, a small town overlooking Grays Harbor. The message was a question which the Pacific Coast amateur was to answer. Two newspapers had exchanged the question and answer by mail previously, but no one else knew the contents of the message. At 12:20 a.m. Maxim sent a pilot message; ten minutes later he dispatched the real message, which had been delivered to him in a sealed envelope just fifteen minutes before. Maxim knew his call was heard by the next relay, 2AGJ, who then tried to contact a station farther west, Mrs. Candler, 8NH, of St. Mary's, Ohio. But the static was so bad she could not hear 2AGJ and in fact, signals could not be heard north of the 40th parallel. Not a single station to the west could be heard. While the QRN (atmospherics) were terrible, the QRM (interference) was nil, indicating that hams asked by post card to stay off the air had rigorously obeyed instructions. Mathews, the trunk line manager at Chicago, said that the weather was at fault: "In ten years of amateur operating, I have never seen anything to equal it." Attempts were made on Friday and Saturday, but again to no avail. The first projected transcontinental relay was a failure.[28]

Activity in transcontinental relaying was not restricted to the ARRL. On January 27, 1917, took place the first real amateur radio communication sent out from a long distance to a definite address. It was accomplished through luck, it was noted, "trusting to any station who may happen to hear it [to relay the message on], and to hand a message to a definite address through definite relay stations." Two brothers named Seefred, 6EA, of Los Angeles, sent two separate messages from the Pacific Coast across the continent on that date. Both communications were addressed and delivered to Maxim, 1ZM, at Hartford while a Mr. Lindley Winser of Bakersfield addressed a message to Tuska at Hartford (the call letters of these two were not given); it too was delivered. These hams had inadvertently created a new trunk line and just hoped that amateurs would send on their messages. As it turned out, from California they all contacted Captain E.A. Smith, 9ZF, of Denver, who relayed the message to Willis P. Corwin, 9ABD, of Jefferson City, Missouri, who transmitted the message to Kenneth Hewitt, 2AGJ, of Albany, New York, who sent the message to its final address in Hartford.[29]

The significance of these achievements lies in the distances the messages traveled. Just yesterday, it seemed, twenty-five miles had been average and a hundred miles the limit of short-wave transmission. On January 17, 1917, the California hams had sent messages received 850 miles away in Denver; from Denver the message went 750 miles to Jefferson City; from there it was sent 1,040 miles to Albany; and from Albany just 100 miles to Hartford. The accomplishment was called by QST "a great event.... In the next issue," it added, "we shall have some interesting details to give about the working conditions." (War intervened; the follow-up was never published.)[30]

But the urgency for a successful transcontinental relay from coast to coast and return in as short a time as possible remained the top priority with Maxim and other officials of the ARRL. By 1917 the probability of war was on their minds. What would be the fate of amateur radio? They felt that the best way to ensure amateur radio's survival was to demonstrate its uses in time of crisis. What better way than proving that messages could be sent by the amateurs' relay system from coast to coast with a considerable measure of reliability? The atmospherics of January 4, 5, and 6 had been unusual, while amateur cooperation to aid the relay had been excellent. There was every reason to believe that another trial might meet with success.

Almost as soon as failure of the ARRL's relay was confirmed, plans were made for another attempt. District managers used the Post Office and telegrams as well as their amateur radios to alert relayers along the designated route. In the early morning of February 6, 1917, the relay was accomplished. The message left New York City amateur station 2PM (call letters) at 1:40 a.m. It passed through relayers in Cleveland, Jefferson City, Denver, and Los Angeles, returning the same way. One hour and twenty minutes later, at 3:00 a.m., the message was received, again by station 2PM in New York. Of the five relayers involved, three—Corwin, Smith, and the Seefred brothers—had participated in the transcontinental relay to a specific address, as mentioned above. "If we had stated in December that in three months time, we would have handled a message out to the Pacific Coast and back in less than two hours, we would have felt guilty of overstating the possibilities," commented QST, "[but] things happen almighty swiftly in this game of ours."[31]

To further prove that the relay from coast to coast and return was not just a stroke of luck, Maxim ran another relay in early March, 1917. The return message was heard by a young ham in Brooklyn who telephoned the note, which was written by Maxim, to the New York Times. H.P.M. informed the Times that the message had crossed the continent and returned in less than three hours and that the ARRL was now conducting such relaying of messages as a normal procedure. The organization had been handling messages dependably for three months, and in a war, he said, it could furnish the government with three thousand radio hams. At least fifty messages had been transmitted across the continent along one or the other of the three routes. In part the headlines read, "Ready for War Service."[32]

In the May 1917 issue of QST, J.O. Smith, manager of the Atlantic Division of the Trunk Lines, pointed out that most progress in radio had taken place within the past five years, and "the greatest strides amateur radio has ever seen were made in the last few months, when the epoch-making transcontinental amateur relays were successfully handled." The month of March had been the best ever for the relays. And another manager, R.H.G. Mathews, Central District manager, looked forward to an Atlantic-Pacific contact with just one relay.[33]

How is this phenomenal breakthrough into long distance contacts explained? Radio's technological advances had been so drastic that by 1918 worldwide contacts were predicted. Vacuum tubes could be used for both sending and receiving and, due to war needs, were about to achieve mass production. They remained basic to radio until the coming of transistors after the Second World War. In 1916 William Dubilier began manufacturing mica condensers that replaced the bulky glass plate—tinfoil condensers. Among many patents issued during these years, the above developments were of incredible importance.

Awareness of the advances being made, as well as the proliferation of true amateurs—those using transmitters—led the editors of QST to query, "Where are we bound?" The relay message was taking hold. If the government continued its paternal encouragement and equipment continued to improve, there was no telling where the movement would go. "Will private citizens be able to communicate with other private citizens without cost? What will AT&T think of such a development? Will we be able to chat with amateurs in Germany or Japan? If the art advances as much in the next ten years as in the past decade, we may confidently expect the reality of overseas wireless communications to be accomplished." So predicted the editors in February 1917.[34]

It is impossible to say what the future of relaying would have been had the movement not been stopped in its tracks, as was all amateur radio, both transmitting and receiving. On April 6, 1917, war was declared by the United States against the Central Powers. The second page of the May QST bore the statement "war measure," and added "we are at war. amateur wireless is at a standstill. our stations are closed." The journal quoted the official order halting all transmitting and receiving. "As we write," said the editors, "the aerials of tens of thousands of us are being lowered to earth, and our instruments disconnected.... In short, the great amateur wireless advance in these United States is stopped."

This was not the end of amateur wireless. Amateurs were urged to join the army or navy, and to cooperate wherever possible with the government. Even their equipment in many cases was contributed to the armed forces. In fact, amateur radio played a significant part in World War I activities both before and after America's entrance into the conflict.

# 3

## Amateurs During the First World War

In the years prior to America's entrance into the First World War, there was no sharp delineation between the activities of amateurs and professionals in radio. Save for a few British operators who had been trained in a telegraphic school, almost all radio operators had begun their careers as amateurs. Many continued their hobby after taking employment as Marconimen. It was not unusual for those hired as radiomen to have with them, whether on ship or ashore, their own sets (transmitter and receiver) as well as the rig they were using in their professional capacities.

Amateurs during these years were occasionally suspected of espionage or seditious conduct, and were always subject to criticism for interference with commercial activities, while they in turn helped expose the undercover actions of the Germans. Until America entered the war, however, government generally left amateurs alone save for the issuing of licenses and minor attempts to enforce the few regulations imposed by law. (The closing of hams in Texas during General Pershing's foray into Mexico is an exception.) As for commercial radio—meaning private companies serving oceanic or Great Lakes shipping—the government may be said to have brought their activities under control by July 1915. All the while the navy was putting together a wireless network from Maine to Puerto Rico and the Canal Zone, up the Pacific Coast to the Aleutian Islands and ultimately around the world.

Amateurs were nevertheless aware of their precarious position in time of war. They did what they could do to counteract situations that might prompt officials to put a stop to their activities. We have already observed the ARRL's attempts at establishing coast-to-coast relays, presenting the achievement as proof of the constructive, helpful part amateurs could play in time of war. Committees, with members including Maxim and Tuska, traveled to Washington to confer with government officials, urging them to support amateur radio.

The ARRL was not alone. In November 1915, a group of amateurs not affiliated with the organization, but possessing considerable ability at public relations, announced creation of the Radio League of America. One of its aims was to aid the United States government in defense of the country should it be attacked. Little more was heard of the Radio League, however.[1]

Another organization, apparently dead on arrival, was announced in 1916. Guglielmo Marconi tried to organize the National Amateur Wireless Association with himself as president. He also knew something about public relations. He sought and received President Wilson's praise for creating the association, which would help the nation in case of war. The group was not heard from again.[2]

War or no war, amateur radio was expanding phenomenally. In January 1917, QST editorialized that in no previous year had amateur wireless progressed as it had in 1916. Instead of a mere handful of stations capable of working five hundred miles, there were now several hundred capable of working close to one thousand miles. A portent of the near future was Lee De Forest's broadcasting from Monday through Friday at 8:00 p.m., from his laboratory at Highbridge (a residential section of the Bronx), a one-half hour concert using phonograph records. A spontaneous movement was also advancing ham radio: the exchange of cards between hams working each other.[3]

The terrible First World War was the culmination of several decades of growing tension. It is usually considered as beginning on June 28, 1914, in the Balkans, the "powder keg of Europe." On that date a Serbian patriot shot to death the Archduke Ferdinand of Austria. Within a few weeks entangling alliances had fallen into place. Germany's declaration of war came on August 3. By August 4, when Great Britain entered the conflict, it was a full scale World War.

Part of the horror of this conflict lay in the realities of new technology. Never before had poison gas, submarines, and internal combustion engines used for tanks, trucks, motorcycles and airplanes, been so widely used. The Maxim machine gun (invented by Hiram Maxim's father) spewed out bullets killing thousands of soldiers who never touched each other flesh to flesh. Still another of the technologies new to warfare was wireless telegraphy. With the outbreak of hostilities the British cut the German transoceanic cables. But how could the belligerents cancel out each other's wireless? How could countries determined to maintain their neutrality prevent clandestine radios from sending and receiving un-neutral messages?

The greatest and most powerful of the neutrals was, of course, the United States of America. It was the principle potential provider of supplies of all kinds for the belligerents. Moreover (and in spite of millions of first and second generation citizens whose mother country was one of the Central Powers), the nation was primarily pro–Ally, pro–British and pro–French.

Nevertheless, President Wilson immediately proclaimed America's neutrality. The concept was not new to the nation: America had proclaimed its noninterference in a treaty with France in 1778, had gone to war with England over the preservation of the nation's neutral rights in 1812, had participated in the two Hague Conventions (1899 and 1907) and had been a participant in the writing of the Declaration of London in 1909, all three of which touched on problems of neutrality. Of particular importance regarding communications was the Hague Convention of 1907 which recognized the existence of wireless and paid some attention to its role in time of war, especially when it involved neutral nations.

The trouble was, the convention's statements simply exacerbated the problem. As stated in its Article VIII, "a neutral power is not called upon to forbid or restrict the use on behalf of the belligerents of telegraph or telephone cables, or of wireless telegraph apparatus belonging to it or to companies or private individuals." This would seem to have tied America's hands: "There are no rulings, no decisions available for reference in questions involving international law," noted an article in the *New York Times*. "We have therefore no precedent for such a question as the right of censorship over wireless dispatches."[4]

From the very beginning it was clear that American policy was in troubled waters. Decisions would have to be made without the backing of precedents. Could a belligerent maintain a wireless station in the United States from which it conveyed information to its warships or submarines concerning ship departures? And if it was allowed to do that, what other sensitive information might it be allowed to convey to its military, naval, and civilian authorities? And what if the station was privately owned?

The two leading nations in radio were Germany and the United States, although France, England, Italy, and Japan were not far behind. The freedom granted radio in America had resulted in the rise of thousands of radio enthusiasts. Ninety percent of them had simply built basic receivers, with antenna, ground, and headphones, and listened to code from wherever it originated. But a small percentage were so intrigued that they mastered Morse code and built or purchased transmitting sets. So good were some of these amateurs that, in the case of a North Dakota ham, one night when conditions were just right, through his earphones came the great German station at Nauen. By 1916 transcontinental contacts were occasionally taking place with no relay in between. If the existence of radios—both receivers and transmitters—did constitute a risk in wartime, then the American problem was acute. There were more hams in the United States than in all the rest of the world put together.

But the commercial stations were the real problem. The Marconi network was the largest, legitimately in contact with ships at sea which were equipped with Marconi apparatus. There were smaller companies; the United Fruit Company, for example, managed its own stations. And there were stations owned by Germany although private ownership was claimed.

Germany troubled American officials from the very beginning because its radio technology was so advanced. It had flaunted its expertise in January and February 1914, heralding the opening of two powerful stations in Germany, one at Nauen, twenty-five miles west-northwest of Berlin, and the other at Eilvese, near Hanover. They had sent messages to stations they controlled in America, which were then conveyed by telegraph, first to President Wilson and next to New York's mayor and some of that city's newspapers. The first message came though on January 27, 1914. It was sent by wireless from Eilvese to the president by way of their powerful station at Tuckerton, New Jersey, a small town about sixteen miles north of Atlantic City. Without a relay, the Germans boasted, the message had crossed 4,062 and ½ miles. But the facts were disturbing. The American station, it was reported, was built "with every effort to keep its whereabouts a secret and to keep the name of the company erecting it from becoming known." More troubling, especially to anyone concerned with matters of national defense, were statements released about how the station could send secret messages by adjusting its apparatus to different wavelengths. Were Americans aware of the potential danger implied by this boast?[5]

Barely two weeks later, on February 11, a thorough test was made of two-way wireless between New York City and Berlin (by way of their stations nearby). The American station receiving these messages was located not at Tuckerton, but at Sayville, a small resort village on the south shore of Long Island. Here the German Telefunken Company, operating under the guise of the Atlantic Communication Company, had erected a high, single tower, with a small operating building near its base. Both the Tuckerton and the Sayville installations communicated with the powerful stations at Eilvese and Nauen.[6]

And then war came. On August 5 President Wilson issued a sweeping order involving commercial wireless; amateurs were not directly involved. To enforce his orders of neutrality insofar as radio communications were involved, his proclamation stated in part "that all radio stations within the jurisdiction of the United States of America are hereby prohibited from transmitting or receiving for delivery messages of an un-neutral nature, and from in any way rendering to any one of the belligerents any un-neutral service during the continuance of hostilities."

Enforcement was placed in the hands of the navy with Assistant Secretary Franklin D. Roosevelt in charge. He announced that censors would be stationed first in the foreign owned radio stations but later in American stations as well. Possibly all amateurs along the seaboard would be asked to remove their roof aerials. The Department of Commerce was to notify the stations of the executive order.[7]

The rapidity with which this executive order was issued arouses suspicion that authorities had been alerted to the possibility of neutrality violations. Both the Sayville station and the Tuckerton station were suspected of violating neutrality regulations. L. Batterman, in charge of the Sayville station, had admitted to reporters that the station had received "about thirty" messages from Germany which it had relayed in code to the German cruisers *Dresden* and *Karlsruhe*, which were in the western Atlantic. An amateur operator on Long Island had intercepted and interpreted messages to the two cruisers; he considered the communications un-neutral. A Columbia University professor of international law, Dr. George W. Kirchwey, pronounced the sending of such information, whether by Germans, British, or French, a violation of International Law.[8]

Thus began months of controversy and indecision as to how to deal with commercial short-wave radio stations, most especially (but not exclusively) Sayville and Tuckerton, which were used by the Germans. Decisions were made and rescinded and made again. The British, determined to eliminate Sayville and Tuckerton, were deeply involved, as were the supposedly private owners of the two stations and their German diplomat defenders.

From August 4, 1914, when England entered the war, until at least the sinking of the *Lusitania* (May 7, 1915), the sure hand of British interests is evident in the enforcement of American neutrality, especially with regard to wireless. Of course they wanted the Sayville and Tuckerton stations shut down. They also wanted stamped out the clandestine stations in every neutral country where such stations might be located, and which were aiding their enemy. The Germans had the edge on them with wireless when the war began, but the British set out to destroy that advantage and to a remarkable extent, they succeeded.

First, unbeknownst to the Germans, the British had captured their three principal codes, but were still confronted with the enemy's ability to transmit that code around the world. As early as August 7, 1914, it was reported that the Germans had "a wireless chain" making it possible for headquarters to contact German warships anywhere on the seven seas. Just how their wireless surmounted enormous distances was a subject for speculation, but the British were certain that clandestine stations on the Chilean mainland, and quite possibly elsewhere in the world, could be used to send messages westward across the Pacific.[9]

The code, which they considered undecipherable, plus worldwide communications by wireless, placed German naval operations on an equal if not superior footing with Britain and the Allies when the war began. When war broke out the German navy was dispersed around the world. She had built an up-to-date fleet of warships, many of which were faster and more heavily armed than British warships. Orders could be, and were, sent from Germany and received by Admiral Graf Spee and his modern fleet anchored at Tsingtao, China. Spee's fleet weighed anchor as soon as war broke out. The flotilla, probably with orders received by coded wireless, headed for a destination known only to the Germans. The British suspected that it was bound for the west coast of Chile. England carried on a brisk trade with Chile and Peru, obtaining nitrates, copper, food, and other supplies now more than ever necessary. Aware of the peril to its shipping, England sent a fleet led by Admiral Sir Christopher Cradock up the west coast of South America. His two heavy cruisers, the *Monmouth* and the *Good Hope*, were outdated, outgunned, and slow; even their crews were fresh and untrained. Cradock knew that if he met Spee's fleet, the most he could hope to accomplish was damage to the Germans while his own ships would probably end in Davy Jones' Locker.

The British had guessed right. It is a long way across the Pacific. The Kaiser's fleet put in for fresh water and food at Easter Island. It was inhabited by a few hundred natives and three white men. There was no radio on the island. The twelve German ships remained there for several days. When they weighed anchor they headed east, just as the English suspected, to harass British shipping along the Chilean coast. Spee was in contact with German merchant vessels that undoubtedly gave him information about Cradock's fleet, as well as from wireless communications from on-shore Chileans who, violating Chilean neutrality and heavily favoring the Germans, gave still more information. The result was a naval battle on November 1, 1914, near the central Chilean port of Coronel. Cradock's weaker fleet was defeated in a brief, deadly encounter. The *Monmouth* and *Good Hope* were sunk and 1600 British sailors were lost.[10]

Spee then ordered his fleet around Cape Horn and into the South Atlantic. Near the Falkland Islands, on December 8, 1914, his armada was defeated by the British under the command of Admiral Frederick Sturdee. Spee went down with his ship.

Confronted with a war in which the enemy's ships, dispersed world wide, were keeping in touch with headquarters in Germany by wireless, the British set out to do all possible damage to the enemy's communications. Through their embassy in Washington, they were quick to notify American officials of possible clandestine stations in, for example, the Maine woods.

Then the British charge d'affaires, Colville Barclay, informed someone in authority that the Tuckerton station was operating without a license. In conveying this information Barclay was emphasizing the necessity of the navy continuing a stringent censorship over the station. In spite of the cogent arguments of the Germans, on the 24th of August, 1914, the government closed the Tuckerton facility. With its eight-hundred-foot-high antenna it was believed to be the only station capable of contacting Germany without a relay.[11]

Early in September both England and Germany agreed to President Wilson's suggestion that coded messages could be sent provided the censors had read and approved them as neutral communications. As the warfare became more intense, however, it was clear that the situation was fluid and bound to change.[12]

American firms, trading legally with one or the other belligerents, needed quick communications with their customers overseas. To close both Tuckerton and Sayville, with their long-distance capabilities, would put a serious crimp on commercial activities involving international trade. What to do?[13]

Wilson's solution was to reactivate the Tuckerton facility, which had been taken over by the navy. The government would use it for transmitting coded messages and would accept, for a fee, transmission of commercial messages. These included coded or ciphered communications, provided they were of a neutral character and had been first examined and approved by a censor. No messages in foreign languages or in unintelligible terms would be accepted. The original owners would be compensated for any profit made by the government through the use of any stations, such as Tuckerton, that might be taken and run by the navy. When the war ended, the government had earned over $400,000 from this business.[14]

Still revealing a fear that the Germans were sending and receiving un-neutral messages, in a few days the authorities changed the regulations still more. The navy men stationed as censors were instructed to paraphrase messages and limit to twenty-five the number of words per message. Official messages either sent or received were to receive priority. The cost was fixed at twenty-five cents a word.[15]

If the British were chuckling over the restrictions placed upon the Tuckerton plant, they lost their smiles rather quickly when the Marconi station at Siasconset, Massachusetts, was likewise ordered to abide by the same restrictions. Officials had proof that the station had received a radio message from the British cruiser *Suffolk* and had relayed it on to New York City. This was in clear violation of instructions issued by the government early in August. On September 24 the navy arrived and closed the station. Hardly two weeks later the Marconi station in Hawaii was closed because, in the absence of the censor, it had sent a message announcing the arrival of the German gunboat *Geier* at Honolulu.[16]

Within two months following the outbreak of the First World War, the United States had shown both concern and understanding, but also indecision, in its actions toward privately owned commercially run wireless stations in the United States. It had placed knowledgeable navy men in charge of the powerful installations and ordered a rather strict censorship. It still allowed commercial and even personal messages to be sent and received, albeit under what seemed strict conditions. Not until April 7, 1917, with war declared, did the government seize all commercial stations, closing many but keeping some operational for use by the navy.[17]

With the outbreak of war the question of amateurs as liabilities or assets arose among government officials. If an illegal station was found, it was likely to be manned by an amateur. After Cradock's defeat at Coronel the British stepped up their campaign to get rid of clandestine or enemy radio stations wherever they might be. They sent letters to key officials offering rewards for information leading to stations transmitting messages to the Germans. They alerted American officials of their suspicion that there was an enemy station in the Maine woods, possibly in Florida, or in Mexico or Chile. They could be transmitting information to the Germans about the movement of warships and commercial vessels and cause enormous harm. They felt that the Germans off the coast of Chile had possessed information about Cradock's fleet that could only have been obtained through stations extending all the way from Maine into Central and South America.[18]

The United States Navy took particular interest in these reports. Secretary of the Navy Daniels, Secretary of State Lansing, and members of the Secret Service discussed the problem at length. They took steps to increase surveillance.[19]

Rumors arose on more than one occasion about a clandestine station in Mexico. One of them was said to be established at the town of Alamo, a mining community sixty miles inland from the Lower California port of Ensenada. Germans were rumored to have landed there from the German cruiser *Nurnburg* carrying with them special equipment which, they said, was for gold prospecting. "The station ... is one of the several stations from which, according to British officials here, German cruisers in the Atlantic and Pacific have been receiving information concerning the whereabouts of English vessels," reported the *Times*. It was said that Mexican authorities had ordered the operation closed.[20]

A hidden station was reported to be somewhere in the woods around Tampa, Florida. The story had Germans operating a powerful station at a sawmill. Three operatives were said to have been landed from the German cruiser *Karlsruhe*. They had represented themselves as timber men from the north, and had even purchased a considerable acreage of forested land. Local amateurs reported hearing unintelligible messages sent in cipher. The rumor was investigated, and the conclusion was that the strange code was that of a boy learning Morse code.[21]

Maine, Florida, Mexico, Central America, Chile: the British and French were concerned that the Germans had a network of stations conveying information to them throughout the Western Hemisphere. So concerned were they that they made formal requests of Secretary of State Bryan that he step in and pressure the Central and South American nations—all of which proclaimed their neutrality—to take active steps to prevent wireless stations from using their countries to help the Germans. They invoked the Monroe Doctrine. Since it prevented them from applying pressure, it was the duty of the United States, they contended, to do it. It was a delicate situation for Secretary Bryan, being aware as he was of the Latin American nations' sensitivity at being told what to do by the Colossus of the North. As would be expected, the countries denied aiding the Germans.[22]

It was suggested that two wealthy brothers who summered on an estate in Maine, Ernesto G. and Alessandro Fabbri, might be guilty of helping the Germans. Alessandro was known to have one of the most sophisticated amateur stations in Maine. Although they had an Italian name, the brothers were wealthy Americans. The town clerk of Bar Harbor reported Alessandro Fabbri as being "pronouncedly pro–German in his sympathies." To determine if there was a clandestine, pro–German amateur station in Maine, manned by the Fabbris or someone else, the Secret Service obtained a list of all licensed amateurs in that state. Twenty-one stations were investigated, of which five were in Bar Harbor. The assumption was that an amateur, whoever he might be, received British messages, then put them in code and sent them by telegraph to German recipients in the United States or Canada and from there the intelligence was conveyed to the German military.[23]

When the war began, the Fabbris came under suspicion because a German liner, the *Kronz-*

*prinzessen Cecelie,* took refuge at Bar Harbor with British cruisers in hot pursuit. Shortly thereafter the ship was moved from her first mooring to one in Frenchman's Bay, opposite the Fabbri estate. The brothers' motorboats were said to have made trips to the German ship, often at night. The captain of the *Kronzprinzessen Cecilie* was entertained by them. They had brought the crew ashore, 250 at a time, served them food and paid their admission to the Bar Harbor Star theater where motion pictures of the crew, taken by one of the Fabbris, were shown; each sailor was given a cigar when he left the theater. The ship's band had played on the estate lawn. Rumors detailed how the brothers, following the ship's arrival, had entered the ship's wireless room, locked the door, and had not emerged for three days. Another Bay Harbor amateur, H.A. Lawford, who was employed by the navy to send coded messages to American naval vessels, said that he was sure the radio aboard the *Kronzprinzessen* had not been sealed. He was certain that the Fabbris could hear messages from Germany, because other local amateurs with less powerful stations had heard them.[24]

The Fabbris, who shortly after the story broke closed their Bar Harbor estate for the winter, vehemently denied the accusations. Ernesto wrote a letter to the *Times* protesting their innocence, complaining of misinformation, of the embarrassment it had caused them, and including supporting letters from officials at the Department of State, the Navy Department, and even from Cecil Spring-Rice, the British ambassador to the United States. All the letters confirmed that they had not been singled out for investigation, that all wireless stations in Maine had been examined. The Fabbris were never cited.[25]

Meanwhile the war at sea was becoming more desperate. With the German U-boat campaign expanding while the British blockade of Germany was tightening its grip, it was just a matter of time before something happened at sea to anger the greatest of neutrals, the United States. It can be argued that the first incident, in which one American was killed, was due to wireless.

On Sunday afternoon, March 28, 1915, the British liner *Falaba,* bound for the west coast of Africa, was confronted just south of St. George's Channel (between Ireland and Wales) by a German U-boat. The submarine had surfaced and its captain had given the passengers and crew a few minutes to disembark before the U-boat sank the *Falaba.* Captain Frederick J. Davis dawdled, all the while urging the Marconiman, named Taylor, to tap out SOS. He was reportedly ordered by the Germans to stop working his transmitter, but he persisted. His message was heard. A ship appeared on the horizon. For the Germans, this was sufficient reason to destroy the *Falaba,* for on the surface U-boats, with their thin armor, were nearly defenseless against even lightly armed enemy vessels.

So the *Falaba* was torpedoed with a loss of over one hundred lives. Among those killed was an American, Leon C. Thresher, about thirty years old, on his way to the Gold Coast where he was employed as a mining engineer. Had the Marconiman not followed his captain's orders, had no ship received the SOS and appeared on the horizon, the U-boat captain would probably have allowed the extra time for the passengers to debark before torpedoing the *Falaba.*[26]

Barely two months later, on May 7, 1915, the *Lusitania* was torpedoed and sank in eighteen minutes. When the torpedo hit, the senior wireless operator, Robert Leith, "bolted from the Second Class Dining Saloon, through the passageways and up the ladder to relieve his junior assistant at the key." The Marconiroom was on the hurricane deck under a mass of wire antennas. Within seconds he had tapped out, almost reflexively, "Come at once. Big list. Ten miles south of Old Kinsale. MSU" (the *Lusitania's* wireless sign).

Leith repeated his call over and over again, even as the ship's electric power was weakening. When the ammeter needle fell to zero, he got up from his chair and switched on the emergency dynamo, powered by its own batteries. His message became more desperate. "Send help quickly. Listing badly." Leith is said to have stuck to his job, sweating profusely, even as the list became more pronounced and water began coming in. He finally gave up and set about saving himself. He was later taken alive from a half-submerged lifeboat. Robert Leith was received at Court in honor of his persistence in manning the transmitter.[27]

One thousand, one hundred ninety-eight lives were lost (or 1,195, or 1,200—the precise count varies), of which 128 were Americans. "It was the Concord Bridge of 1915, a rallying cry like the 'Alamo' and the Maine," according to one history of the disaster. Certainly it helped cement negative American attitudes toward Germany. The tragedy has been the subject of almost as much controversy as the *Titanic* disaster.[28]

Was wireless involved in any way in the torpedoing? Two respected historians, Thomas A. Bailey and Paul B. Ryan, describe a book by John P. Jones entitled *The German Spy in America* in which Jones suggests a scenario that, write Bailey and Bryan, "is irresponsible fantasy." Jones wrote that a Sayville operator picked up a message sent to the Admiralty concerning the *Lusitania* and in an admittedly twisted and rather bizarre fashion, sent orders that ultimately were received by the *Lusitania's* skipper, causing him to veer into the area near Old Head of Kinsale, where two German submarines lay waiting to torpedo the passenger liner.

Probably Jones' explanation of the torpedoing is fantasy, but for Jones' detractors to call it ridiculous reveals a lack of knowledge of the miracle of radio. When conditions in the ether are just right, messages that theoretically can be heard just a few hundred miles away are often picked up a thousand or more miles away. Wireless *probably* had little or nothing to do with the fatal meeting of the *Lusitania* with Kapitan Leutnant Schweiger's *U-20*. The possibility, however, no matter how remote or bizarre, should not be ruled out.[29]

What was the status of the Tuckerton and Sayville stations as of April 1915? The high powered Tuckerton facility had been taken over by the navy, but the powerful Sayville station, which was being used by the German embassy, continued to operate, albeit with navy censors on duty there. In April 1915 the Telefunken people quietly began increasing its power. Three 500-foot towers were being assembled and the wattage was being increased from thirty-five to one hundred kilowatts. All the new equipment, possibly even the towers, had been manufactured in Germany, transported to Rotterdam, placed on board a Holland-America liner, carried to Hoboken, and from there moved by rail in six large freight cars to Sayville. Because static was so bad in the summer, it was felt necessary by the Telefunken officials to increase the power of Sayville so that it could overcome the static and be in constant touch with the station at Nauen. The station did not yet have a license for its increased power, but it was expected to receive one.[30]

Certainly those searching for reasons of the fatal meeting of submarine and passenger ship saw the transparent force of wireless as a key factor. As early as May 10, just three days after the tragedy, suspicion was being aimed at the Sayville station. As a result of a meeting of three experts, including Frank J. Sprague, a former president of the American Institute of Electrical Engineers, a call was made to President Wilson requesting that every message sent in the past few weeks either to or from Germany be turned over to the president, translated if in code, and studied by his experts. They called attention to the German embassy's warning in the New York papers about the risk of sailing on the *Lusitania*, and suggested that the Germans had been made aware of sailing time and the route of the great liner.[31]

In London it was said that the Germans must have known of the inevitable torpedoing of the great passenger liner. If not, it was asked, how else could they have distributed "bulletins with huge headlines" so soon after the disaster? "There is no doubt," it was said, "that the German Admiralty received a wireless message from New York, giving the date of the *Lusitania's* sailing, and that a number of submarines were ordered to torpedo her at any cost."[32]

Proof was lacking, but the Sayville station became a source of mistrust, especially after its increased power and new towers became operational. Someone reported that Professor Jonathon Zenneck, a German wireless expert, was staying at a Sayville hotel and was suspected of advising the Sayville staff on the station's operations. The press found him a rather mysterious person. He was variously described as being a captain in the German submarine service, a member of the German marine reserve, a German Army man, and as a radio expert here to testify in a court case involving the Marconi Company. It was even said that he was here when the war broke out and had not been able to return to Germany. The professor was quite well known in America. Zenneck's *Wireless Telegraphy*, priced at $4.00, had been advertised in *QST*.[33]

Enter now Mr. Charles P. Apgar, an adult radio ham who possessed state-of-the-art equipment and also indulged in experiments. By sheer chance he knew W.J. Flynn, chief of the Secret Service. Apgar, who lived in New Jersey, showed Flynn his rig, which included a "loud talker-horn" (a loud speaker) and a phonograph using round wax cylinders to record code, music, or voice. Flynn immediately saw this as an opportunity to study in depth messages being sent to Germany from WSL, the Sayville station. "Could you listen in every night for a couple of weeks?" he asked Apgar, "recording everything you hear on the phonograph cylinders?" As Apgar recalled in an interview at the N.B.C. Museum in 1933, he willingly complied with the suggestion.

For two full weeks starting on June 7, 1915, Apgar tuned in every night at 11:00 p.m. and listened until WSL was through sending messages, sometimes as late as 3:00 a.m. He filled 170 cylinders with messages sent out by the Sayville station. The Secret Service, analyzing the recordings, found direct evidence that to get past the censors the Germans were using coded messages as well as messages in English. For example, orders were placed for items that were easily obtainable in Germany; or it was requested that messages be sent and re-sent, until a pattern appeared. "It was really his absolutely faithful records of all of the signals sent out from Sayville that caused the United States to seize the famous station," Flynn said.[34]

Seize it the United States did. The legal issue that gave the government the right to take over Sayville was the application for a new license to cover the more powerful station. Dr. Zenneck's employment at Sayville was also noted, although at the time no accusations were made. The Providence (Rhode Island) *Journal* was far more positive in its accusations than the *Times*. It came right out and said that Count Von Bernstorff and Captain Boy-Ed, the naval attaché at the German embassy, had used code to convey all kinds of un-neutral information to Germany. The *Journal* had likewise kept track of messages, had turned them over to federal authorities, and noted that "the documents prove conclusively that the Government has been consistently fooled by the German Embassy at Washington and under the very eyes of the censor the German ambassador has violated every obligation of neutrality."[35]

With the Sayville acquisition on July 9, 1915, the navy proceeded to operate it as it had been operating Tuckerton, sending commercial messages at a fixed price overseas. The new operators found the station so powerful that it could send messages by day as well as by night, and in a short time it was sending out three times as many messages as had been sent when it was under Telefunken control.[36]

A final example—and there were certainly others—of hams helping search out German operated wireless stations occurred after America's entrance into the war. Mention has already been made of rumors of a station in Mexico being closed early in the conflict. In 1917 the British were sure that another pro–German wireless station was operating there. This led naval intelligence in San Francisco to ask Harden Pratt, a young man employed as an "Expert Radio Aide" at the Mare Island Navy Yard, to try to find the whereabouts of the station. Pratt was an amateur and also a professional. He had worked as a radio operator on ships. With two naval officers assigned to him, he constructed four direction-finding receiving sets. One was stationed at Point Loma near San Diego, one in a barn on a bean farm near Los Angeles, one at Phoenix and one at El Paso.[37]

In conducting their search, Pratt and his aides at the bean field station had an unexpected call. One of the navy men had obtained a Western Electric amplifier from the Wilson Observatory laboratory at Phoenix. "It squealed loudly when you did not want it to," Pratt recalled, "and on the second night about four in the morning we were ordered out with our hands up at the point of a gun." The bean field farmer had become suspicious of those strange noises. Of course, explanations were accepted and the army men and the radiomen all enjoyed a good breakfast in the local town—but the farmer was not invited.[38]

Through triangulation, Pratt was able to determine that the wireless was operating in Mexico City. In a letter to the Mare Island officials, the American ambassador to Mexico gave details of the station, and its closure. He said that he had watched it being built in Chapultepec Park while on his morning horseback rides. Later it was found that C. Reuthe, the German manager of the Sayville station, and his Polish engineer had fled to Mexico the day before the United States declared war on Germany. The rig had been smuggled through the British blockade on a German ship. Mexican authorities had subsequently ordered the facility closed.[39]

Although amateurs were not involved, a wireless message in code intercepted by the British in January 1917 and released to the United States government in February certainly hastened America's entrance into the war. This was the Zimmerman Note, a foolish message, sent in code from Neuen, offering Mexico a return of what it lost to the United States in the Mexican War if Mexico would declare war on the United States. The Germans' use of Neuen to send code that they did not believe had been broken was a serious mistake on their part.[40]

In April 1917, the month in which the United States declared war on Germany, *QST* was able to announce that in February it had sent a relay from New York City to Los Angeles and return in just one hour and twenty minutes. Fearful that all radio might be closed down, the editors suggested that amateurs' appetites would just be whetted if they were forbidden to operate. It also noted that there were 5,425 licensed first class grade and special station operators in the United States. Several articles in the April issue, which was printed prior to the outbreak of war, were written with the ominous sense of impending conflict and the probable closing down of all amateur stations.[41]

Came the Declaration of War, April 6, 1917. In the May issue *QST*'s leading article began by quoting in entirety the order from the Department of Commerce, "To All Experimenters," directing them to the

> immediate closing of all stations for radio communications, both transmitting and receiving, owned or operated by you.... I direct that the antennae and all aerial wires be immediately lowered to the ground, and that all radio apparatus both for transmitting and receiving be disconnected from both the antennae and ground circuits and that it otherwise be rendered inoperative for both the transmitting and receiving any messages or signals, and that it so remain until the order is revoked. Immediate compliance of this order is insisted upon and will be strictly enforced.

All efforts to convince the government of the usefulness of amateur radio in time of war, with relays, cooperation with government officials, and donations of equipment to the military, were of no avail. Lightning had struck, as *QST* described it. "All plans for improvements are cancelled. All the plans of our manufacturers are in mid-air." The editors had thought about it a lot during the previous months, they wrote, "even to the holding of special Directors' meetings, drawing up resolutions and sending a Committee to Washington to appear before the Secretary of the Navy."[42]

Readers were also informed that Maxim and Hebert had been called into consultation with authorities who asked their help in securing operators and equipment for wireless listening facilities along the coasts. The lead editorial urged amateurs in good health and free from dependents to join up. Both the army and navy had made the situation for wireless operators "more agreeable than any military service ever known.... To those who do not [enlist] it will be one of the big regrets of their lives in the years to come."[43]

By June the initial shock of the total closure of amateur activity had worn off. A bit of sarcasm and criticism crept into the editorials and the Old Man's comments. "If anyone wants to know the candid opinion of Yours Truly about this amateur close up business," he wrote in anger, "let him be advised right here and now, that he considers it ROTTEN.... Do you suppose Mr. Wilson knew the awful results that would follow...? Ten to one he never thought of it, but just went ahead and did what some sore head said he ought to do." The Old Man went on to ask, what harm would we do listening in?[44]

He emphasized the ban's absurdity by relating the discussions at the regular meeting of the local radio club. (Many of his essays were placed at a local ham club where heated discussions took place.) If the American flag was hung from the mast, did that make the aerial legal? Because the aerials and grounds had to be dismantled, how about hitching the radio to the electric door bell system? That worked with one of the hams. Another hooked his rig up to the electrical wiring in his house, and that worked also until night came on, and his family needed light. Careful, though! One of the club members had discussed the radio ban with the local chief of police who warned that the gendarmes would be after them, as well as the Secret Service. Better dismantle and seal and put away for the duration! The one good aspect of the ban was this: the local radio clubs were more determined than ever to stay together so that when the war ended and victory was won, they would be ready to operate again, with the latest available equipment.[45]

Sarcasm and suggestions for getting around the ban did not prevent QST from recruiting amateurs for the armed services. One full page of the June issue urged "bugs"—as amateurs were being called at that time—to join up. The journal was especially strong for the navy, whose naval reserve concept, in which the amateur signed up solely for the war's duration, carried great appeal. The lead essay in the July issue bore the title "WANTED: BY UNCLE SAM: 2000 AMATEUR WIRELESS OPERATORS." "You must not fail [Uncle Sam] in his hour of need," it stressed. It then described the navy's radio schools, one at Harvard, one at the Brooklyn Navy Yard (later moved to Columbia University) and one in San Francisco at Mare Island. Even more to the point, QST included an application blank to be filled out and taken to the amateur's nearest recruiting station. The August issue's lead article was submitted by the Navy Publicity Bureau: "What the Naval Reserve Offers Men of the ARRL." Space was also given to letters from amateurs in the service describing their experiences at the Great Lakes Naval Training Station and the nature of their subsequent assignments. It was noted that "even our lady wireless wizard, Mrs. Candler of 8NH, St. Mary's, Ohio, has enlisted in the Naval Reserve, and is daily hoping that she be assigned as operator to one of the Great Lakes Stations." The army's Signal Corps was also mentioned, although the editors acknowledged that they did not have as much information on it. No question about it, QST did genuine patriotic service in encouraging amateurs to sign up.[46]

In retrospect, the government's orders to take down all aerials, remove all grounds, and seal all transmitters and receivers seems utterly ridiculous. It struck H.P.M. that way and thousands of amateurs when the order was issued, and it certainly seemed that way by the early 1920s when radio receivers were being sold by the millions. When Maxim wrote that some "sore head" had got to President Wilson with the absurd idea, he meant it! His disgust crept into his writings several more times in the remaining months of QSTs 1917 issues.

In the last issue (September) before QST ceased publication he expanded on his beliefs. In discussing German espionage using wireless in America, the Old Man suggested that hams could ferret out clandestine stations. "Why isn't there a place for us stay-at-homes with the flat hands and square feet?" he asked. "Did it ever occur to any of you bugs that maybe we could be a big help in locating enemy wireless if we could satisfy the Government that we could be trusted?" He then boasted about amateurs' receiving sets, some of them better than the navy's best. "The junk [the amateurs used] would make a Government official sick to his stomach, but it works in our hands," he boasted. "We know how to use a match on an audion bulb or where to locate a magnet and how to monkey a motley array of home made tuners and variables until wonders and miracles are performed.... Not one ordinary paid operator in a hundred will scratch as hard as most amateurs to read faint signals.... Maybe the trouble is that the Government don't [sic] see any way to trust us. If that's the obstacle, isn't there some way we could be sworn in, or give bonds or oaths or some darned thing which would register us?"[47]

What is equally remarkable is how stringently the ruling was enforced, at least initially. Many an amateur believed that, with his aerial and ground removed, he could continue to experiment, but it was officially stated that "radio apparatus cannot be used for experimenting at all." Doctors could continue using induction coils for X-rays, but "all experiments involving inductance and capacity and such power and voltage as would justify the experiments being called 'radio telegraphic experiments' are forbidden.... In short, all experiments usually performed by amateurs must cease." Thus wrote a lieutenant commander for the director of Naval Communications.[48]

The journal editorialized unhappily that "the lid is on tighter than we supposed.... The facts are that we must not touch any radio apparatus. All we can do is to read radio books and think radio thoughts. Until we become Germanized, we at least have these liberties," the editors wrote. "Of course, all radio inventing must stop if anything more than a lead pencil and a piece of paper are involved."[49]

In the atmosphere of extreme patriotism and xenophobia that characterized the war effort of 1917–1918, the reporting of radios said to be in attics or basements or closets reached proportions one usually links with a dictatorship . Neighbors spied on neighbors and no one was above suspicion. Policemen, sheriffs, and secret service agents were informed of suspicious activities and found themselves knocking on doors where someone had just installed an electric door bell, or where a teenager had a buzzer connected with a telegraph key and was practicing Morse code.

In Buffalo, New York, the sheriff received orders from the governor to dismantle all radio stations in the county. He assembled his sixty deputies, enlisted the police force also, and set about sweeping the city of aerials and rigs. On April 16 a wireless station was set up at the federal building in Buffalo to catch amateurs foolish enough to be active under the circumstances. By April 17 it was reported that more than three hundred radio stations had been dismantled. As sort of an afterthought, it was suggested that hams fly Old Glory from the top of their masts, in lieu of their aerials, until peace came.[50]

How thoroughly the order was enforced throughout the United States is open to question, but probably obedience was widespread. In New York City, where an estimated three thousand radio stations were said to be in existence, authorities set to work immediately and by April 9 were said to have already dismantled eight hundred stations. Statewide, there were believed to be as many as four thousand stations. A few days later it was reported that of 1,028 stations found, just eighteen had resisted the sealing of their rigs. In St. Paul, Minnesota, an ex-governor, John Lind, protested but subsequently allowed his radio to be dismantled.[51]

Throughout America's involvement there were occasional reports of secret radio stations discovered and seized: one at Norwich, Connecticut; the discovery of a radio outfit in the tower of a building on Broadway in New York City; a seizure in Columbia Heights, Brooklyn; and the arrest of a German as a "wireless suspect" at his farm near Englishtown, New Jersey. On May 29, 1917, Professor Jonathon Zenneck was arrested and interned for the duration of the war. His arrest caused considerable interest.[52]

This was the same Professor Jonathon Zenneck who had been found occupying a hotel in Sayville, who was said to be an officer in some branch of the German military, and who was such a brilliant radioman that one of his books was sold by QST. "The Government looks upon him as one of the most dangerous German subjects in this country," noted the Times. He was described as "one of the enemy's most skilled wireless experts." After warning of the dangers inherent in the enemy's use of wireless, the Times suggested that the government should lay out a plan for utilizing amateurs in the detection of espionage.[53]

QST's Old Man just couldn't keep from writing a column about Zenneck. "Our Department of Justice would never have selected Dr. Zenneck to occupy a front seat in the cooler for the rest of the war if the old gent had not been monkeying with the oscillatory business somewhere," T.O.M. speculated. "Zenneck," he added, "has been one of the world's authorities on radio. He was the big boss at WSL, Sayville, it seems to me." Maxim then stated matter-of-factly how the Germans from Sayville "sent plain war dope to their German ships in the Atlantic.... We amateurs all noticed it." Especially Apgar, who "gave the biggest lift an amateur ever gave the Government. Apgar, the old sleuth, smelled something rotten just about the time the rest of us did." T.O.M. speculated that Zenneck had developed a system of sending messages to Germany without American knowledge." Hmmm. "After we make up again with old Z, let's ask him into the ARRL. He's probably OK. His trouble is with the Kaiser."[54]

Meanwhile hundreds of young men were enlisting, attending the navy radio schools. QST even ran a few letters from hams in the service. Bill Woods, 9HS, wrote, "It is 1:00 a.m. and I am sitting—as I did almost every evening last winter—at my wireless set with the phones on. However, in this case it is not my set, but is the Marconi station at Manistique, Michigan, recently taken over by the Government, call letters WMX. I am a 1st Class Electrician ... and on the midwatch. That is, from 12 midnight to 6 a.m."[55]

It is estimated that eighty percent of amateurs served in the armed forces during the First World War, most of them in France. That figures to about four thousand men. On November 11, 1918, the Armistice ended the war. Amateurs looked forward to a renewal of activities. And of the amateurs who had served in the armed forces, a high percentage looked forward to practicing their hobby of amateur radio in peace time.

But ... not so fast! Trouble was brewing.

# 4

## Amateurs Between
## War and Peace

The war was over, but peace still had to be made. President Wilson decided to leave the United States and attend the peace conference in person. A German ship, captured during the war and renamed the *George Washington*, was to carry the president and his entourage to and from France. To ensure the president's contacts with Washington, officials installed on the vessel a powerful state-of-the-art transmitter and receiver. Thirty-five men were considered necessary to send and receive messages and keep the equipment operational.

The radioman placed in charge was all of twenty years old. His name was Don C. Wallace, at six feet four inches a hefty, handsome young man, born in Minnesota but raised in Long Beach, California. From childhood Don had shown an interest in football and ham radio. His interest in electronics, and radio, had been whetted when he found a physics text in a box of his father's college books. His curiosity brought him to the offices of the United Wireless Company, about a half hour's bicycle ride away. The personnel were friendly to the boy, helped him master Morse code and showed him their transmitters. In 1909, when just eleven years old, Don built his first receiver. He used an oatmeal box coil and a coherer created from a glass tube from a chemistry set. It was filled with iron filings, with a brass plug inserted at each end. With it the lad could hear stations five or six miles away. Young Don Wallace was hooked. He remained a ham—arguably the outstanding amateur in the United States—until his death in May 1986, six weeks prior to his eighty-seventh birthday. He will be mentioned several times throughout this book.[1]

In 1914, when just sixteen, he obtained his commercial operator's license, second grade. To qualify for his first class license he had to prove experience over a twelve month period. He prevailed, and in 1915 obtained his first class operator's license. As soon as he graduated from high school he picked up good experience as radio operator on a ship working the West Coast trade. When autumn came, Don matriculated at Hamline University in St. Paul. He played center on the Hamline team that won the conference championship that year, and was elected president of the freshman class. He also met, fell in love with, and married Bertha Lindquist. Then came the war. Don joined the navy, was radio operator on a submarine for a short while, and, when it was found that he could send and receive messages at forty-five words per minute, and could service a 40-kilowatt arc transmitter, he was placed in charge of the radio crew on the *George Washington*.[2]

This was quite an assignment for a twenty year old. Among the men Wallace chose was Fred Schnell, who for many years would have a distinguished relationship with the ARRL. All thirty-five of the men accepted could send and receive code at great speed. All were proud of their assignment, and the work of the radio crew functioned well. They sent messages on one hour and received them on the other hour. Much of the time during the receiving period they serviced the transmitter, which was extremely powerful and subject to burnouts. One day President Wilson sat down to talk on the radiophone to Secretary of the Navy Josephus Daniels, and the president suffered mike fright. He just sat there with nothing to say. Don cleared the room and waited. Finally, after about ten minutes, Wilson found his voice and carried on the conversation. The president was intrigued by the rigs but enjoyed even more the company of the young radiomen. He often came to the shack, as the room was called, and conversed with them rather than with his advisers and Very Important Men who were a part of his entourage. When the mission was ended, young Mr. Wallace left the navy and returned to Hamline University.[3]

The First World War had ended at 11:00 a.m., October 11, 1918, but this was not a peace treaty. After nearly three years of bitter controversy over the Treaty of Versailles including the Covenant of the League of Nations, Congress by joint resolution simply declared the war with Germany at an end as of July 2, 1921. Treaties with Germany, Austria, and Hungary were not signed until the following October. Failure to sign a peace treaty theoretically allowed Secretary of the Navy Josephus Daniels to postpone his department's relinquishing of total control of American radio for nearly three years, if he so wished. Secretary Daniels did so wish. His stubborn refusal to allow amateurs to renew their hobby, and his determination to push through Congress legislation granting power to the Navy Department over all wireless, kept radio, and especially amateur radio, in limbo until the fall of 1919.

To fully understand the opposition to amateurs, it is necessary to review the essence of Congressional hearings since 1912 on the general subject of radio regulation. This takes us back to January 1917, prior to the outbreak of the war. With the incredible progress that had been made in radio, the 1912 law was clearly in need of revision. Naval officials were the proponents. They hankered not only for needed new regulations, but for complete control of radio itself, and their most active supporter was Josephus Daniels.

Apart from the time-honored tendencies of bureaucracies to expand, the navy felt that it had sound reasons for being placed in charge of all wireless activities. Until the explosive appearance of commercial radio in the 1920s, the greatest contribution of the new communications technology lay in the safety it offered to ocean and Great Lakes shipping. In an appendix to the January 1917 hearings is a listing by year, beginning in 1899, of "Wireless as a Safeguard to Life at Sea." In the span of seventeen years, to and including 1916, more than 270 shipping mishaps are listed in which wireless helped prevent or allay tragedy, more than seventy incidents occurred in 1916 alone.

A portant of coming troubles for amateurs took place in January 1917, before America's entrance into the war. Hearings on *A Bill to Regulate Radio Communication* were held by the House Merchant Marine and Fisheries Committee. The transcript of the hearings ran to more than four hundred pages.[4]

Much of the space was given over to repetition. The navy and its supporters repeated over and over again their reasons why the bill should be passed, and the opposition expressed their views over and over again as to why the legislation should be defeated. Naval experts would come back insisting, among other arguments, that the bill "left amateurs just where they are now."[5]

All of the positive statements, the reassurances for amateurs of their continued status under the Radio Act of 1912, were capped by a letter to the committee from none other than Hiram Percy Maxim, writing as president of the American Radio Relay League. His approval of the bill placed him in the distinct minority of amateurs, possibly for the only time in his distinguished career. One speculates that the interference caused by the young fellow with the crude spark transmitter so bothered Maxim's enjoyment that he testified in hopes that bill would eliminate that pesky kid.[6]

Others testified that the bill was drawn up to satisfy "the greed for power of a few Navy officers." Even David Sarnoff, the future CEO of RCA (Radio Corporation of America) and NBC (National Broadcasting Company) testified against the service.[7]

Representatives for amateur organizations besides the ARRL appeared before the committee, categorically opposing the bill. The first of these men to testify was Charles H. Stewart, representing the Wireless Association of Pennsylvania, and by special request, the amateur associations of Germantown, Pa., the South Jersey Radio Association of Collingwood, and the Atlantic City Association.[8]

Stewart referred to certain paragraphs in the proposed bill which, he believed, would change the status of the amateurs or at least make their situation susceptible to change. "I cannot understand his [Maxim's] attitude in favoring the bill, because of the possibilities," he said. "We are fearful of what might follow." He said that interference was actually being brought under control and that amateurs, restricted as they were to 200 meters and less, were making remarkable progress in overcoming the obstacles. As for the young amateurs, adverse regulations would prevent them from improving themselves, "because wireless starts the train of thought among the young men and is a good influence on them generally.[9]

Stewart's testimony was followed by that of Frank B. Chambers, also representing the Wireless Association of Pennsylvania. He vigorously defended the amateurs, those "little boys of 14 and 18" who were working their rigs primarily with cheap, usually home-made apparatus. "A lot of them find old junk around the house and put it together to make their sets with," he said. They made "sliders on their tuning coil out of a tomato can and their sliding rods out of strips from the stairways.... Those little fellows can sit down there and do work that surprised the men."[10]

*Scientific American* editorialized against the bill on behalf of amateurs. "The good the wireless amateur accomplishes is soon forgotten," reminded the writer. Readers were reminded of how England without an "army of amateurs" had been forced to train thousands of them when the war began. The work of the ARRL with its relays was mentioned favorably. "Should it be destroyed through adverse legislation?" the editorial asked.[11]

As the hearings were coming to an end, Congressman William Stedman Greene of Massachusetts vented his opposition. The committee, he said, could hardly be expected in a few days to make a decision on so important a subject.

"We could not expect it, we could only hope," replied Commander Todd, who had just testified.

"You could not have much hope," Greene suggested. "You will have to go a long time before you get that, and you will get a chance to rest for quite a while."[12]

So nothing came of H.R. 19350, but Secretary Daniels did not rest. On April 19, 1917, H.R. 2573 was introduced, again requesting total navy control, and not mentioning amateurs by name. The war had just come to America, however, and the bill died.

When the war ended amateurs had every reason to believe that the government would lift the ban on all amateur radio activities. They were in for a surprise. Secretary Daniels dragged his feet on relinquishing navy control over radio. Not until April 1919 did he allow receivers to be reactivated. He still held tenaciously to his power over transmitters, insisting that they would only be allowed when a peace treaty was declared. In April 1921, Congress passed the resolution ending the war; it was signed by President Harding on July 22, but not until October were actual peace treaties signed with Germany, Austria, and Hungary.

Part of Daniels's attitude surely had to do with the exemplary record in radio activities chalked up by the navy during the eighteen months of American warfare. He boasted of the navy's accomplishments. More than ever he was convinced that Congress should pass legislation granting the navy the power to regulate all radio. Just two months after the Armistice H.R.

19959, aimed at Congressional approval of navy control of radio, came up for consideration before the House Committee on the Merchant Marine and Fisheries. Hearings lasted five days beginning on December 12, 1918. Meanwhile the ban on amateur radios continued.

When Secretary Daniels appeared before the committee as the first person to testify in the bill's behalf, he was reinforced in his arguments by the navy's many accomplishments in the radio realm. He cited major steps taken by the navy to control interference. He boasted of the smooth running of the high powered stations at Annapolis, Sayville, Tuckerton, and New Brunswick. He stated that all the executive departments favored the bill and that Central and South American countries likewise desired complete government control of radio. In sum, Daniels was as dead set on government monopoly of all radio activities, assigned to the Navy Department, as he had been nearly two years before with the earlier bill. Forced to acknowledge that the bill would create a monopoly, he informed the committee several times that it "ought to be governmental owned and operated, for the good of the whole people, and for the national need."[13]

The status of amateurs was soon raised and amateurs did not like it. So strongly did they oppose the bill that a "Board of Direction" of the ARRL had sent a "Little Blue Card" strongly protesting the bill to former subscribers, urging them to sign the cards and forward them to the committee. One major criticism was its lack of any mention of amateurs. In reply, the sponsors of H.R. 13159 said that they had considered amateurs under a clause defining experimenters and specifying regulations under which they could operate. Clearly, this did not satisfy the amateurs, and in deference to them, early on the sponsors suggested an amendment specifically mentioning amateurs and the regulations by which they must operate. Amateur stations were defined as "being used for private practice or experiment in radio communication and not operated for profit in either receiving or sending signals." An operator to be licensed should be able to receive seventy-five letters a minute which works out to about fifteen words per minute based upon an average of five letters per word. It also cut power to one-half kilowatt within one hundred miles of major bodies of water and one-fourth kilowatt within five miles of government receiving stations. Amateurs would be allowed to use wavelengths up to 250 meters.[14]

The first witness to testify in opposition was none other than Hiram Percy Maxim. He appeared before the committee in the capacity of president of the Maxim Silencer Company and president of the ARRL. Committeemen looked upon him as a respected businessman, a distinct asset. He announced his opposition not just to the original bill but also to the proposed amendments involving amateurs.

H.P.M. began his testimony by defending the 1912 Radio Act, which amateurs viewed as being imminently fair in its stipulated regulations as they were enforced by the Commerce Department. He estimated that when America went to war the licensed amateurs numbered as many as 7,000, that twenty-five times that number had possessed receiving sets, or as many as 175,000. Manufacturers of the amateurs' equipment numbered about twenty-five and did substantial business with them: he estimated investments in amateur radio as of the outbreak of the war (April 1917) at $10,000,000. He also emphasized the relays which, he said, were just gaining perfection when President Wilson ordered all radio sealed for the duration of the struggle.[15]

He mentioned the improvements made by amateurs, citing especially Captain Edwin H. Armstrong's regenerative circuit. "We amateurs wonder what the Secretary of the Navy would have done without amateur Armstrong's circuit," Maxim said dryly. "There must be thousands of inventions in the making and which may be of priceless value, and which you will kill if you prohibit American citizens from operating wireless stations," he said.

Although teenage boys seemed impressed upon the public mind—and on the Congress— Maxim went out of his way to state that "the amateur is not always a boy in short trousers." He named a number of wealthy men with rigs costing up to $5,000, and included himself as a man who became interested at age forty and now had a rig valued at about $2,000. But most amateurs, he added, are of ages twenty to thirty; they represent "the mentally keener among us," are fired with "the American spirit"; and radio appeals to them "profoundly, probably because it represents the profoundly difficult." Parents, he stressed, are proud of these young men, most of whom come from families of modest means.

He stressed how a boy could learn Morse code with a push button and a door bell. Representative William B. Bankhead, who was from rural Alabama, asked Maxim how much a rig would cost "if, for instance, a country boy in my district wants to experiment with wireless?" Maxim replied with a detailed description of the boy setting out to build his rig. First he obtained baled hay wire for his aerial. He used a tree at the back of the house for one end and the hencoop for the other. ("I cannot say why, but he always selects a hencoop," Maxim added.) Then the boy went to town and purchased exactly the length of copper wire he wanted, and with it he wound cardboard cylinders, one inside the other, to make his receiving tuner. From a friend or a friend of a friend he acquired a piece of galena. Then he obtained a little piece of the finest wire he could find, and used it for the "cat's whisker" to tap the galena for a station. His one big purchase would be headphones, whose cost began at about $3.50. The boy had saved up for the purchase, ten cents at a time, "and when they come in to pay for them they will have a handful of dimes, which they usually dump out of a little bag."

The remainder of those testifying in behalf of amateurs opposing the proposed bill essentially repeated Maxim's arguments. A recurring theme was the experimentation hams carried on. "Take care of amateurs and wireless will grow," said the representative of the Hoosier Radio Club of Indianapolis.[16]

When the hearings ended it was clear that several arguments against the bill had dominated the sessions. The most consistent reason, although it was often carefully shrouded in patriotic declarations, was nevertheless the suspicion of naval control. The navy was (and is) a military organization. It does not issue orders democratically; those orders must be obeyed; recourse in case of protest by a non-military enterprise even in a peacetime navy could be difficult. Amateurs were wary of their art being regulated by such a tough agency. Opponents also felt that naval control would end experimentation; that the navy would grab patents and refuse to pay for their application; and that couched in the bill's terminology was the power to eliminate the amateur forever. Again and again, the Commerce Department was praised for its application of the 1912 Radio Act. Truth was, Commerce had indeed been an "easy boss," clamping down upon occasional violators, but for the most part leaving the art alone. As an executive agency concerned with the peace-time affairs of American business, amateurs considered Commerce the proper, most understanding executive department to regulate them.

In reading the hearings, it becomes quite clear that most members of the committee were hostile to HR 13959. One reason was the Congressional distaste at enlarging government. This was emphatically a grab by the navy for a facet of communication that was a growing commercial activity. As the manager of the United Fruit Company informed the committee, "If legislation such as this is to be an outcome of the war, then the United States will have been made unsafe for business."[17]

A final comment should be made about the hearings. At no time did anyone, whether favoring or not favoring the bill, sense what would be happening to radio within four years. None of those testifying, none of the Congressmen asking questions, envisioned a time when commercial AM radio would burst upon the American scene. Receivers, which some still thought should be licensed, were going to be found in almost every home in America. And no one, *no one*, had the foresight to envision this commercial and communications revolution.[18]

H.R. 13159 was never brought to the floor of Congress.

The hearings ended on December 19, 1918. The Armistice had been signed more than two months previously, on October 11. Amateurs, including the thousands who restricted their operations to receivers, were still denied the privilege of unsealing their rigs, erecting their antennas, fixing their grounds, attaching their headphones, and listening to whatever was being sent by wireless. This power to crush all private radio, granted to the navy during the war, was now being used as a method of forcing passage of the legislation the navy so desired—total control of all radio.

Meanwhile, amateurs were reactivating everything but their radios. In April 1919, Maxim and a few of his aides reached into their pockets and came up with enough money to send former subscribers a "midget issue" of QST, labeled "The American Radio Relay League Special Bulletin." QST admitted that no one knew when the navy would lift its ban on radio save that it would do so when a peace treaty was signed. Speculation was that the treaty could be signed as early as May "provided the anarchists and Bolshevists have not got the upper hand." Those who had been in uniform were encouraged, "as they heard Atten—SHUN for the last time, to exchange that order to ——·— ··· – [QST]."

What about QST and the ARRL—how to restart the journal and the relay movement after a two year hiatus? The solution, as announced in the April 1919 midget issue, was the floating of a bond issue at five per cent, payable semi-annually, and to be retired in two years. The amount needed was just $7,500.00. Would amateurs dig in their pockets and support the suggestion?

The main article in this mini-issue was on "New Developments." Indeed, in the two years that had elapsed since amateur radio was silenced, impressive progress had been made. Most significant was the development of the vacuum tube. At first known as audions, bulbs, or lamps, the originals lacked uniformity, manufacturer to manufacturer; needed careful adjustments of the filaments and plate voltage; and were simply too delicate for "the Service." American engineers got to work and soon produced a standardized tube with a base having four pins to plug into a socket. These tubes automatically connected the filament, grid, and plate leads. "They are steady, efficient, and fine oscillators," the article stated, and "it is probable that ... manufacturers will arrange to put these tubes on the market within reach of the amateurs' pocketbooks."

Bulb transmitters—as apart from receivers—soon to be referred to as tubes also—were likewise greatly improved. It was predicted that in a short time they would be the coming thing among amateurs. For receiving, the loop antenna—one loop or several loops of wire—which could be placed in the same room as the radio or in the attic, did away with much of the static and, being somewhat directional, could be used by amateurs to minimize QRM (interference). Probably even the editors did not dream of how rapidly these developments would become available at affordable prices.

Of course the Old Man (Maxim) published his column. He radiated joy and optimism. "Can it be," he wrote, "that the old junk is at last to come down from the attic, and that the old bits of wire are to be straightened out and pieced together and a whole litter of new holes hammered in the plaster again and the lightning switches, kick back preventers, hot wire ammeters and mechanical miscellany generally to be worried back into place under the old table once more? It makes the cold shivers of delight oscillate up and down my spinal ground lead to think of it.... It would be sweet music to get the [head]phones on again and hear the boys over in Illinois threatening each other with Wouff Hongs, Ugerumfs, Rettysnitches and what not and chewing the fat about QRM." (Maxim loved to create new words: others were Flitchgobber, Oofhogger, and Zabsnifer.)[19]

QST did not publish a May 1919 issue, but when June arrived, the old QST in form and layout was back in print. Meanwhile, Assistant Secretary of the Navy Roosevelt had announced the lifting as of April 15, 1919, of the ban on *receiving* sets; transmitting sets would remain sealed, however, until a treaty of peace was signed. At least some progress had been made.

QST's June issue reflected the enthusiasm of radio fans with their new freedom. The magazine announced that it would be issued monthly from then on. Articles stressed how amateurs everywhere could be observed up on roofs, trees, and hen houses putting up new aerials, or digging trenches in which to bury copper wire for grounds. A list of long wave stations that could be "worked" was included. The prohibition on transmitting, it was predicted, would soon be lifted.

In anticipation of the end of the transmitter ban, the June issue looked ahead to the reinstituting of the coast-to-coast relays. It listed the names of those in charge of the six districts and predicted a bright future for the ARRL. The directors saw in the end of the two year hiatus an opportunity to upgrade relay quality. "It is hoped that all officers and members of the traffic department will have the courage to refuse the 'Greetings-by-wireless' type of message," suggested QST. "There will be enough of the 'Will arrive tomorrow' kind to keep us all busy." Membership blanks were available from headquarters. Stations were to be established close enough to one another that daytime contacts could be maintained. Progress in radio technology had reduced problems of decrement and static. Small wonder that there was considerable enthusiasm.

The June issue also noted one of the tragedies of the times. Recorded in a black-bordered notice was a death by pneumonia as a result of influenza: William D. Woodcock, ex–8SK, had passed away at the Great Lakes Naval Training Station on September 28, 1918. In a letter expressing her son's love of amateur radio, Woodcock's mother enclosed a check to help the ARRL on its way. The pandemic had attacked amateurs as well as everyone else.

The July issue dealt with the troubles involved in starting up receivers. Many had been stored for two years in uninsulated attics subject to freezing cold in winter and excessive heat in summer, or they had been stored in damp basements. The Old Man found his rig "as full of tricks as a flivver with a pint of water in her gasoline." Advice was given as to cleaning hard rubber panels that had become green from age and exposure to light and air. (The procedure included the use of Bon Ami—a popular kitchen detergent of the time—and water, grain alcohol, and rubbing with a soft black cotton cloth that had been dampened with castor oil.)

The bond issue was going well, with half the $7,500 already raised. It was reported that QST's founding editor, Clarence D. Tuska, had stepped down to be replaced by Kenneth B. Warner. The July issue included an expanded list of long wave stations including, besides American installations, those of British, French, Italian, German, and "Miscellaneous—Mexico City, Petrograd, Stavanger [Norway] and Java." A portent of the near future was the request by readers for schedules of radio-telephone stations. QST reported that none of them had regular schedules by radiophone, although experiments were being carried on. Activity at this time still dealt almost exclusively with dots and dashes.

Even as amateurs were reading the July issue, worrisome events were taking place in Washington. Josephus Daniels and the Navy Department just refused to give up. On April 18, 1919, censorship was removed on all cables except those addressed to or from German territory, and on July 23, even that restriction was lifted. Daniels noted that this resulted in an enormous increase in communications, the cables being so backed up with messages that radio communications were also being hard pressed to meet the demand. On July 19 the secretary sent a letter to the Speaker of the House including a short, terse bill (House Document No. 159) simply granting the navy the right to use stations under its control for commercial purposes. "It is exactly what its title intimates," noted QST, "and contains no possibilities of complexities from our standpoint."

Then on July 28 Daniels sent identical letters to Vice President Thomas R. Marshall and Speaker of the House Frederick H. Gillett (House Document No. 165) requesting the appointment of a committee to investigate radio, and again pleading for the navy's total control of wireless. He wanted a naval monopoly on all commercial communications by radio. At the same time Senator Miles Poindexter was chairing a subcommittee of the Naval Affairs Committee, making a similar investigation. His committee requested that the navy "tender a draft of their idea of adequate legislation ... as set forth in Document 165." This did not materialize immediately, but would be the subject of ARRL concern within a couple of months.

Amateurs were little worried by these statements while the newspapers took a dim view of Daniels's proposals. "Congress should turn a deaf ear to this request," editorialized the *New York Times*. "Not for any temporary and not for any permanent cause, or merely assumed cause, should the Government be allowed to put its bungling and paralyzing hand upon private business.... At the very moment of the removal of the censorship the plan of making the Government, under a new authority, a transmitter, and hence a potential muddler [sic] of business messages, seems peculiarly inappropriate.... Commercial interests are best served by commercial interests."[20]

Stubbornly, even though cables were once again open, and radio receivers were allowed, Daniels refused to lift the ban on transmitters. He kept insisting that it would be lifted when the peace treaties were signed.

Subscribers received the August QST—late. This was because rumor had the navy lifting the transmitter ban as of August 1, and the editors postponed mailing QST until they could tell the good news. It was, according to a QST editorial, even "unofficially announced" by the service that "the lid would go off August 1st." Then came news that the secretary of the navy had disapproved the measure. Amateurs must wait until the war was officially over, when the navy by law would automatically lose its control over radio.

QST speculated on the reasons why the navy still denied transmitting rights. An ARRL committee had traveled to Washington and expressed before naval officials their concern, not only involving the failure to lift the ban on transmitting, but also about the proposed legislation that, in leaving out mention of amateurs, made the way clear for a committee, without amateurs represented, to literally destroy the hobby. Officials denied that the refusal was in any way linked with the navy's request for legislation granting it power to regulate all radio. The League's representatives could not confront Mr. Daniels, who was in Hawaii, but Assistant Secretary Roosevelt informed them that Daniels had disapproved of the order and he (Roosevelt) did not know why. "Apparently Mr. Daniels is personally responsible," commented QST. "The whole proposition is so basically unjust, so uncalled for, that we do not believe it will long obtain." QST called upon amateurs to contact their congressmen and senators to demand the early release of the transmitting ban. Whatever the reasons, the ARRL delegates explained to the navy brass that this refusal was creating a "bad odor" and an explanation was necessary "if the suspicion with which the amateur world regards the Navy Department is to be eliminated."[21]

Traffic Manager J.O. Smith, editing QST's "Operating Department," which ran reports from the relay's district managers, put in his two cents' worth. "Possibly no one has informed the Honorable Secretary that the war is over," he wrote sarcastically. Daniels was clogging the legislative machinery "with his personal, pet theories while thousands of radio amateurs are prohibited from exercising the privilege of operation of their stations."[22]

September was passing by. Late in the month Secretary Daniels, apparently home from Hawaii, reiterated that restrictions would be removed "just as soon as a state of peace exists." He regretted that there was a misunderstanding on the part of the amateurs as to the attitude of the Navy Department, which "appreciates their service to the country during the past war." He even granted that they were responsible, as developing scientists, in making valuable advances in the radio art.[23]

Daniels must have been feeling the heat. The ARRL delegation had called upon Congressman William S. Greene who, as a member of the Committee on Merchant Marine and Fisheries, had defended the hams. They expressed to him their frustration with the navy, explained that they could not understand Daniels's refusal to lift the ban, and asked for his help. Greene's questions at the hearings in January 1917 and December 1918 had made quite clear that he was not favorable to the navy's grab for power. Now he introduced House Resolution 291, referred

Secretary of the Navy Josephus Daniels at his desk in 1918. He was forced by Representative William S. Greene to lift the ban on amateur transmitting (courtesy of the North Carolina Office of Archives and History, Raleigh, North Carolina).

to the Committee on Merchant Marine and Fisheries, which inquired of Secretary Daniels why he had refused to lift the ban. At about the same time the Navy Department presented the Senate Committee on Naval Affairs with another bill designed to give the navy a monopoly over radio. Amateurs saw in this bill, even when (and if) the ban was lifted, a threat to their continued existence. True, it did mention amateurs, but it placed decisions in the hands of a committee to be made up of representatives from the various executive departments who, in star chamber seclusion, could determine all kinds of rules and regulations concerning amateurs. The hams, as they saw it, had no representative on the committee and would have no power to protest decisions made by it—decisions not set forth by existing law. The league made plans to fight this bill. "We are out to oppose any despotic and un–American scheme by which we amateurs can be swept off the map by a brush of the hand whenever some unfriendly Government officer or committee takes a notion to do so," editorialized QST.[24]

When Daniels ignored H.R. 291 Greene came right back with House Joint Resolution 217, "to direct the Secretary of the Navy to remove the restrictions on the use and operation of amateur radio stations throughout the United States."

Daniels obeyed, one suspects, with great reluctance.

The October issue of QST, already going to press, inserted an additional unbound page—a supplement—announcing the good news. The director of the Naval Communication Service announced that as of October 1, 1919, "all restrictions on amateurs and amateur radio stations are removed." In a final paragraph the statement reminded all amateurs that their licenses must be renewed. Applications should be mailed to the Commissioner of Navigation of the Department of Commerce.[25]

"The Biggest Boom in Amateur Radio History.... WE'RE OFF!" headlined QST.

In the supplement the ARRL gave itself credit for forcing the issue. According to Jan Perkins in his biography of Don Wallace, another possibility for Daniels's defeat could have been President Wilson, who had enjoyed talking with the radiomen aboard the *George Washington* and had been impressed by their enthusiasm. Could he have intervened? No one knows. As would be expected, QST's November 1919 issue devoted additional space to the lifting of the ban. It ran a photograph of Congressman William S. Greene of Connecticut, the one who had inserted the resolutions requesting the secretary of the navy to explain why he refused to lift the ban on transmitters. The second Joint Resolution (H.J. Res. 217) had resulted in Daniels' lifting the ban.[26]

Maxim, in the November issue's lead article, stressed how organization had made the difference between success and failure in dealing with Congress. Even the Radio Act of 1912, he wrote, might have put an end to amateurs, but the work of a few local clubs, and especially of the Pennsylvania Wireless Association, had ensured amateur continuity at that time. He cited as an example of the organization's value the more recent work of the ARRL "when legislation was attempted by the Navy Department, which many of us think would have made easy the total elimination of amateur radio." So successful had the organization been, urging amateurs from coast to coast to write to their congressmen and sending delegations to Washington to confront navy officials and seek Congressional help, that the bills had died in committee—had not even been brought to the floor of Congress. "When an amateur asks the old time question, 'What do I get out of joining the ARRL?' the answer should be, PROTECTION," Maxim stressed.

Amateur activity was not long in coming. As a league historian described it, "Glorious old sparks! Night after night the mighty chorus swelled.... Character: Nervous impatient sparks ...

clean, businesslike sparks ... good natured sparks that drawled lazily and ended in a throaty chuckle as the gap coasted downhill for the sign off...." The rush was on to obtain licenses. Thousands of new hams, having served in radio during the war, were applying. They would be an element in American life working for the good of all.[27]

Daniels had been forced to rescind the ban on transmitters and so far his attempts at legislation granting regulating power to the navy had failed, but he was not through. On March 8, 1920, Senator Poindexter of Washington introduced a "Statement to Accompany the Bill (S 4038)..." which was a report from the Navy Department on its achievements in radio during the late war and the reasons why its power should be continued. Hearings on the proposal did not take place until October. QST took notice of the proposal, and opposed the bill.[28]

Maxim, who was unable to attend in person, sent a letter in protest which was inserted in the hearings. The proposed legislation, he wrote, was the strongest yet proposed for control of radio. He described it as "wholly

Representative **William S. Greene.** From the Collection of the **Fall River Historical Society.**

un–American in many of its provisions and would provide us the amateurs no assurance of continued operation." It was feared that its sponsors would "put it through with a rush during the short session of Congress" which was to begin December 6, 1920; in other words, it would be "railroaded" into law. The bill, Maxim explained, provided for the formation of a "National Radio Commission" with "authority to formulate and promulgate regulations for the operation of all classes of stations, and to change those regulations as they see fit: The bill says simply that the Commission 'shall have full power to regulate radio communications in the United States.'" Representatives from the Navy, War, Commerce, and Post Office departments were to comprise the committee, but its secretary was to be a naval officer of the line, appointed by the secretary of the navy, "thereby giving practical control ... to the Navy Department." The bill was never brought to the Senate floor. Daniels was frustrated once again.[29]

On March 4, 1921, the Wilson administration came to an end. Until 1933 Daniels was in private life; then President Franklin Roosevelt appointed him ambassador to Mexico. While serving there Daniels came into conflict with the greatest radio quack of all time, Dr. John Brinkley. By the mid–1920s this smooth talking charlatan had grasped the potentialities of radio. He ran station KFKB from the little town of Milford, Kansas. Through radio advertising he brought thousands of impotent men to his hospital for a ludicrous fifteen minute operation in which a slice of the gonads from a Toggenberg goat—the goat chosen by the patient from a herd in the back yard of the hospital—were inserted into the scrotum. When his station was closed by authorities, the doctor moved to Villa Acuna, a Mexican village just across the border from Del Rio, Texas. From there he broadcast not only his useless nostrums, but also the tirades of Bible thumpers, mystics, stock jobbers, real estate get-rich-quick drummers, and others. His station, XER (later XERA), was the most powerful in the world. It interfered with the legitimate stations of mid–America.

The United States mounted a campaign to get rid of Dr. Brinkley, but he was a shrewd wheeler-dealer, and Mexicans were wont to embarrass the Colossus of the North. Deeply involved in this effort was none other than Ambassador Josephus Daniels. He failed, although eventually Brinkley was driven off the air.[30]

In his memoirs, Daniels concluded his statements about his attempt to silence Dr. Brinkley's station:

> Naturally, as Secretary of the Navy, I had early been deeply interested in wireless communication, and during World War I, when its operation was a government function under the Navy, many of the important developments had been made by experts under my direction in researches financed by the government. I had felt that radio should remain a government function, and the part played by Brinkley at the time did not change my mind.[31]

And now amateur radio was entering an exciting, challenging, and dangerous world. It would thrive in the next three decades, but it would have its troubles, too. BCL's (Broadcast Listeners) would protest those dots and dashes coming into their AM radios. But, as we shall see, amateur radio prevailed.

# 5

## Surviving the Broadcast Revolution

One day in February, 1924, Forrest Bartlett's father went into town to visit the editor of the now defunct *Boulder News-Herald*. He wanted his nine-year-old son to receive some publicity for his achievements. Forrest, with a simple Crosley receiver, had heard President Coolidge speak from station WEAF, New York City. Moreover, his dad informed the editor, Forrest had "brought in over eighty stations" from such distant places as Havana, Cuba, Schenectady, New York, Seattle, Washington, and other faraway points in all directions. The editor was impressed. On February 24th he ran a story of Forrest's achievements with the headlines "BOY 'RADIOPHAN' SAID TO BE ONLY ONE HERE TO GET COOLIDGE TALK."

The Bartletts were business journalists. They lived in the country northeast of Boulder in a house that still exists at 37th and Jay streets. Relaxation in those days often took the form of quiet walks around the countryside. Often Mr. Bartlett chatted with nearby farmers, one of whom, named Harrop, had a precocious son named Clarence. While Forrest's dad talked with the father, Clarence escorted Forrest to the well house where his father had let him construct a radio. It was there that Forrest heard his first radio signals, and from that moment on, he was hooked on radio.

Supportive parents of young amateurs appear throughout the history of ham radio. Forrest's dad was no exception. When he returned from interviewing trips to Denver he brought radio magazines for his son. This only whetted Forrest's interest. Then a relative died and left money for the Bartlett children. Forrest knew from an advertisement in one of the magazines just what he wanted with his share: a one-tube Crosley radio which he ordered by mail. Well ... it was the basis for a radio anyway: It arrived without tube, batteries, earphones, aerial and ground accessories. Also it was in kit form, unwired. His dad purchased locally the parts still needed while Forrest went about erecting the aerial and providing a ground connection. Neighbor boy Clarence took the kit of parts back to his dad's farm workshop and carefully followed the wiring diagram to connect and solder the components together. When he brought the little one-tube back, Forrest anxiously plugged in the tube and hooked up the batteries, aerial and ground wires. The set worked. This was, of course, just a receiver, but the joy of tuning in the far-off stations sufficed for a few years. Forest was good at it.

To adequately describe the history of amateurs in the 1920s it is necessary to place it in the milieu of the times, with emphasis on the growth of commercial radio. The 1920 census gave the nation 106 million inhabitants; the 2000 census estimated the population of the United States at more than 280 million. In 1920 Colorado had just short of 940,000 residents; in 2000

## IT'S COMING, FELLOWS !

**This cartoon from *QST* is self explanatory. Published in the April 1919 issue of *QST*, p. 3.**

it had more than 4 million. Boulder, city and county, boasted about 10,000 residents in 1920; in 2000 it had more than 273,000. These statistics alone emphasize how the nation has changed. An old man's memory of the past he experienced already embraces a world so different from today's that it is now studied by historians.

People like Forrest or his parents were witnesses, whether they realized it or not, to the greatest material changes in the world's history. Between 1900 and 1925 more innovations took place in transportation, warfare, communications, illumination, and energy than in all the years preceding to the dawn of history. When Forrest's parents saw the first sputtering car in 1900— they would have been eight years old at the time—there were about 8,000 automobiles registered in the United States; by 1925 there were more than 17 million. The Wright brothers first flew in 1903; by 1925 mail was being carried by air on a regular basis. Barnstormers flew from a pasture outside of one town to a pasture outside of another, charging $5.00 for brief rides over the countryside. Medicine had a long way to go (it still has), but already bacteria had been identified and vaccinations were being developed to curtail the devastation of more diseases. A better understanding of the needs of pure water and good sewage treatment were bringing typhoid fever under control. The wide use of x-rays improved the setting of broken limbs.[1]

And then there was radio. In 1922 radio sales were estimated at $60,000,000; in 1929, $842,548,999. By the end of the decade one in three homes in the United States had a radio, networks had been established, and radio personalities had become celebrities. Next to the mass sale of automobiles, radio spurred the good times. The radio boom—the "great awakening" as some have called it—occurred in 1922. Factories could not keep up with demand. The only temporary loss of listeners occurred in the middle of the decade. This was because there was so much chaos in broadcasting that listeners could hardly hear one station without suffering interference from another. Radio's businessmen, reversing custom, actually asked for government intervention.[2]

Innovations—the regenerative receiver and the development of the vacuum tube—have already been mentioned, but other minor improvements all added up to radio as a much more sophisticated art at the end of the First World War than it had been when the war began. There was consolidation of small companies into bigger ones. Marconi was bought out and the American Telephone and Telegraph Company (AT&T), General Electric, Westinghouse, and the just-created Radio Corporation of America (RCA) entered the picture. All were vying for power. To clear up some of the patent difficulties, they resorted to the policy of cross-licensing, whereby companies pooled their patents in exchange for a portion of the anticipated profits.

They were great corporations, but at war's end the concept of radiotelephony as something entering the homes of millions of people had not penetrated the entrepreneurial or inventive psyche to the point of critical mass. Radio, as perceived well into 1920 or even 1921, was still considered a method of communication intended for a specific person or receiver. It was a device improving the safety of ships at sea. Money was made by the sender levying fees for the service. That is how the telephone and telegraph worked. In the money-making sense, radio was envisioned as a similar means of communication.

In retrospect, it was inevitable that a "critical mass" would take place somewhere. Suddenly the concept of radio broadcasting as a communications device would take hold. It would be recognized as a source of entertainment, information, news, propaganda, profit, and as a channel for marketing everything from religious sects to soap.

Although Detroit station 8MK vies for the credit, KDKA of Pittsburg is usually considered the first licensed commercial station. The story involves Dr. Frank Conrad, an amateur in his own right but also a researcher for the Pittsburgh based Westinghouse Corporation. Conrad was an avant-garde ham experimenting with a radiotelephone station, 8XK. Twice weekly at a fixed time he broadcast music, news, sports announcements and trivia to amateurs in the transmitting area. Soon they were listening regularly to his scheduled broadcasts. Newspapers mentioned his service. As the broadcasts became known, stores began selling radio receivers for ten dollars and up. People wanted to hear the broadcasts. A form of unpaid advertising was also taking place: Conrad played phonograph records in exchange for mentioning the stores that had lent them to him.[3]

Westinghouse officials experienced an epiphany. Radiotelephony was not just for specific listeners such as Marconimen aboard ships or with corporations, being charged by the sender for the service. Instead, radiotelephony should be free to everyone and anyone who possessed a radio receiver. The company's decision makers still believed the profits were to come from the manufacture and sale of radios, as Sarnoff had envisioned in 1915. Westinghouse, tooling up for mass production of radios, would be on the ground floor of a new industry. The company moved quickly. It set up a radiotelephone station on the roof of its East Pittsburgh factory, obtained from the Department of Commerce a commercial license—it was to be radio station KDKA— and began broadcasts on election night, November 2, 1920. Returns of the Harding-Cox presidential campaign were interspersed with gramophone and banjo music. (Detroit's 8MK also broadcast the results but without a commercial license.) In 1921 the Dempsey-Carpentier heavyweight boxing championship fight was broadcast, as was the World Series.

Westinghouse was manufacturing "radio boxes"—soon to be dubbed simply "radios"—as fast as possible, but it could not keep up with demand. The company soon had plenty of competition. A.C. Gilbert, maker of Gilbert Erector Sets, entered the field selling "Radio Outfits"— kits for both transmitters and receivers. Magnavox, Crosley, Atwater-Kent, General Electric, Zenith, and literally hundreds of other companies began manufacturing radios for the ordinary citizen. To us today, pictures of those jet black horns (loud speakers), of radios with batteries, aerials, grounds, and several dials, look incredibly ugly, yet, like pictures of old cars, they are strangely fascinating.

It was not until March 1922 that QST ran articles indicating recognition of the AM commercial radio boom. "With Our Radiophone Listeners" gave simple instructions for erecting a good aerial and promised more informative articles to come. Note that the article referred to "our radiophone listeners." This implies that they were placed in a different category from hams, but being a journal devoted to radio, QST was willing to help these citizens who were tuning in to commercial radio. Amateurs were encouraged to work with them as changes took place that might threaten the amateurs' freedom. It is clear, however, that there was as yet no comprehension of the coming explosion of radio broadcasting. Whatever developed, hams who had enjoyed near autonomy prior to America's entrance into the World War wanted their status kept that way.[4]

Until about 1923 the public considered anyone with a receiver, such as a simple crystal set, a ham, an amateur. Their numbers proliferated. Receivers far more sophisticated than crystal sets soon appeared on the market. Placed in living room–acceptable cabinetry, the batteries discretely hidden below the radio, but still with that ungainly out-of-place speaker in the form of a jet black horn atop the cabinet, the family sat down to face the radio and listen in. At the very least, the owner of a multi-tube set would have several dials to adjust. There was the radio frequency amplifier, the detector tuning dial, and the regeneration control. He or she had to adjust one or more rheostats to control the filament current to the tubes. Volume was adjusted by "detuning" either the radio frequency or regeneration dials or adjusting the audio amplifier

# Which would you choose?

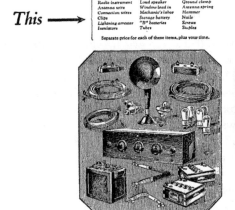

*This* ——→  ←—— *or this?*

## What "complete self-contained" means as in De Forest

IT is the best of fun, we admit, to hook up a radio set, to string your antenna from tree to house, to connect your ground wire—at least it is fun if you are mechanically minded.

If, however, you want principally to use a radio set, there are two things of primary importance—first, that its tone and quality shall be absolutely pure, non-metallic and accurate; secondly, that it shall be as little

fuss and bother to you as is humanly possible. This means De Forest D-12 Radiophone—the leader in the field—bearing the imprint of Dr. Lee De Forest, the man whose great invention paved the way to radio broadcasting.

As to tone—it is impossible to describe the clean and natural quality which this instrument gives. You simply must hear it

and judge for yourself. And as for convenience, remember these important things: it is self-contained and complete in one unit—usable within five minutes after it enters your home—easily movable from room to room because it does not need to be attached to either antenna or ground.

When you find the De Forest agent in your vicinity you find a man who knows

radio—a man who has given us his word that he will see that every instrument he sells is thoroughly inspected and properly serviced after the sale.

Avail yourself of his help. He desires, as we do, that you should get the fullest enjoyment and satisfaction from your instrument.

DE FOREST RADIO COMPANY, JERSEY CITY, N. J.
*Also makers of De Forest Tubes, The "Magic Lamp" of Radio*

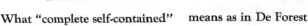

## DE FOREST RADIOPHONE       DE FOREST RADIOPHONE

As early as January 1925, the more affluent could purchase such radio receivers as the De Forest Radiophone, which absorbed the many parts of a receiver within a single cabinet.

Crosley Radio advertisement.

tubes' filament rheostat. Some sets had what was called a "tap switch," which selected the number of turns on a "vario-coupler" or "variometer" used in conjunction with the tuning and regenerative controls to give the strongest signal. There were also single tube sets, like the Crosley Pup, that could be tuned with a dial and a couple of knobs; and crystal sets, of course, just needed the cat whisker. Real amateurs, persons with transmitters as well as receivers, aware that an aerial and a ground and batteries and the tuning with several dials were all necessary, felt that radio was just too complicated for the average citizen. They were wrong. Moreover, technology advanced so rapidly that within just a few years one dial turned on the set and controlled the volume, one tuned the stations, and a third determined the tone.

Even when compared with today's rapidly changing world of computers and the World Wide Web, the speed with which broadcast radio spread over the land in the early 1920s is astonishing. QST compared the boom in radio to the rush to the East Texas oil fields. Westinghouse quickly established 500-watt transmitters in its factories in Newark, Chicago, and Springfield, Massachusetts. Plans were for stations in San Francisco and Dallas. General Electric, RCA, and AT&T were not far behind in exploiting the new bonanza. Marconi was establishing "phone broadcasting stations" in Montreal and Toronto. It seemed that just about every facet of society wanted a licensed radio station. Religious sects, retail stores, manufacturers of radios and speakers, colleges, and medical quacks all entered the field. Within twenty-five months after KDKA's first broadcast, the Commerce Department had issued 670 licenses. By the end of 1922 Secretary of Commerce Hoover estimated that there were at least 600,000 receivers in American hands, and possibly as many as a million.[5]

Programming had been born. Westinghouse was broadcasting a digest of the day's news given it by a Newark newspaper. It also gave government weather forecasts three times a day and time signals from the navy's Arlington station. Every Tuesday and Friday at 7:00 p.m. the company broadcast "Man-in-the-Moon Fairy Tales for Children." Election returns, bulletins of sporting events conveyed to the studio direct from the playing field, and lectures helped fill the broadcast day.[6]

Yet most programming was abominable. Commercial radio's first few years were characterized by all kinds of experiments to fill up time on the air waves. Here is a sampling of programs amateur 1ZE listed for "What We Hear Tonight":

W GUY (WGUY)—Beantown, Mass. (360.5 Meters)
    8 p.m.—Boston Homicidal Statistics
KAY WHY TRUBBLE YOU (KWTU)—Windburg on the Lake (360 meters)
    6 p.m.—Dramatic Reading of Sears, Roebuck catalogue by John Smith
    7 p.m.—How to Roll Your Own—recital By Prince Albert
    8 p.m.—Electrical Cat Chorus by the Loose Connection in the Set
WOP-PHILADELPHIA (360–567 Meters)
    8 p.m.—Boiler Recital of the Baldwin Locomotive Wks. Quartet, 1st sledge,
    2nd hammer, bass-bellows, baritone-riveting machine. Very selective.
    9 p.m.—The Dangers of Smoking—by Bluenose Mary of the U.S. Weather Bureau
CAL-LOSANGELES (361 Meters)
    11 p.m.—Samples of Indoor Lightning by Steinmetz
    12 p.m.—Hawaiian Song "Put Coconut Shells on Your Feet and Clatter Home"
KOW-DALLAS (412 Meters)
    10 p.m.—Chinese Ballad—"No Tickee, No Shirtee"
    12 p.m.—"Bull I Have Thrown," by a Prominent Cattle Man
OWL-ATLANTA (400 Meters)
    3:52 a.m.—Chimes Announcing Breakfast and the Regular Sunrise Program.[7]

Another comment, again suggesting that commercial radio would soon collapse, was the remark that there was "a growing tendency among broadcast fans to regard the call of the phone station as the best part of the program."[8]

Was all this enthusiasm just a fad, a flash in the pan? In those early years, 1921–1927, the hysteria of radio resulted in all kinds of problems. Much of the programming was awful. Interference was terrible. Stations broadcast on the same wavelengths as other stations in the vicinity. Aimee Semple McPherson, the Los Angeles evangelist, maintained a powerful station that constantly switched wavelengths. The 1912 law was obsolete and new legislation was necessary.

Amateurs did not help the situation. When people began listening to music and speech over radios, some hams, as mentioned above, switched to phone rigs and began broadcasting programs on their own. Until regulations were issued in January 1922, putting a stop to it, not just "broadcast listeners"—people purchasing receivers—suffered. So also did some of the licensed amateurs. How, asked QST, could dot and dash amateurs be heard over the din of radiophone operations at the same wavelength? "And as to the awful stuff that most amateur phones put out from ten cent records on $1.89 phonographs with a supply ripple like a threshing-machine and a wave like the Atlantic—we are glad it is gone and hope it never comes back! Amateur radio is decidedly better without it!"[9]

QST thought the BCLs (broadcast listeners) would soon tire of radio and turn to other leisure-time activities. Of course, they were wrong on both counts. Programs got better and gradually federal regulation eliminated much of the interference. Still more significant, however, was the creation of nationwide networks. These had the resources to broadcast higher quality programs and thus keep listeners sitting at their radios.

The cross-licensing of patents was but one aspect of the consolidations that were taking place in the radio industry. Two powers emerged in radio: the Radio Group, consisting of General Electric, Westinghouse, and RCA; the other was AT&T. At first they worked against each other, AT&T asserting monopoly rights over the telephone lines, denying the Radio Group access to them. Conversely, AT&T was denied the privilege of manufacturing radio equipment. Tortuous negotiations between the two parties resulted in major agreements in 1926. The upshot was the creation of the National Broadcasting Company (NBC). It was owned by General Electric (50 percent), RCA (30 percent), and Westinghouse (20 percent). The National Broadcasting Company would pay AT&T for use of its telephone lines. Thus was created the first company organized specifically to run a radio network. On November 15, 1926, NBC ran a four-and-a-half-hour broadcast coast-to-coast, with twenty-five stations participating. Because it was charged with having monopoly powers, in 1927 NBC divided into two networks, the Red and the Blue; in time, the Blue became the American Broadcasting Company (ABC). In 1928 William Paley created the Columbia Broadcasting System (CBS) and in 1934 the Mutual Broadcasting Company came into being. However, the networks still represented a minority of radio stations. Most remained locally owned, operated, and programmed.[10]

As commercial radio expanded in the 1920s with overwhelming rapidity, the question arose as to who controlled it. Sidney Head, in his excellent textbook *Broadcasting in America*, calls the American system permissivism, in which the Federal Communications Commission is empowered to maintain radio (and television) in the public interest while the government itself does not participate in broadcasting or televising.[11]

With strong opposition commercialism entered in. The first advertising broadcaster paid a "toll" for using the radio. This was a realtor who was charged fifty dollars by New York Station WEAF for ten minutes lauding his apartments. The concept of charging to advertise over radio took hold with incredible rapidity. Virtually accepted today, the concept of advertising met at first with strong disfavor. Even the radio entrepreneurs expressed an attitude of idealism, hoping that the medium would be a culturally uplifting service to the people. Opposition to advertising was strongly expressed at each of the four radio conferences called by Secretary of Commerce Hoover.[12]

T.O.M. (Maxim—the Old Man) reflected true amateurs' anger and frustration over the proliferation of radiophone broadcasting in a diatribe appearing in the May 1922 issue of *QST*. Amateurs yearned for the good old days. Here is part of what T.O.M. had to say:

> Say, son, what are we coming to, anyway? If this daggone broadcast stuff keeps on increasing something is going to bust up. The air is so chock full of jingles and jazz and foxtrots and speeches and advice as how to peel potatoes and bedtime bunk that it isn't fit to breathe anymore. Darned if I don't think folks will be going crazy pretty soon.
> ... From 300 to 400 meters it is one grand smother of stuff they call music and speechifying and what-not, all tangled and snarled up until if you listened to it long enough the bats would begin to show in your belfry, as sure as hellsamantrap.... When the whole blooming country starts yapping and yowling and hollering, and all of them trying to bawl their heads off at 360 meters, it just simply unseats a man's reason.... If they don't go easy pretty soon, not only will the great American public degenerate into a lot of snickering imbeciles but 360 meters will get worn out.[13]

True amateurs, as reflected in the same May 1922 issue, still did not know what to make of the explosion. Were the million new radio listeners potential candidates for membership in the ARRL? Should they subscribe to *QST*? In March 1922, the journal started a new department, "With Our Radiophone Listeners." The May issue included a photograph of the broadcasting station at Radcliffe College with coeds working it with headphones on. It also carried a listing of sixty-seven stations, their cities and call letters, licensed by the Commerce Department as of March 10th, 1922.[14]

*QST* predicted in July that with the summer static, all of a sudden people would lose interest, fly-by-night radio stores would go out of business, and the craze which had cleared shelves of radio products in previous months would end. Programming continued to be abominable. That and static, it was predicted, would drive private radio out of business. Really serious radio fans would turn to code, and, anticipating this transition, *QST* ran a number of articles on mastering Morse. Still not comprehending the BCL revolution, even the radio industry at large continued to consider the receiver-purchaser an amateur. RCA issued a thirty-five cent booklet, "How Radio Enters the Home," in which it called the purchaser of a receiver an amateur. Another book, "Radio Telephone for Amateurs," at $1.50, also continued referring to anyone purchasing a receiver as an amateur.[15]

In the February 1923 issue of *Radio News*, editor Hugo Gernsback expressed negative thoughts about the future of ham radio. "All thinking men, and most intelligent amateurs themselves," he began, "had long come to the conclusion that, ever since the broadcasting popularity started, the radio amateur was indeed doomed." It was, he wrote, the amateur who doomed himself! "Of what real use is the amateur today?" he asked. "His utility is microscopic.... The real usefulness of the American amateur in the United States is practically nil. This does not sound very nice, but it is the whole, unvarnished truth."[16]

Gernsback was clearly off base, but, indeed, amateurs did have a new fear. It was no longer the navy threatening their freedom, it was the man on the street, the average citizen who had purchased a "radiophone," risked his neck to raise an antenna, installed a ground, hooked up batteries to a receiver, sat himself down in his chair with headphones in place and attempted to master the several dials on the control board to bring in a "radiophone" station. When he heard static peppered with Morse code, he blamed the amateurs and demanded they cease transmitting.

The ARRL was well aware of the danger posed by BCLs. *QST* repeatedly emphasized the need for good public relations. The new purchasers of receivers, it sarcastically stated, knew little about radio and virtually nothing about amateurs and the Morse code. When they heard interference they complained. The one culprit they could identify, because he or she was in the neighborhood, was the amateur, so all the troubles were blamed on amateurs. Remember, warned league officials, these complainers are stalwart citizens. Their usual demand is that the amateurs be eliminated. "For you see, men, the novice listener doesn't yet know that there are others besides himself that amount to anything in radio, and at the present time he wants *all the air*, the same as we used to have all of it for ourselves."[17]

"Won't you let us help you?" the hams should ask the BCLs. Amateurs were reminded of the knowledge they possessed. Tell them that "we have radio clubs in every town and we want you to feel welcome to come around and get acquainted.... We would like you to know that when you hear our dots and dashes that it isn't 'the American small boy' playing around, but an organization of thousands of young men who are about a more or less serious business, engaged in mastering a complex art." Explain to the BCLs that interference comes from many sources, from leaky electric light lines, elevators, x-ray machines, violet-ray machines, welding machinery, transmitters on board ships and commercial and government stations—and that all of them can, and do, cause interference. Emphatically, it was not just amateurs. "Please be fair to us, won't you?" was the hams' plea.[18]

In the spirit of cooperation, by April 1923 the board of directors of the ARRL "earnestly call[ed] upon each and every amateur to refrain from transmission of any kind, for the good of our game, between the hours of 7:30 o'clock and 10 o'clock, local standard time, every night of the week." This was known as the Rochester Plan, whereby amateurs in communities with strong QRM (interference) voluntarily refrained from operations during those hours.[19]

Even as BCLs were raising more and louder protests again them, amateurs during the 1920s increased their services to, among others, the Department of Agriculture, the army and the navy. To some degree their contributions checkmated the criticisms they were receiving from BCLs and the regulators and politicians who were receiving their protests, their demands to do something about those "damned amateurs."

As for transmitting market reports, the weather, baseball scores and the outcome of professional boxing fights, amateurs by 1923 were already performing these services. Beginning on April 15, 1921, the Radio News Service of the U.S. Bureau of Markets (which was in the Department of Agriculture), which had already experimented with wireless reports within a three hundred mile area covered by WWX in Washington, D.C., expanded its activities to include three other radio stations: WWQ in Bellefont, Pennsylvania, KDEL in St. Louis, and KDEF in Omaha. The system functioned in cooperation with the Post Office Department, which at the time had a department called the Air Mail Radio Service. It sent agricultural and weather statistics gathered from the Middle West to the East Coast to the Bureau of Markets in Washington. There reports were prepared of prices, conditions of leading fruits and vegetables, livestock and meats, grain, hay and feed at important national markets. At 5:00 p.m. and 7:00 p.m. a Radio Marketgram was sent in code to the above named stations. They in turn relayed the information to hams within their region. It was estimated that there were 2500 amateurs within the 300 mile range of the four stations. As many hams as were willing were urged to participate in the program. Taking code at 15 words a minute, it took 10 to 15 minutes for an amateur to receive each report with a more thorough Marketgram taking perhaps 20 minutes.[20]

Hams interested in participating in this service were urged to contact the United States Bureau of Markets, which would furnish them with blank forms in which they could jot down the reports in pen or pencil. Once this was accomplished, amateurs were urged to furnish the information to state bureaus of markets, county agents, farm bureaus, and individual farmers, or to place the information on bulletin boards in post offices and give the information to telephone operators who could then convey it to those desiring it. How well the system functioned is not known. Ham stations at several universities participated. Within a year or two the system was made obsolete because broadcasting was now linked nationwide by telephone lines, which furnished the information to local radio stations. Nevertheless, it demonstrates the wide uses that amateur radio was being challenged to fill.[21]

Both the army and navy paid attention to amateur radio in the post-war years. Army plans called for an amateur station at every Army Corps area headquarters. National Guard units began installing their own sets, built by their own men prior to the receipt of government issue apparatus. An Amateur National Guard net was being developed by 1923. These activities were not well coordinated until the Army Signal Corps approached the ARRL in August 1925 with a plan to affiliate with the league. Described by league personnel as "the biggest opportunity that has ever been offered us," the proposal called for an amateur station in every unit of the National Guard and Organized Reserve throughout the country. To handle army traffic it was estimated that three thousand amateurs would be needed. Chief Signal Officer Major General C. McK. Saltzman said that "the amateur's participation will be a concrete and positive answer to those who question his right to continue as such."[22]

The plan was implemented by November 1, 1925, with the league named as the amateurs' representative. Immediately representatives were selected for the nine army corps areas. They were to act as advisors to the Signal Corps officers of each corps area and to make appointments in each territory. Hams accepting the invitation were obliged to work their station just one day a week. Within six months the plan had been realized, with more than a hundred certificates issued in the First Corps area alone. All nine corps areas were reporting good progress.[23]

The Signal Corps went beyond this. It established the Signal Corps Training System at the Citizens Military Training Camp stationed at Fort Monmouth, New Jersey. At this facility, where the War Department carried out its radio research, the corps set up summer camps for radio enthusiasts. It was divided into four parts: Basic, Red, White, and Blue. Each section took a full summer encampment to complete. After four summers, those graduating from Blue were eligible for a second lieutenant's bars in the Signal Corps.[24]

The army-amateur affiliation worked well. Hams within the program learned to work what were termed the Army Nets, enciphering and deciphering messages, contacting officials from governors to local administrators as well as mastering the army system. The military looked upon amateurs, and the cooperative ARRL, as a major source of expertise in time of war. Only amateurs—not telephone and telegraph operators—were trained to service and work the low power portable transmitters and receivers the army used.[25]

The navy was perhaps even more aware of the usefulness of amateurs and amateur radio. Fred Schnell, the head of the league's traffic department and a Naval Reserve officer, did liege service during the summer of 1925 aboard the U.S. Navy ship *Seattle* in the Pacific; he served at the navy's request. Schnell did such excellent work with low wavelength, low power short-waves, out performing navy standard equipment using twenty times the power, that he received a letter of commendation from Admiral R.E. Coontz, commander in chief of the United States fleet. He received even more than that. In a letter to ARRL headquarters, Admiral E.W. Eberle, chief of Naval Operations, described Lieutenant Schnell's services to the navy as "invaluable. It is considered that largely through his efforts high-frequency radio is now definitely in the Navy, both ashore and afloat." The admiral then stated that Schnell had been recommended for promotion to lieutenant commander in the U.S. Naval Reserve.[26]

The navy approached Maxim, as president of the ARRL, for its cooperation in assisting the navy "in the organization of a top-notch A-Number-One Radio Naval Reserve." The service contemplated a reserve force of six thousand skilled amateur operators. The plan was unusual in that it demanded no drilling or cruising, unless requested by the reservist—in which case he might be called for a two week's cruise in the summertime.[27]

Progress involving the army, navy, and amateurs continued. By 1929 the navy had established the Volunteer Communication Naval Reserve Class V3, "which included outright enlistment, rating of amateurs and their training, so that in case of a national emergency they would be available immediately, and in turn be assured of a good berth. Hiram Maxim and other ARRL officials (including Schnell) were active in the program's administration."[28]

From 1919 on, the Forest Service was experimenting with amateur, short wave radio. Amateurs were also cooperating with the Bureau of Standards Radio Laboratory. By 1921 the Radio Lab's activities occupied a two story building. Many of its activities were of interest to the amateur. It worked to ensure the accuracy of wavelength, capacity, inductance and resistance; helped standardize the definition of radio terms and symbols for radio instruments and apparatus; kept an up-to-date file of standards; and maintained a comprehensive list of abbreviations. It already had over ten thousand references to pertinent articles in periodicals and radio literature. It had become a clearing house for radio information.[29]

Other ideas involving cooperation cropped up in these years. Examples of collaboration by amateurs calling their hobby Citizen Radio began appearing now and then in the pages of *QST*. By the early 1920s most cities had police radios at four hundred meters. Amateurs were urged to help them. When the police broadcast descriptions of stolen cars, for example, amateurs in nearby towns were urged to listen in and then notify their own law officers with details of the stolen vehicle. The cooperation occasionally paid off. Richard Frank and William Michael, 2TK, of Union Hill, New Jersey, listened in on the NYPD's radio, which broadcast descriptions of stolen vehicles every evening at 7:30 and 10:30. One night they heard "Broadcast Alarm 1068" which described a stolen "two-ton truck, painted green, marked [on] both sides "GILLEN BROS., 24 TWELFTH STREET, BROOKLYN"; the truck was carrying "ninety-seven boxes of oranges." The two young amateurs spotted the truck, reported it to the Union Hill police, and the vehicle was found.[30]

The *Radio News* had been emphatic in advocating the end of the spark transmitter. As it caused far more interference than the vacuum tube, its elimination was expected to end some

of the complaints registered by BCLs. Again, this had been recommended ever since continuous wave transmitting became practicable with the coming of the vacuum tube. By 1924 *QST*
estimated that less that one per cent of hams were still using the spark transmitter, and shortly
thereafter, following recommendations at the Third National Radio Conference, the Commerce
Department made its use illegal. Of course there was nostalgia for it, like the yearning for the
plaintive whistle of steam driven locomotives when diesels took over. One experienced ham
described the sounds of the sparks:

> And of the older radio men amongst us many yearn for that sweetest sound of the old days, the
> musicale of the sparks beginning with a rumbling basso, their voices coming with a rapidly increas-
> ing volume, and ever rising inflection to high clear notes like those of silver bells. Alas, those days
> have gone. No more clear notes come winging through ether at eventide and at midnight, telling
> us that some brother is searching for communion with another soul—a friend whom he has never
> seen but whom he knows well. No more do we hear those beautiful tones that sounded like organ
> pipes playing softly in a great cathedral, from the deep booming bass to the shrill flute-like sta-
> catto.... [Now] we sit and listen to what is left, the noise like an army of skeletons on the march.[31]

In early 1923 the league created a public relations department. It established friendly relations with scores of radio editors (or columnists), mailed bulletins to more than five hundred
newspapers, and appointed a publicity editor for each of the league's divisions. A noticeable
decline was noticed in BCL protests; they were more likely to blame other sources. A test conducted by the Bureau of Standards confirmed ham radio innocence. Its statistics on the "relative magnitude of obstacles to reception of radiophone broadcasts" showed that only about nine
percent of the time was reception hampered by some form of man-made interference "exterior
to the fabric of broadcasting itself." This included "funny noises, commercial spark sets, ships,
government stations, and amateurs." So hams were responsible for but a small percentage of
that nine percent. "The crisis in our young lives when the slogan of the general radio public was
'Damn the Amateurs' has passed," observed *QST*.[32]

In retrospect, Hugo Gernsback's pessimism, his tirade about ham radio being useless, was
without foundation. Amateur radio by 1922 was on the verge of its most vigorous decades. The
commercial radio revolution did not destroy it. Some real battles took place involving litigation
at the local level as well as legislation at the national level, but the amateur consistently prevailed. Yes, the ham had to accept some restrictions concerning wavelengths and times of operation, but in the final analysis, the restraints were minimal.

Troubles began at the grass roots and extended to the halls of Congress. BCL anger over
interference, consistently blamed on amateurs, resulted in local ordinances against ham radios.
Such conflict was inevitable. Atchison, Kansas, prohibited amateur transmission within its limits, and other Kansas communities, including Kansas City and cities elsewhere including Seattle, discussed similar actions. The editors of *QST* assured their readers that such legislation was
almost certainly unconstitutional, yet dangerous. It urged hams to attend citizen meetings and
do everything in their power to promote understanding. Convince novices (another name for
BCLs) that amateurs have rights too![33]

The editors were correct in stating that local ordinances were unconstitutional. Atchison,
Kansas, learned of the futility in such legislation when its city attorney, Orlin A. Weede, ruled
that "a city ordinance cannot supersede a federal statute" and the Atchison city commissioners
were clearly in the wrong. Amateurs could transmit within the city.[34]

This did not stop other municipalities from attempting to regulate amateurs. Portland and
Salem, Oregon, and Minneapolis, attempted to regulate them but either withdrew their motions
upon legal advice, or their ordinances were successfully contested. The ARRL's general counsel
was Paul M. Segal, 9EEA. At about the same time the small community of Wilmore, Kentucky,
placed a one hundred dollar per year tax on any amateurs within its city limits. Mr. Segal con-

tested the ordinance. Judge A.M. Cochran of the District Court of Kentucky handed down an opinion that municipal ordinances designed to limit or regulate amateur radio transmitting stations were unlawful and unconstitutional and could not be enforced.[35]

While hams were concerned about actions taken against them, the chaos in commercial radio broadcasting was becoming a major national concern. The 1912 Radio Act was obsolete. By the spring of 1922, there were at least a half million receivers in the United States and sixty broadcasting stations. Rumors were that the Commerce Department had another five hundred applications pending. Secretary of Commerce Hoover reacted by calling four national radio conferences to discuss the problems. The first took place during the eight days from February 28th to March 7th, 1922. Hoover had selected a blue ribbon committee which included Hiram Maxim, who was to represent amateurs. At the time it was described as the "most important radio body that ever sat."[36]

The first day corporations took the stand. They received considerable criticism for their patent quarrels and failure to furnish top-rate, state-of-the-art receivers. The corporations also had to defend themselves against charges of monopolistic practices. On the second day the amateurs, represented by ARRL officials, presented their case. Paul Godley, well-known for his part in a trans–Atlantic relay (see chapter 6) protested the proliferation of propaganda flooding newspapers about "the small American boy" who was accused of interfering with just about everything pertaining to radio. The adverse publicity had been so consistent, he suggested, that many amateurs thought it was "inspired propaganda from unfriendly interests." Suggestions floating about that Secretary Hoover wanted to curtail amateur activities prompted Hoover to make a strong denial. Godley included it in his testimony:

> I would like to say at once that anyone starting any such suggestion that this conference proposes or had any notion of limiting the area of amateur work was simply fabricating. There has never been any suggestion of the kind.... The amateurs were asked to be represented in the conference and they are represented here today, and the starting of that sort of information is one of the most treacherous things that can be done. So I wish to sit on that right at the start—that the whole sense of this conference has been to protect and encourage the amateur in every possible direction.

The conference was in executive session for most of the eight days. It came up with suggestions that, for the most part, amateurs approved of; in fact, the principal recommendations were very similar to what the ARRL had proposed. Amateurs were to have 150 to 275 meters and the secretary was to divide these meters between beginning amateurs, more experienced ones, and radiophone operators. Moreover, again as suggested by the ARRL, amateurs were to be allowed to police themselves, choosing deputy inspectors. The one provision of which they disapproved was the decision to allow broadcasters to use 350 meters, which was so close to the upper limits granted amateurs that interference was a certainty. Amateurs would have preferred broadcasters granted space from one thousand meters up. One unusual committee recommendation was that broadcasting stations be absolutely prohibited from running advertising.[37]

In July 1922, radio bills that were considered the outgrowth of the conference were introduced in the Senate as S3694, in the House as H.R. 11964. While in agreement with many of the bill's provisions, amateurs were unhappy to discover that they were not mentioned by name or given status. The bills made no headway before Congress adjourned, but the legislation was reintroduced in the Congress convening in January 1923.[38]

Hearings were then held on the proposals. Hiram Percy Maxim in his capacity as president of the ARRL presented a superb defense of the amateurs. He boasted of their recent accomplishments and mentioned the league's campaign to have amateurs honor quiet hours from 7:00 until 10:30 p.m. In order to end ham opposition he also suggested a few corrections. Hams wanted the word "amateur" to appear where justified. They wanted an assurance that no one person or committee should have absolute power to control or destroy the art. The bill passed in the House but never came up for a Senate vote. The chaos in radio increased. Because something had to be done, Secretary Hoover called members of the Radio Telephone Conference to meet again on March 20th to discuss ways still open to the secretary, under the still valid Radio Act of 1912, to combat the turmoil.[39]

At the second conference the primary enemy of amateurs was not the BCLs, but commercial radio. The power of its entrepreneurs in Washington had greatly increased. They dominated the conference and demanded just about all of the ether. Nevertheless hams came out of the sessions feeling pretty good. Amateurs were denied access below 150 meters because some vested interest wanted it saved for the future. "For tugboats and millionaire's houseboats?" asked the editors. Amateurs were granted 150 to 220 meters, however, with some special licenses granted higher meters. Moreover, they could range through their meter span instead of being on a crowded 200 meters. Hams did not like their voluntary "quiet hours" policy being made into a government regulation with Commerce declaring hams off limits from 7:30 to 10:30 p.m. local time, with the regulation to be rubber stamped on all licenses. The measure did, however, standardize the quiet hours system throughout the nation. Hams concluded that they could have suffered much more, and they could live with the new restrictions.[40]

Their situation improved when, in July 1924, Commerce relented on the "nothing below 150 meters" rule. It permitted the use of several other meter bands: 75 to 80 meters, 40 to 48 meters, 20 to 22 meters, and 4 to 5 meters. Amateurs working these categories did not have to observe quiet hours. The Bureau of Navigation (the branch of Commerce regulating radio) urged hams to make quick use of their extended rights because another radio meeting was coming up in September, commercial groups might be vying for those meters, and amateurs would need good ammunition in order to hold onto them.[41]

Lacking passage of a congressional act regulating radio, Hoover found it necessary to call the Third National Radio Conference, to be held in Washington from October 6th to 10th, 1924. More than ninety representatives from all phases of radio attended. Again, amateurs emerged from the conference in good shape. The bands below 150 given them by the Bureau of Navigation were officially recognized. "Not only were five bands, comprising collectively a range of eighteen-plus megacycles per second, allotted to amateur operations, but in his opening speech Secretary Hoover referred to the importance of amateurs' work and the desirability of having that work preserved and encouraged within its own proper limits. Indeed, it was said, the Secretary's remarks were as justly appreciative of the best activities of the amateur that one might almost suppose that during his leisure moments he was a radio amateur himself. (His son was an amateur.)[42]

Although hams were pleased with the conference results, they were also made acutely aware that they were no longer the exclusive occupants of the short wave region. Commercial radio was allotted meters down to nearly 200, alerting amateurs that they now had competitors on both sides of their bands. They knew that this would create new problems.[43]

Most of the suggestions of the Third National Radio Conference failed to be implemented. Chaos simply got worse, so in December 1924, Hoover introduced by way of Congressman White a short bill that gave the secretary of commerce extraordinary powers to control radio. It was opposed by the ARRL and broadcasters because of this. As the president of the Zenith Radio Corporation said in his dissent, "I ... would be in favor of putting this tremendous power into the hands of the Secretary of Commerce on one condition, and that is, that Hoover give to the radio broadcasting industry a guarantee that he will live for one hundred years and that he will serve as Secretary of Commerce for that hundred years. In other words, Hoover, we don't know who your successor is going to be." The bill did not pass.[44]

The Fourth Annual Radio Conference was called for November 9–11, 1925. More than seven hundred attended, and every indication was that a congressional bill regulating radio, with authority placed in the Department of Commerce, would be forthcoming. The tone of the meeting was markedly different from earlier conferences. At earlier meetings, attended heavily by commercial broadcasters, the attitude toward hams had been one of disgust and surprise. "What? You amateurs here? How did you get in?" This time the atmosphere had changed. Hams were hailed as the folks who made high-frequency radio possible. They were being approached in a friendly way because they were considered experts on shortwaves.[45]

Amateurs were worried that their 150–200 meter domain would be infringed upon, but such was not the case. Secretary Hoover defended them, informing the committee that he opposed reducing the meters allotted to hams. And he went out of his way to compliment the hobby: "Radio in this branch [150–200 meters] has found a part in the fine development of the American boy, and I do not believe anyone will wish to minimize his part in American life."[46]

Again, amateurs emerged from the conference essentially unscathed. Spark transmitters were ruled illegal, but hardly one percent of amateurs still used them. Radio telephone users were granted meters from 83.3 to 85.6 in addition to the 170–180 meters they already were allowed. Broadcasters were the real recipients of changes because there were too many of them and they infringed on each other's wavelengths. They pleaded for the secretary to step in and enforce stricter regulation.[47]

Following the Fourth Radio Conference, Congress again concerned itself with radio legislation. The bill favored by the Coolidge administration was subsequently passed by the House while the Senate bill called for a five man commission responsible to no one, and its decisions would be final. Although it was not liked by amateurs or, for that matter, by the entire radio industry, it passed the Senate. However, there was not enough time left before Congress adjourned to iron out major differences in a conference committee between the House-passed bill and the Senate-passed bill.[48]

This failure left the radio regulatory situation chaotic, and during the slow congressional processes something happened that exacerbated the situation. A court decision involving commercial radio denied the secretary of commerce most of the powers he had assumed for regulating radio. *United States versus Zenith Radio Corporation and E.F. McDonald* involved a complex situation involving the rights of radio stations in the Chicago area in the use of wavelengths, kilocycles, and hours of operation. Federal Court of Appeals Judge James H. Wilkerson handed down his decision on April 16, 1926. The gist of the court's opinion was that all the regulations that had been set up and were being enforced by the Commerce Department were illegal, not being provided for under the obsolete, but still valid, 1912 Radio Act. The court, editorialized *QST*, "disclosed a large and husky Senegambian who had been carefully concealed in the Commerce Department's wood pile." The attorney general, requested by Hoover to clarify Hoover's remaining powers, stripped the secretary of almost all powers the Commerce Department had previously held with regard to regulating radio. The secretary was obliged to grant any broadcasting station a license upon demand, and that station could then use practically any wavelength and power and operating hours it so chose. "Government regulation of radio had broken down completely," editorialized *QST*.[49]

What should amateurs do under the circumstances? They were complimented for continuing to adhere to the regulations that were in place, even if the regulations were no longer enforceable. They were to stick to the wavelengths allotted them by the Fourth National Radio Congress. Radio legislation was inevitable, and amateurs did not want to be discriminated against. *Don't give our enemies an opening.*[50]

Chaos in broadcasting or not, the wheels of government continued to run—slowly. As of January 1927, Congress had not acted. Commercial radio companies were blamed for the impasse, which centered around government radio control being placed in the hands of a single agency.

It was also speculated that politicians did not want to see Mr. Hoover, secretary of commerce, receive the credit due him.[51]

On February 23, 1927, President Coolidge signed into law the bill creating the Radio Act of 1927. It created the five man Radio Commission, each member to come from a different geographical zone of which five were designated for radio purposes. The first chairman was Rear Admiral William H. Bullard, U.S. Navy retired; in previous years he had testified against amateurs and in favor of total navy control of radio. QST's editors predicted chaos, and said they would not serve on the committee for the combined salaries of all its members. All licenses were to be renewed. It was suggested that the sixteen thousand amateurs be exempt because they were working smoothly under provisions set down by the Fourth National Radio Conference. It was predicted that there would be so much litigation involving licenses to commercial stations that previous conditions would seem like "tiddlywinks."[52]

The chaos did not materialize. The Bureau of Navigation was moved out of the Commerce building and replaced with the Federal Radio Commission, directly under the supervision of Secretary Hoover. On March 15th the commission extended all amateur licenses with the same force and effect as though it had issued new permits. The continuance was indefinite, meaning that at some time in the future, under new regulations, amateurs would be expected to renew their licenses. All of this meant that for the present the regulations set down by the Fourth National Radio Conference would remain in force. Amateurs had bands at 150–200, 80, 40, 20, 5, and ¾ meters. From March 29 to April 1 the commission held hearings in which League Secretary K.B. Warner testified, defending the amateurs' bands, especially the range from 150 to 200. He explained that the nearly seventeen thousand hams were also experimenters who had made many advances in the uses of this region. Fortunately neither broadcasters nor radio manufacturers wanted an extension of commercial radio into these wavelengths because to do so would make obsolete thousands, possibly millions, of radio receivers. Warner said that 200–545 meters and 550–1500 kilocycles were plenty for commercial radio. One reason for his defense of 150–200 meters was that this band was the one used by radiophone amateurs. For the time being, hams appeared to be safe within their meter limits.[53]

Yet a danger did exist. Unless amateurs made vigorous use of these allocations, most especially 20 meters, other interests were waiting to pounce on them. Hams were urged to make more use of 20 meters. It was pointed out that when the 1912 law gave them 200 meters and down, amateurs tended to work around 240 meters. They had almost succeeded in working at an honest 200 meters when the great potential of short waves became known. Now they were "allotted" specific bands—¾, 5, 20, 40, 80, 150–200 meters, and it was up to them to use these bands to the extent that no one could deprive the amateur world of them.[54]

It is something of a miracle that amateurs came through unscathed from these four radio conferences and the final legislation in 1927. Complaints against them were on the rise during the very years in which the conferences were held and legislation was being considered. In 1925 Commerce sent the ARRL. a letter emphasizing the rising number of BCL complaints against "code interference." If nothing was done, quiet hours would have to be imposed on the lower wavelengths and the hours extended. League members were urged—as they had been urged many times before—to get acquainted with BCLs and help them solve their problems. "Rig up a wave-trap for them and show them how to use it; point out the too-long antenna; recommend the right kind of equipment to them; make over their single-circuit tuners into loose-coupled receivers.... Listen to their interference and tell them what it is—most of the time, it won't be an amateur."[55]

Even more drastic action was suggested. In each league district where there was a big city, the district director was to appoint a vigilance committee. It was to consist of three reliable hams, a BCL, and a public relations person, preferably a newspaper reporter. The director was to announce the committee's existence and invite complaints. When the cause of the interference was determined, designated hams were to call on the BCL-ers to determine what was wrong, and if indeed it was an amateur, then the committee was to approach that ham and get things straightened out. Clearly ARRL. officials saw this "code interference" as a very serious threat to amateur activities.[56]

That the Commerce Department meant business was clear when it was announced that it had rescinded the sending licenses of more than a hundred amateurs.[57]

Forrest Bartlett was well aware of the problem. Although nearly eighty years have elapsed, he still recalls the family's experience with the problem:

> We were still living in the country. On Saturday afternoons in the fall it was discovered that a radio station was broadcasting direct accounts of what I would guess were University of Colorado football games. Dad was intrigued by the play-by-play action. He would sit by the radio with phones on his ears and a pad of paper so he could diagram the plays as they were described.
>
> And then would come a series of raucous crashes. He would yell, "Forrest, come fix the radio."
>
> I didn't know what these crashes were so consulted with Clarence. He said this would be code sending by engineering students using spark transmitters and no, there was nothing you could do to quiet the noise because they were radio signals just as much as the broadcast station.
>
> Dad was of course miffed but to me this was a new and fascinating side of radio. I wondered what these code signals were saying. It probably became a subject of conversation around the supper table. On an interviewing trip a short time later, Dad bought a "Learner's" telegraph set. Mounted on a polished piece of wood were a telegraph key and a telegraph sounder. A lithographed plate spelled out the code symbols and a pair of binding posts were where a battery was connected. It operated off an NR6 drycell which is what energized our telephone and was also the battery supply for the filament of the D12 tube the Crosley receiver used. I'm sure I knew how to borrow either a battery from the telephone or disconnect the one for the radio. Hooking it up to the telegraph set, operating the key would cause the sounder to click.
>
> I knew it didn't sound like what I heard on the radio but I stuck with my practice until I was familiar with the dots and dashes for each letter of the alphabet. But the sound was so different from what I heard on the radio that [the] code was still a mystery. It wasn't until we moved into Boulder and I met Avery Lamont that I found that I had learned the wrong code. American Morse was different from the Continental code used in radio.

With all the fuss stirred up over radio, it must be remembered that amateur radio was and remains a hobby. The men and women who call themselves hams or bugs or just amateurs are devoted to it for the pleasure it brings them. They are busy with relaying, DX-ing, and helping in times of crisis such as floods and hurricanes. Many indulge in "rag-chewing"—maintaining contacts with ham friends throughout the world. Most belong to both national and local organizations, from the ARRL down to the local amateur society. Besides these interests, many are also experimenters—just born experimenters. "Would this work? Let's find out!"

Some of these ham activities are the subject of the next chapter.

# 6

## Relays, DX-ing, Strange Languages, and Clubs

Forrest Bartlett was just one of hundreds of thousands of "radiophans" who spent long hours at their receivers listening for the most distant broadcast stations. Some even sent QSL cards to stations they had heard. Booklets could be purchased listing station locations, call letters, power, and place on the dial. One could buy blank booklets formatted to list stations heard. While the hobby continued well into the 1930s, the real amateur world of hams working short-wave transmitters and receivers ceased considering people listening to commercial broadcasts as participants in their hobby. As for Forrest, his interest in listening for long distance broadcasts declined as his knowledge about shortwave radio and Morse code increased. By 1928 he was teaching himself code. Not until 1930, however, did he receive his amateur license, W9FYK. During the years of the 1920s while he was growing up and gaining knowledge of radio, the amateur world expanded with incredible rapidity.

The concept of message relay had been Hiram Percy Maxim's idea when he founded the American Radio Relay League. Efforts to create a system by which amateurs could convey messages to people across the country had challenged thousands of radio hobbyists. Their aim was to achieve successful relays. It thrilled a ham when he or she knocked on a neighbor's door to deliver a message from relatives several hundred miles away, a message received by the ham in code via shortwave. The relaying of messages was a mastic that brought hams from all over the nation together with a common purpose. The challenge was constant. Successes and failures and how better to work the system were subjects of discussions at amateur club meetings. Relaying reminded hams of their continent-wide fraternity, that there were thousands of others out there with the same interests, the same goals.

Just how much the leadership looked upon relaying as a useful public relations ploy, elevating the image of amateur radio from a hobby of teenage boys to the status of a valuable national asset, is evident from the enthusiasm with which ARRL executives promoted the relay idea. This is made clear by an examination of the complexities of the system established by league headquarters. That QST devoted several pages in every issue to activities of the various relay divisions into which the United States (and Canada, shortly after lifting of the wartime ban) were divided is an indicator of the enthusiasm for the system not just by league headquarters but by thousands of hams.

It comes as no surprise that reestablishing the relay system as soon as the First World War transmitting ban was lifted was, along with reissuing QST, one of the league's top priorities. By September 1919, the Operations Department, which was the information unit for the relay system, could list the trunk lines agreed upon and request the division managers "to arrange accordingly." District managers were urged to have made all necessary arrangements so that activity could begin just as soon as the ban was lifted. Distances between stations were to be close enough for daylight transmission, although regular relay work was to be between nine and twelve p.m.[1]

Headquarters' Operating Department—which tended to be called the Traffic Department—handled the activity. In the field the system was subdivided into DMs (division managers), ADMs (assistant division managers), DSs (district superintendents), CMs (city managers), and ORSs (official relay stations). Each division was in the charge of a DM, each state in the charge of one or two ADMs, the states being further divided into districts in the charge of a DS. In cities of 25,000 or more there was a CM. The real work, it was explained, was done by the ORSs, each of whom kept count of the number of messages handled each month. This station reported to his DS or CM who reported to the DM through the ADM; the ADM in turn sent his report to the TM (traffic manager) at headquarters, who inspected and verified the reports and had the results published in the Operating Department of QST.

By July 1919, preparations for start-up were progressing well throughout the United States. Division managers and district superintendents were hard at work recruiting hams interested in the process. The aim was to have at least two district superintendents in each state. Once recruited—and a good many had held the positions prior to the ban—they were to enlist hams to participate in the relaying process. The many independent, local clubs were also urged to affiliate with the ARRL. "Secretaries of Radio Clubs," challenged the officials, "we invite your cooperation!" Although the system's start-up was rather slow, by 1923 it was functioning well. In July of that year, ordinarily a poor radio month, 25,546 messages were registered as being sent.[2]

Obligations for participating in the relay system were simple enough once mastered. The ARRL furnished an official blank for relays which it sold at cost. On these forms the amateur filled out the office of origin, the date the message was filed, the number of words in the text, the name and address of the person to whom the message was to be delivered, the text itself, and the name of the party sending the message. The participant was urged to keep a log, a record book of all transactions. All good amateurs have always kept logs: Forrest's dates back to 1930. The left column of each page is for the exact time of the communication. Additional columns vary but always have the call letters of the ham contacted.[3]

Fairly elaborate rules and regulations for the handling of traffic were established, printed, and sent to each member of the ARRL. Newcomers to the hobby were advised to be in touch with an experienced ham and learn the process from the "old timer." "Messages could be filed by anybody with any amateur who was then to transmit and relay the message to its destination without charge," stated F.H. Schnell, the traffic manager. The amateur's pay for the service was the satisfaction he or she derived from doing it. Sometimes the message was sent across the country in an hour, sometimes from Chicago to Indianapolis in weeks, and sometimes it was never delivered—a skyhook of some sort stealing it on the way.

Besides abiding by the somewhat intricate instructions printed by the Traffic Department and mastering the on-site directions of the experienced amateur, the beginner needed to know the abbreviations widely used by hams. There are so many, and they are used so often, that by 1923 the Department of Commerce was already publishing a list of them. Moreover, the beginner would soon learn of the ARRL news broadcasts every Saturday and Sunday nights beginning at 10:30 p.m. Over a hundred hams cooperated in these broadcasts, which we presume were sent in code. (This was in 1923.) He or she should listen to these informational programs to keep up to date with amateur activities.[4]

A hypothetical example of the transmitting of messages as part of a relay was described by Traffic Manager Schnell. He placed his novice at the rig of a hypothetical TOM in Kansas City.

(This is confusing: TOM has nothing to do with Maxim, who wrote as T.O.M.) He had the amateur make contact with 1ZE, which were Maxim's call letters. On the radio the novice hears 1ZE. He calls him three times: "1ZE 1ZE 1ZE de TOM TOM TOM." Assuming 1ZE answers, the novice asks 1ZE if he has anything coming his way. But he doesn't spell this out in code, instead he taps out the letters "QTC?." (Or "K," which means the same thing.) Then IZE sends: "TOM de 1ZE: GE OM QRK? QTC QRV? K." IZE has said, "Good Evening old man, how do you receive my signals? I have a message for you, are you ready to receive? Go ahead." The novice then taps out the abbreviations meaning that IZE's signals are good and he (the novice) is ready to receive the message: "IZE de TOM QRK QRV K." The message comes in: "TOM de IZE Nr 1 fm Marion, Mass. IZE to Mathes radio 70E I have been working WNP nearly every night since he left Wiscasset sig IZE."

The ham should then acknowledge receipt of the message: "IZE de TOM R nr 1 K," meaning that the message number one has been received and go ahead with the next one. IZE has nothing more and sends "TOM de IZE TKS NM cul 73 GN," meaning "Thanks, nothing more, see you later, best wishes, good night."

The ham should then acknowledge receipt of the message: "IZE de TOM R nr 1K," meaning that message number one (K) has been received and to go ahead with the next one. IZE has nothing more and the ham copies "TOM de IZE TKAS NM cul 78 GN," meaning "Thanks; nothing more, see you later, best wishes, good night."

But the novice is still not through. He is in the relay business; he must try to send the message on to station 70E. Here he refers to the call book which contains all the licensed amateurs in the United States. He finds that 70E is a ham in Bremerton, Washington. At this stage of radio development, he does not dream of attempting to contact 70E directly. He must relay the message; that is, he must contact a reliable ham located on the way to Bremerton. He listens in for ten minutes or so until he hears a ham located in that direction. He contacts him in much the same way as he contacted 1ZE, transmits the message and expects that station—(9AMB in Schnell's description) to relay the message on towards its final destination, to 70E in Bremerton, Washington.

Bright men and women, some disgustingly young and some quite old, participated in this game, for they looked upon it as such. They carried out the instructions down to the final requirement of registering their work. Some could prove more than 400 completed relays handled in a month. By the autumn of 1923 the relaying of more than 26,000 messages a month was not unusual, although, it has to be admitted, the delivered total was less.[5]

Even as relaying flourished, it was changing rapidly. The vacuum tube, which made CW transmission possible, regenerative receivers and constant experiments were moving operations into shorter wavelengths (higher kilocycles). Reliable contacts over expanding distances were becoming commonplace. On January 11, 1925, IXAM at South Manchester, Connecticut, worked 6TS at Santa Monica, California. The two hams were using 20 meters with less than one kilowatt of power at each end. "It isn't an accident, it isn't a freak, it is a beginning," they insisted. And they were right.[6]

The trunk line idea died. "When you no longer have to string out your stations in a chain, but any one of them is able to work all the others, what have you?" queried QST. "You have a 'net,'" and that is what the ARRL traffic handling system became—what the government services call a "free net." The message was to be passed on to the next ham in the right direction "without benefit of schedule, clergy, or trunk line." The dividing of ham radio into four frequency bands further fragmented the ham fraternity. The result of these changes was rather chaotic. Something had to be done.[7]

That "something" was to be known as the Five Points System. Under this modus operandi each ham station would arrange schedules with four other stations, one in each direction. If the amateur was one station's western connection and another station's eastern connection, then that ham had to find just two more hams—one north, one south—to round out his obligation. Besides being a highly workable system, it had the added advantage of helping the amateur make new friends, indulge in "rag-chewing" and experimenting.[8]

Prior to America's entrance into the First World War, programmed relays had been staged for publicity purposes. A little over a year after the ban on transmitting was lifted the ARRL announced another "Washington's Birthday" relay. It was to be run on the night of February 21, 1921. Described as "the biggest free-for-all relay ever," more than 7,240 hams sent in reports. The thirty-word message was obtained from president-to-be Harding: "May the spirit of Washington be our guide in all our national aspirations and may the current year mark the return of tranquility, stability, confidence and progress thruout [sic] the entire world."[9]

The contest worked as follows: fourteen words of the message were sent by a designated ham from the Atlantic coast to the odd-numbered districts; fourteen other words were sent by a designated ham from the Pacific coast to the even-numbered districts, and the remaining two words were sent by a designated ham from the Mississippi Valley. When the participating hams had received all the words, they were to deliver the message to mayors, governors, and if possible, to President Wilson; they were also to notify league headquarters of their success. Fourteen governors received the message. One Southern governor was roused out of bed to receive the message, which was delivered to him by a Catholic priest who was also an amateur. The governor thought it was a MSG (message) from night riders, but the priest replied, "No, it's the Night Radioers." Two hundred and forty-seven mayors were disturbed in their sleep to sign for the message, twenty-two United States senators, thirty-five state senators, plus over five hundred chiefs of police, selectmen, city councilmen, sheriffs, postmasters, and news editors. It was said that one enterprising ham even gave a copy to President Wilson.[10]

Seventy-eight prizes donated by companies in the wireless business were distributed on the basis of judges' decisions, the judges consisting of three prominent radio men. Awards were based upon "speedy and correct reception," with additional credit for precise wireless reports about transmitting and receiving conditions. One prize was won by a lady ham, Miss Winnie Dow of Tacoma, Washington. Enthusiasm for the Washington's Birthday Relay prompted suggestions that it become an annual affair. It was noted that it also brought amateur radio favorably to the public's attention.[11]

About a year later, during the First National Radio Conference (February 28–March 7, 1922) the ARRL sponsored the "Governors-President Relay." The object was to relay a brief message from the governors of the forty-eight states, carefully written on official ARRL blanks, to President Harding. These were to be delivered to him by an amateur chosen from the Washington, D.C., area. Three days were allotted for the transmission of the messages. Forty states responded, which was considered a successful number. The only trouble was that President Harding was in Florida! (When he returned he sent a message of thanks.)[12]

A number of Governors-President and Washington's Birthday relays were run throughout the 1920s, as well as occasional unusual relays. For example, in July 1922, the league sponsored the Police Chief's Relay. The amateurs were to deliver a brief invitation sent via short wave by Chief August Vollmer, head of the organization. The message invited all police chiefs to attend the International Association of Police Chiefs convention in San Francisco during the week of June 19th. The relay was deemed a great success. The brief message was received by hundreds, possibly by thousands of hams, who recorded the information, delivered it to the local police chief, and sent a copy, duly signed by the chief, to ARRL headquarters for confirmation.[13]

By January 1923, the league was rewarding the most diligent relay hams, those that had worked dozens of stations, with membership in the Brass Pounders League, a sort of fraternity of outstanding hams. All active participants, in those early years, had the number of stations they had worked and reported to headquarters listed in a forthcoming issue of QST.

The success of the relay system tells us a lot about the character of radio amateurs. They had to master the sending and receiving of Morse code. They had to know code abbreviations, of which there were many. Once this was done, they had to learn the ethics of amateur operating, the keeping of a log, and the registering of relays. If they were relay enthusiasts they had to discipline themselves to be at their rigs rather consistently at given times on certain days of the week. Ham radio, especially ham radio involving code, demanded intelligent, dedicated, stable membership. Call them the nerds of yesterday (and today) if we will, they were and are major contributors to the intellectual wealth of America.

Not all hams were members of the ARRL, nor were all of them participants in the relay idea. Yet it can be said with confidence that a majority were at least members of local amateur clubs and that indeed they did participate in occasional traffic handling (relaying). Others were interested only in DX-ing—working stations far away. A considerable number found the most enjoyment in contacting other stations, exchanging signal reports and discussing everything from the local weather to a new circuit design seen in QST. Forrest Bartlett, W9FYK, fell into this group for his on-the-air activity but he was also constantly experimenting with new circuits and new ideas. He kept regular schedules with amateur friends—although Forrest was proud of working Tasmania and never turned down a request to relay a message. Communication was usually by Morse code although as the years passed there was an increasing amount of phone use. Forrest used—and still uses—both.

Hiram Percy Maxim was certainly no typical ham, but his habits probably drew nods of agreement from many a dedicated amateur. He related working his rig so late into the wee hours that he had grown dark circles under his eyes and "had aroused the suspicion of our Chief of Police that I have a hidden stock of booze in my cellar.... At this date I have all the symptoms of the wood alcohol brigade, when in reality it is two weeks come Tuesday since that last near beer went the way of things earthly." He always had the cat with him, which he spat upon (so he wrote) and, in preparation for several hours at his rig, had several corncob pipes filled with tobacco and lined up to be smoked as needed. "How do others stay up so long?" he asked. "Don't their stomachs go back on them?"[14]

Maxim (T.O.M.) did a lot of complaining about "young squirts," as he called them, with cheap spark transmitters. They cluttered the air with their QRM. "Scratch and rasp and squawk your blame heads off. That seems to be what you enjoy most, so have your fill," he complained. "Dear little boys, go on and bang-whack the ether with your miserable squeak coils on any old wave, any old decrement, and [sic] old power, and any old business that will make a scratchy sound at any old time of day or night."[15]

This was a bit too much criticism for some hams. One wrote a sharp rejoinder:

> Who in tarnation and thunderashun is the wild galoot from the west who is always hollering "Rotten"? By heck, this bewhiskered old sun-of-a-gun has got my horned animal, or to be brief, explicit, and to the point, my goat. For the last five hectic and suffering years all I've heard him yell is "Rotten." Tell him to go take a walk, take a bath or a shave. Perhaps he can take a drink (if he can get it).[16]

The spark transmitter was rapidly disappearing with the appearance of the vacuum tube and continuous waves, and the natural but unpleasant resentment of the experienced ham with a good rig against the novice with a poor transmitter disappeared before any serious cleavage took place. Maxim did, however, continue in his inimitable style—he really could turn a good phrase—storming about poorly rigged sets and lackadaisical obedience to the government-established rules and regulations.

Among the increasing society of amateurs were women. Many had become telegraph operators, but were also hams, and had mastered both American Morse and Continental codes. Even before the '20s decade, a few women had mastered Morse code and were active hams. Probably the best known was Mrs. Emma Candler, wife of the St. Marys, Ohio, high school principal, Charles Candler, call letters 8ER. In January 1915, she became the first woman in the United States to be issued a second grade commercial license, 8NH. According to information at the Auglaze County (Ohio) Historical Society, by October 1916 she had been heard in thirty states, including Denver and Lewistown, Montana. QST reported that she had also worked ships off Key West and Cape Race, Newfoundland. Her antenna was said to be eighty-seven feet long and fifty-eight feet high, with six wires and a feed-in from the center. According to the Historical Society, during the First World War she and her husband, also a ham, assisted the government in communications. "Their activities," it was reported, "were 'hush-hush.'" Mrs. Candler is mentioned several times in QST for her part in pre-war relays. When the ban was lifted she renewed her interest in ham radio.[17]

So apparent was it that some technologically-inclined women were becoming amateurs that hams were warned as early as August 1917 that "the Ladies Are Coming." When the ban was lifted after the war, it was predicted, there would be several hundred of the fair sex scattered throughout the nation. "We will have to be careful where we use OW," warned the writer. (OW stood for Old Woman, as hams referred to each other as OM for Old Man.) The women would not take kindly to being referred to semi-affectionately as Old Woman, hams were warned, although they might not mind being referred to as OLs for Old Ladies. How about referring to them as DGs for Dear Girls? (Their designation actually became YLs for girls, XYLs for wives, YF for young wives, and occasionally OW for old woman if the ham had to sign off to do a chore demanded by his wife. More than their designations, hams were warned that amateur language had to be cleaned up. No more saying, "Keep out, you big ham!" It was predicted that the female presence would result in a general uplift throughout the fraternity. A glad hand was extended to them.[18]

By September 1920, women were making their existence known in amateur circles. "Well, where are we now?" asked "the Old Woman," the pseudonym one of them—otherwise unidentified—used in writing an article entitled "Beginning at the End." She was one of the women who had taken lessons from a military unit. "Now what?" she asked. "I sort of expected to find a corner for us in QST, with a prize for the best use of a bent hairpin, or how to keep your number forty-two copper wire combed and brushed and ready for use. There were fifty of us when I began, and there must be hundreds [of female hams] now." She said that it had mortified her to have to explain that she was not Mrs. Candler, "that energetic and enthusiastic hameff."[19]

By April 1920, QST, which just a year before had consisted of a bulletin distributed by Maxim and his colleagues at ARRL headquarters, filled ninety pages and had a circulation of ten thousand. In January 1923, the previous month's relays reached fifty thousand. By June of the same year there were more than sixteen thousand licensed amateurs. They continued running "transcons"—attempts at relaying across the continent. As K.B. Warner of the ARRL wrote, "There is something really very theatric in the line-up for a big Transcon test.... The wonder of it never palls. There is a genuine fascination for us on any old night when we marvel of our ability to sit in a half-darkened room before a little collection of instruments, with the audions dimly glowing and hear the messages from our friends come buzzing in through the night." The January 1921 tests were considered highly successful, with one relay making its way from coast to coast and return in just six and one-half minutes. Annual transcons were planned.[20]

If amateurs could work across America, why not across the oceans? Could a one kilowatt station working at 200 meters communicate with hams in the British Isles? Already there were claims that American hams had been heard there. In late 1920 a Scot amateur named George W.Z. Benzie of Aberdeenshire sent a letter to American radiophone ham Hugh Robinson of Keyport, New Jersey, informing him that he had heard his transmission. However, although it seemed valid, the contact was not officially confirmed.[21]

Marshall and Loretta Ensor (brother and sister) lived on this farm near Olathe, Kansas (near Kansas City). Their 90-foot-tall antenna was raised in 1922 (courtesy archives of the Ensor Park and Museum of Olathe, Kansas).

This station is located at our home 6 miles south of Olathe, Kansas. The transmitter runs 1 KW to pair of 822, modulated by pair 822 Class B. The rondaroom 30 watt speech amplifier. The receiver is a Super Skyrider SX 16 - Dumont Oscillagraph. The aerial is a horizontal Hertz, zep fed, 246 ft long, 75-100 ft high. 73's. Marshall N. - Loretta Ensor, oks.

Hams often exchange information beyond the air waves. Here's a picture that the Ensors sent to many fellow enthusiasts depicting Marshall and Loretta with their equipment. Note the descriptive caption giving the station's stats (courtesy of the archives of the Ensor Park and Museum of Olathe, Kansas).

This is a picture sent to the Ensors from a couple identified only as 9CTJ and 9DBU from Edgar, Nebraska. Note their collection of QSL cards in the background (courtesy of the archives of the Ensor Park and Museum of Olathe, Kansas).

Loretta Ensor was one of the first YLs in the Midwest. She was licensed in 1923 with the call 9UA. In 1926 she became the first American woman to send her voice across the Pacific, and was one of the founding members of the Young Ladies' Relay League (courtesy archives of the Ensor Park and Museum of Olathe, Kansas).

In February 1921, the league announced plans for attempted contacts. The first trials failed because the Operating Department had not had sufficient time for preparations. Even so, amateurs during the test were heard by ships on the far side of the Atlantic. Officials of the league decided to try again. They boasted in print that if a "dyed-in-the-wool American ham could be sent across 'the pond' with a good American regenerator," signals from America could be copied. This idea was the genesis of one of the great milestones in the history of amateur radio, short-wave, and indeed, of world communications.[22]

Detailed preparations were made for the tests on this side of the Atlantic and in the British Isles, where the assistant editor of a British publication, *The Radio Review*, coordinated preparations. At first the league restricted the tests to powerful, selected American stations. A competition was held to select hams who had proven contacts of one thousand miles and were known, through their participation in the relay system, for their reliability. From seventy-eight entrants, twenty-seven were chosen. They were each given fifteen minute periods, specified by the radio districts in which they resided, in which to attempt contact with England. The tests were scheduled from December 7th to 16th, 1921. In addition, possibly as an afterthought, a free-for-all period was designated in which any amateurs could attempt contact with England, and they were encouraged to do so.[23]

So desirous of success were some amateurs that a half dozen leaders, among whom was the inventor E.H. Armstrong and all of whom were members of the important Radio Club of America, determined to build a state-of-the-art transmitting station that was most likely to be heard

Marshall Ensor received "Radio's William S. Paley Prize in 1940 for his feat of teaching nearly 10,000 regular listeners to his W9BSP radio training station over a ten year period," enabling them to pass the FCC test for an amateur radio license (courtesy archives of the Ensor Park & Museum of Olathe, Kansas).

in Great Britain. Station 1BCG was set up hurriedly at Greenwich, Connecticut. Six men, reported *Radio News*, spent two weeks without sleep and very little food in a forsaken little shack far from the main road during one of the coldest periods of the winter. By the time of the trials they were ready, although adjustments were still being made. Not only did 1BCG work Great Britain; subsequently it worked many other far-away stations.[24]

At the ARRL convention in Chicago league officials had decided to send the most qualified ham available to England, with American equipment, to help make the experiment a success. They said they wanted to compare American equipment with the British equivalents. Another reason sounds a bit strange. English hams were not such "boiled owls" as Americans, meaning that they refused to stay up late at night listening to their rigs. If an American was present, quite willing to burn the midnight—and early morning—oil for several nights in a row, chances of success would be increased.[25]

The man chosen was no ordinary amateur. Thirty-two-year-old Paul Forman Godley, slim, of average height with black hair and a narrow face, had adapted the Armstrong regenerative circuit to shortwave use and had originated the variometer regenerators which made possible such wonderful shortwave DX work. He had spent his adult life in radio, working for companies involved with wireless, and in 1913 had helped Brazil establish an "Amazon to the Andes" radio service. His own station 2ZE had established distance records for both day and night transmission. Godley was sometimes called Paragon Paul for the trade name of his radio devices. Officials gave him an impressive send-off on the *Aquatania*, and he was equally well treated in the British Isles. While there he had the pleasure of meeting both Dr. Fleming and Marconi.[26]

It is fortunate that Godley was still a young man in good health. He had planned on working from England but with a severe cold he found the only comfort there in bed with a hot water bottle at his feet. London's fog, so thick he could barely see twenty feet ahead, also convinced him to go elsewhere. His next choice was hardly better. His assignment was carried out in Scotland in chilly, rainy, inclement December weather. He fixed on a small fishing and shipbuilding village southwest of Glasgow named Androssan, found a vacant field nearby, persuaded the owner to let him use it, and there Godley set up his equipment. The field had been fertilized with seaweed, and was a mass of muck. His equipment was carried there in a one-horse wagon which stuck in the mud so that the teamster and Godley and his aides had to get behind the wagon and push. (A Scot District Inspector for the Marconi Company, Mr. D.E. Pearson, "stood a constant watch with him during the tests and verified the reception of every signal.") The 1300-foot-long special antenna was known as a Beverage Wire. It rose twelve feet above the muck and was constantly in need of servicing because the poles seldom remained upright. His 10' × 18' tent blew away on its first rising, but was finally anchored satisfactorily; boards were placed for a floor and boxes the equipment came in were used as chairs and tables. The tent was warmed more or less—mostly less—by a "two dollar oil stove"; there was also a lamp and, of course, Godley's state-of-the-art superheterodyne receiver with batteries.[27]

Godley did not have a transmitter. Arrangements had been made with the very cooperative Marconi company to file a daily message at the company's Carnarvon Radio station, MUU. The report was to be sent at 7 a.m., which was 2 a.m. Eastern Standard Time, "sent slowly by hand so that the amateur world could copy it direct and so get first-hand word from Godley at the earliest possible moment." The Radio Corporation's station receiving the message at New Brunswick, WII, was to repeat, slowly, Godley's messages so that American amateurs "could get the dope instantly." Godley later expressed his wish that he could have been permitted by the British powers-that-be to have his own transmitter, so that he could instantly have contacted his American colleagues.

An atlas will help understand Godley's ordeal. The little fishing village of Androssan is somewhere between the 55th and 56th parallels. Follow them west and you go through the center of

Hudson's Bay. In December the sun rises late and sets early. In addition, howling winds and cold, drizzling rain poured down upon him and Pearson. Godley had to have a good constitution, for he was conducting experiments with a cold so bad that he coughed and coughed and felt weak when he was through. He complained of pains in the back, sore muscles, headaches, and a very stiff neck. And he recalled the "lack of enthusiasm on the part of both Pearson and myself to drag ourselves out of a warm corner by the open fire in the lounge at the hotel, in order that we might don rubbers, overcoats and raincoats, and march out into the awfulness of the Scotch night, only to sit on a hard wood box in a very drafty tent." He wrote that Pearson, a good Scot, suggested that he (Godley) had not consumed enough "Scotland Honeydew."

According to his own narration, it was December 7th at 1:33 a.m. that Godley heard the first American amateur, 1AEP (later verified as 1AAP—name not given). "That this was an American ham there was no doubt!" he wrote. "I was greatly elated, and felt very confident that we would soon be hearing many others! Chill winds and cold rains, wet clothes, and the discouraging vision of long vigils under most trying circumstances were forgotten amidst the overwhelming joy of the moment—a joy which I was struggling to hold within!"

Super station 1BCG at Greenwich, Connecticut, was heard on December 9th at 12:50 a.m. "Signals from 1BCG were steady and reliable," Godley recorded. "Remarkable performance." 1BCG had sent the first private radiogram across the Atlantic, and Godley heard it. In the span of the tests he heard thirty more American stations while eight British amateurs also made contacts with American hams. In one night Godley received eighteen stations, but in six other nights he did not receive one. This was attributed to atmospherics caused by a full moon, which appeared on the fifteenth, and possibly by a severe storm that crossed the Atlantic and struck Scotland toward the end of the tests. (Yet he believed that initially the storm may have helped along his contacts.) The atmospherics led Godley to cease operations on Friday the 16th, one day short of the ten days planned.[28]

A British amateur, W.R. Burns of Sale, Cheshire, heard seven American stations and was awarded a British prize of equipment valued at $600. The leading French amateur, Leon Deloy of Nice, heard seventeen American amateurs during the Godley transatlatics. Maxim reported on the French reception to the *New York Times*. The French were puzzled at some of the cables Godfrey sent back to the States. By previous agreement certain code words were chosen for letters of the alphabet. Thus "one boy cast George" was really the powerful station IBCG.[29]

Godley returned to the United States as a hero. "The scientific world is startled at our ARRL's achievement," crowed the editor (K.B. Warner) of *QST*. "In the most graphic way, we have demonstrated the high radiation efficiency of shortwaves. To put a message across the Atlantic on less than one kilowatt! *It was done!*" Warner, it was predicted, would be measuring his head for a new hat, "old man Maxim will carry a face split from ear to ear!," and the six dedicated hams at 1BCG "will go around with chips on their shoulders and chests stuck out."

By February 1923 American amateurs were reporting many contacts with British hams. By March of 1924 it was reported that two-way transoceanic communications were becoming so commonplace that they no longer bore mention. A "Royal Order of Transatlantic Brasspounders" had been created. Italian and Dutch stations were being heard and, on the other side of the world, rumors were flying that Australian amateurs were contacting America. Most were being worked at between 115 and 150 meters. Within months—hardly years—amateurs were DX-ing throughout the world. Their first contacts with far-off places are noted in amateur periodicals. Hams contacted Hawaii, the Philippines, Japan, China, New Zealand, Australia, Chile, Argentina, South Africa, and European nations that allowed ham activity. By the mid–1920s, it can be said, amateur radio contacts were only restricted by government regulations. The world had become their oyster.[30]

Worldwide contacts brought forth a need for more "intermediates"—meaning letters in call numbers signifying countries. When the system was initiated in 1923 the number of countries with amateurs was considered minuscule. Within two years the number with hams—"some officially recognized, some operating in anticipation of government permission and some—well—bootlegging"—had proliferated. Conceivably two hundred or so countries would want identifying letters, but how could they do it with a twenty-six letter alphabet? C stood for Canada and Newfoundland, so what should Chile's designation be? The solution was for some countries to identify themselves with two letters. Chile's became CH and Finland, which lagged behind France in amateur communications, had to settle for FN. More changes would be necessary as hobbyists of more countries entered the ether.[31]

With the rapid expansion of long distance amateur communications, QSL cards—the means by which hams acknowledged contact with each other—proliferated. They had originated years before out of the desire of one ham to prove that he had worked another. With no official encouragement these notifications had standardized into a fixed format. They were (and are) postcard size, usually printed with the call letters in red or black and at least an inch high. In smaller print are the address of the station sending, the time and date at which the other station was heard, possibly a few words of technical information, and the wavelength used. The amateur's name appears inconspicuously on the card. Most hams tacked every card they received on the wall behind their rigs. Some thought this was unaesthetic: why not just tack up a card from each continent worked?[32]

Terminology and language posed problems. "One boy cast George" was just the beginning of French vexation. They were as interested in contacts with America as the British, but the terminology used—the "QST-English" as it was sometimes called—simply baffled them. For example, one of Godley's messages sent to the ARRL read as follows:

> Heard one ram unit two fox pup two boy mike love stop Code words of these three verified. Courtesy stop. Also heard ... one spark one able ram yacht one boy dog tare two boy kind two dog man three dog pup also following contin wave one able ram yacht one boy cast george one boy dog tare one boy george fox one yacht king one x-ray mike two fox dog two easy have eight able cast for eight away vice stop Strong and reliable—Godley.

A French radioman quoted a correspondent who wrote him: "I'll be damned if I understand anything of this mystery where rams, dogs, foxes and even x-rays play such an important part! Might this not be a code?" And indeed it was. The first letters of each word indicated one of the numbers or letters in the contact's code station. "One able able yacht" was the designation for American amateur 1AAY. But even with amateur phone communication there were difficulties. How were English speaking hams to understand French, German, Spanish, Russian, or any other languages? How were the foreigners going to understand English, including QST-English with its dozens of abbreviations?[33]

Suggestions for solving the problem were considered. Although as many as a hundred attempts had been made at universal languages in the past two hundred years, by the 1920s just two dominated the field. One was called IL, or Ido, or ILO, or IAL—an International Auxiliary Language. It had first appeared in 1907, developed by several linguists. A man named Oscar C. Roos publicized IL and sold IL (or ILO—the "O" stood for singular) grammars for twenty cents to a dollar fifty. He claimed that studying ILO just two hours a week for six months, involving twenty-six lessons, would suffice to bring the pupil to "very advanced grade." Elsewhere he boasted that the language could be mastered in eighteen months. "Think this situation over," he asked readers. "The need of a special means of international communication between radiomen for business and progress is upon you. It has come like a flash from a blue sky. What are you going to do about it?"[34]

In a boxed statement preceding Roos's article, the editors of QST also posed the question of how hams could communicate with fellow radiomen speaking different languages. The ARRL

had been considering the problem for a year. If amateurs all over the world could adopt an auxiliary international language, it was suggested, the problem would be solved. Two such projects
seemed feasible. Roos's ILO was one of them.

Roos noted that numerous international language societies using ILO were being formed
in Europe to promote the idea. They called themselves RAILS, an acronym for Radio Auxiliary
International Language Society. The Russian communists, he said, had adopted ILO and made
it compulsory in their propaganda. It was emphasized that the new idiom was not to replace
national languages but was to meet the needs of technology and business. It "was a tool, not a
paint brush"; it could not be used to write poetry. The Crosley broadcast station in Cincinnati,
WLW, was teaching ILO by radio, and other stations in the Northeast had given talks in English
and ILO. To read Roos's arguments, one would believe that ILO would solve all the world's problems. It was as good as the development of sliced bread.[35]

The other proposed international language was Esperanto. It had been created in 1887 by
a Polish scholar, Dr. L.I. Zamhof, and by the 1920s had thousands of adherents. It was said that
delegates crossing the ocean to a convention in Europe could pick up enough Esperanto while
on board to talk with anyone at the meeting. Fifteen Esperanto conventions had been held
between 1905 and 1924 involving up to five thousand delegates from forty-three countries, all
speaking Esperanto. This, it was strongly suggested, should be the language of radio amateurs
throughout the world.[36]

So concerned were league officials that they queried foreign amateur radio organizations
about the needs for a universal language. Most of them suggested Esperanto. In September 1924,
the league endorsed it, urging all amateurs to purchase texts. At the same time the league recognized that few hams, who after all were hobbyists, would take the time to master it. If, however, an international language came into being, it was almost certain to be Esperanto or based
upon Esperanto. So, league officials argued, the amateur who learned it would be ahead of the
game. It even gave the address of where texts could be purchased.[37]

None of the international languages succeeded with amateurs. In time it became clear that
amateurese, or QST English, worked quite well throughout the world. The long list of abbreviations used by hams, much of it adapted from telegraphers, became in itself a kind of international language. "Q" signals, for example (QST, QSL, QRM, QRN, QSA, QRV, QRO, etc.), are
known the world over. Other abbreviations are also familiar: WX, for example, always means
weather, so temperatures are easily given using WX plus the suffix F or C for Fahrenheit or Cen-
tigrade. Exchanging names has become rather standard: "hr name Jim." Descriptions of equipment do not need much foreign language to be recognized. The mastering of amateurese is an
additional task not often mentioned when a person proposes to learn code, but it is a very important part of becoming an expert amateur.[38]

The discovery that high frequencies (kilocycles) of shortwave radio could work the rest of
the world inspired league officers to contemplate the creation of an international amateur organization. Just as hams need protection both by and from the federal government, so do amateurs
in other countries need as much help as possible to keep suspicious governments from restricting and sometimes forbidding ham radio. In fact, one of the saddest aspects of amateur history
is the paranoia revealed by so many nations in relation to ham activities. Restrictions range from
total denial of amateur privileges, as occurred in pre–World War II Czechoslovakia and Switzerland, to minor restrictions even in Canada and Great Britain. An international organization, it
was thought, might help in the fight for the freedom of amateurs as well as coordinate the hobby
in numerous ways.

Hiram Maxim was aware of the possibilities. With league backing, its officials made arrangements to meet with European amateurs "in a common effort to coordinate free amateur radio
communication between the private citizens of the various countries of the world, represent ama-

teurs in international conferences, encourage international fraternalism, etc." On March 12, 1924, Maxim dined at the Hotel Lutetia in Paris with delegates from nine countries—Great Britain, France, Belgium, Switzerland, Italy, Spain, Luxemburg, Canada, and, of course, the United States. (Denmark could not send a delegate but was represented by a letter strongly supporting any organizational plans the group might suggest.)[39]

French delegate Comte du Waru called for an immediate organizational meeting which culminated two days later with another dinner at another Paris hotel. At this meeting Maxim was elected president with Dr. Pierre Corret named secretary of the not-yet-created society. Thus began the movement for the International Amateur Radio Union. It was further planned that during the Easter half-days of 1925 there would be held another meeting at Paris. In the interim the ARRL was requested to draw up a constitution to be presented to a temporary committee which would study it, make suggested changes, and present it to the meeting for approval on those Easter half-days. Maxim ended his report to the league with the optimistic statement that "there seems little doubt that the spirit of brotherhood that pervades amateur radio in America can be made to spread its mantle of good fellowship over the civilized nations of the earth."[40]

The meeting did materialize. In the peaceful days of the mid–1920s international meetings were easily orchestrated. Representatives of twenty-three nations registered at the conference. Their work proceeded smoothly. On April 17, 1925, the International Amateur Radio Union came into being. It adopted a constitution which had been formulated by ARRL officials; officers had been elected; and four national sections has been recognized. Any country with twenty-five or more amateurs qualified for membership. The influence of the ARRL was apparent when Hiram Maxim was elected its first president and K.B. Warner of the league's staff was chosen international secretary-treasurer; QST was named the union's official organ. When the meeting adjourned the IARU had one hundred and twelve members. Cost of membership was one American dollar.[41]

Warner wrote of the American delegation's pleasant surprise to discover that the French amateurs were not, as they believed, all elderly scientists wearing beards, men who would never be brass pounders. "This was all wrong," he wrote. "The French amateur, and all the rest of them, are just like ourselves, a noisy, happy bunch of keypushers of our own age, tooting whistles and discussing circuits, and talking QST English, bless 'em."[42]

The organization flourished. In 1928 it drew up a new constitution. Previously members had been individuals from countries with few amateurs, but rapidly the hobby proliferated. Most countries began having amateur associations. The new constitution specified that the IARU was to be made up of societies rather than of individuals. On October 30, 1928, the new constitution had been ratified, by which time there were fourteen national societies.[43]

Through depressions, revolutions, civil wars, and two world wars, the International Amateur Radio Union has continued. It warrants a department in QST and provides a strong link of friendship throughout the world. Through it members were able to coordinate activities in defense of amateur radio at the international radio conferences that took place in the 1920s and 1930s. Maxim noted that just two years after its creation, the IARU had been helpful, that because of it some of the other countries involved had learned at least something about amateur radio.[44]

The IARU came into existence when amateurs were busier than ever DX-ing. Those amateurs hooking up with transoceanic contacts were experiencing new sensations, communicating with amateurs in foreign countries more than ever before. They were learning new expressions, new abbreviations, and new styles. "A new kind of fist is coming into being—a heavy, firm style that will make each dot carry across separately—the transocean fist," noted K.B. Warner of the league. A new society, the Royal Order of Transatlanic Brasspounders, had its first members initiated at an ARRL convention.[45]

As the decade of the Twenties advanced, and with the rapid growth of all kinds of wireless communication, regulations, restrictions, and standards agreed upon internationally were necessary. A number of meetings were held, some just regional, some planning meetings in preparation for another huge conference. Amateurs were present, defending their rights, at all of them. That they emerged from them virtually unscathed is something of a miracle.

The most important was the International Radiotelegraph Conference held in Washington, D.C., in October and November 1927. Participating governments submitted a mountain of suggestions which the Department of State published as a book of proposals. It was described as being the size of a Sears Roebuck catalogue and, wrote K.B. Warner, it "made a little bit of light summer reading." Pessimistically Warner reported that amateurs were going to suffer devastating restrictions; at least that was how things looked during the meeting. Of fifty-two nations attending, just four were favorable to amateur radio: the United States, Canada, Australia, and New Zealand. (On one issue Canada was opposed and on others the Italians sided with the Americans.) Other countries (Czechoslovakia, Switzerland) advocated the outright denial of any rights, while some favored restrictions that would seriously hinder the art (Great Britain). Clearly, paranoia is a worldwide problem.[46]

The majority had several reasons for opposing amateurs. It was said that hams violated a nation's monopoly over communications; that a government would be unable to control amateurs; that countries wanted to preserve all wavelengths for themselves; that amateurs would undermine the security of the state; that they indulged in war talk; that nations were reluctant to take on the administrative work incurred by having amateurs in their midst. Even though none of them had many amateurs—the United States with more than sixteen thousand had more than all the rest of the nations combined—they came into the conference determined to slam down amateur radio.[47]

It must have been one of the last of the old line international conferences. Accompanying the delegates were their interpreters, commercial interests, manufacturers, technicians, cranks, and other assorted members of humanity with some interest in radio. Warner described the plenary sessions as colorful affairs, "with frock coats and silk hats, uniforms and brass braid, and plumes and medals and a hushed air of dignity in the main hall.... Many languages in the corridors.... Mimeographed literature and minutes. Tons of it. A busy, busy scene."

"Fellows, they don't like us. We're a nuisance," Warner wrote.[48]

Yet, when, the agreement was signed on November 25, 1927, by 200 delegates and 178 special representatives from 55 countries and 23 dominions and colonies, amateurs came out almost unscathed. The Senate ratified the document on March 21, 1928, it was signed by President Coolidge on March 28, 1928, and it was to become operational on January 1, 1929.

This was a significant victory for amateurs the world over. Warner, representing both the league and the International Amateur Radio Union, and Charles H.S. Stewart, vice president of the ARRL, both put in the better part of two months lobbying in Washington. Marconi, though not a delegate, was quoted in the New York Times as defending amateurs and giving to them the credit, along with American affluence, for the great strides made by radio in the last five years. Another reason for the good fortune was attributed to amateurs' cooperation since 1919 with the army and navy. Major General Charles McK. Saltzman, chief signal officer of the army, Lieutenant-Commander T.A.M. Craven, U.S.N., and Captain S.C. Cooper, U.S.N., representing their military branches at the conference, did liege service in defense of hams. Commander Craven was singled out for his diplomacy, persistence, and resistance when demands harmful to amateurs were made. "It may be said," wrote Warner, "that he is personally responsible for the successful negotiating of the wavelength agreements embodied in the Washington Convention of 1927." From the conference the radio amateur emerged with international status and recognition, with rules for his conduct set forth, his rights and privileges defined. "The privileges," Warner editorialized, "in most respects are entirely adequate. We have achieved a great victory."[49]

Highlights of the agreement included amateur bands retained at 160, 80, 40, 20, 10, and 5 meters, with amateurs in every country allowed to use those bands. Countries retained control over their amateurs, however, with power determined by each nation and—a negative—each nation was permitted to permit or prohibit amateurs as it desired. Each nation could withhold amateurs from any or all bands. International message traffic was to be allowed only by special arrangements between nations. Finally, a new system of amateur calls to indicate nationality was introduced. "International Intermediate" was replaced with simple "de," (meaning "of").[50]

"The unyielding world has yielded, and we amateurs are safely written up in the greatest communication document of the age, the Washington Convention of 1927," Warner concluded.[51]

When the 118 page document was available in English, amateurs, although still pleased with their victory over overwhelming opposition, sobered up. They realized that they were going to have to meet new demands. Available bands were going to be reduced, receivers had to be sharper than ever, transmitters had to be more stable and frequency determinations more accurate. Ninety-eight percent of American hams did not have such sophisticated equipment. The league set out to help, developing the Technical Development Program. Schematics of the more sensitive equipment appeared in QST and, by January 1929, when the document became effective, to a remarkable degree American hams had achieved the necessary goals.[52]

In the summer of 1929 the European Radio Conference was held in Prague, but it hardly affected amateur radio. In October an "International Consultive Committee on Radio Communications" met at The Hague. Such a meeting, every two years, was provided for by the Washington Conference. The State Department assigned K.B. Warner to attend as a technical expert. At this conference, which lasted two weeks and was attended by 180 delegates representing 48 nations, efforts were made to draw up an identical list of regulations for amateurs, but the attempt failed.

By the end of the Twenties decade, amateur radio can be said to have established a permanent niche in the wide field of radio and in the national psyche. With nearly twenty thousand licensed hams, with a system of relaying in place, DX-ing an active pursuit of many of its adherents, and experimentation encouraged, it offered the radio enthusiast a broad spectrum of activities. It had the ARRL as a staunch supporter. A well edited journal, QST, kept adherents informed of technical, social and political events affecting the art. League leadership, especially that of Hiram Percy Maxim and K.B. Warner but including other league officers, kept an eye on developments and adequately defended amateur rights. Secretary of Commerce Herbert Hoover's influence was also vital. (During his administration he sponsored a "Herbert Hoover Cup" awarded annually to the ham with the best, primarily homemade rig.) "We believe that he [appreciates us] to a much greater extent than many of our government and military authorities who are not engineers," commented QST. Federal legislation and the rules of the Washington Radio-telegraphic Conference recognized amateurs specifically, and while regulating the art, they also served as protectors.[53]

The league helped maintain a high plain of amateur activity by sponsoring contests, many of which gave prizes donated by manufacturers of radio equipment, and some of which honored the winners by listing their achievements in QST. Brasspounders received monthly mention as did DX-ers working stations of members of the International Amateur Radio Union. Forrest Bartlett in the early days was determined to have his radio signals sound superior. In April 1931, the league sponsored an "All-April CW Contest." The purpose was to pick the most consistent and reliable radiotelegraph signals from the nine United States districts and Canada. To be eligible the signals had to be "high quality." In September 1931, QST published the results. W9FYK, Forrest's station, was listed with three asterisks indicating the number of times the call was reported. It was a real victory for Forrest, sharpening his desire more than ever to be an above-average ham.[54]

Early on officials of the ARRL recognized the advantages for all amateurs if their local clubs could be somehow linked to the league. It encouraged this, urging officers of the relay system to proselytize the advantages of associating with league headquarters. The process simply involved registration; not even a pennant was given the clubs. But the league did begin a department with its April 1920 issue, "With the Affiliated Clubs" (clubs associated with the ARRL). In its first appearance the editors pointed out how amateur radio's future depended upon a strong federation of interests, and predicted that amateurs would "weld themselves into an impregnable body" that in their hour of need would "be able to stand shoulder to shoulder."

What about the amateurs, hams, or bugs as they were sometimes called? Can we personalize these teenage boys, young men, older adults, and women? That they were in the mainstream of America can best be shown by describing one of their gatherings. The Midwest Convention, as it was officially called, met at St. Louis on December 28, 29, and 30, 1920. Amateurs from small town and country America were awed at the beautiful Hotel Statler, headquarters for the convention, and at the big city itself. Even more were they impressed by the presence of hundreds of fellow hams whose badges, imprinted with their call letters, were automatic invitations to shake hands, get acquainted, and talk shop. An aura of the Twenties permeated the activities. This was a regional convention, but others, whether of a state's hams or radio districts or a national assemblage, had and still have a similarity one to the other. They were and are much the same as other group meetings, whether they be automobile dealers, historians, or garbage haulers.

Although experienced, older hams often seemed to have forgotten the "kid with the little spark transmitter," it was clear that some teenage boys and many very young men constituted the majority of those attending. Singled out and asked to give a talk was a brilliant seventeen-year-old radio technician named Mr. Haddaway. His family was barely middle class; there were no funds available for ham gadgets. The young man had a spark transmitter, a CW transmitter, and a radio telephone, all of which he had built himself. Everything: "storage batteries, B batteries, telephone transmitters, amplifying transformers, chokes, filters, condensers, and bulbs. Yes, indeed, bulbs [tubes]. The only thing he did not make was a pair of pliers.... Makes his own grids, filaments, and plates, pumps his own vacuum, and by heck, builds his own vacuum pump to pump his own vacuum." When one burnt out, he just went down to the kitchen and made another one. One adult ham said he had tried some and found them better than a good many he had bought.[55]

Mr. Haddaway so impressed his audience that they—if T.O.M. is to be believed—predicted that this "good-looking lad with the appearance of a scholar and a reader" would one day be a radio engineer. Amateur radio, the City of St. Louis, and the United States of America would be proud of him. Even T.O.M. praised the young man.[56]

One evening during a profound discussion a Mr. Benson asked for recognition from the floor. He explained that "some kind of nut" was outside the room and wanted to show a new static eliminator. Even as he spoke, the "nut" walked in pushing down the center aisle a hand truck upon which was a crate. He unloaded it in front of the speaker's platform. The moderator, a man named Benwood Bill, began asking questions of the "nut." Benwood Bill explained that there were booths outside the room for the demonstration of apparatus, but not *here*. Bill's temper rose. Then someone suggested that since the machine was already there, the audience should watch the demonstration. Hiram Maxim was chosen to open the crate.

Just as he got his hands on the "blamed thing ... it busted open, and a young lady dressed for anything but winter weather hopped out." She began singing, hugged Maxim, then circled the audience singing how she loved everybody "but was not getting a crumb of satisfaction ...

then the damsel gave a squeal and disappeared through a back door." Now that the "static eliminator" had been eliminated, the session continued.

The high point of early amateur confabs was the first National Convention, held in Chicago on August 31–September 3, 1921. Headquarters were at the Edgewater Beach Hotel but the radio exhibitions were installed in the Broadway Armory and big meetings were held at the Nicholas Senn High School Auditorium. Twelve hundred out-of-towners plus several hundred local hams attended, along with 50 exhibitors. More than 300 sectional clubs affiliated with the ARRL were represented. The convention reporter's record came to 563 pages. QST's report was fourteen pages long with a separate seven page article on "Some New Apparatus at the Convention."[57]

A pattern of national activities had been established. Conventions were being held in radio districts, states, and sometimes in regions put together by the hams. Annual confabs became an integral part of many an amateur's year. The meeting was planned for, money saved for, anticipated with great enthusiasm, and when attended was rarely a disappointment. The ARRL was expected to have a representative at these meetings, giving them an aura of legality within the hobby.

Forrest Bartlett was present at a number of these meetings. He recalls how he, Harvey, and Marvin pooled their meager resources and set out from Boulder to attend their first real ham convention. This was the Midwest Division Convention (the Ninth Radio District) scheduled at Grand Island, Nebraska, March 26–27, 1932. They rode in Forrest's recently acquired single-seated Model A Ford. There was a problem. Fumes from the engine compartment were so bad that they had to drive with the windows open, in chilly March weather. After passing through Fort Morgan, Colorado, the three decided to do something about the problem. They determined that the fumes were coming from the oil filler cap. One of the boys took the rag Forrest used to check the oil level, removed the filler cap and stuffed the rag inside the pipe. For a few miles they drove in comfort with the windows up, then a loud banging came from the engine and the Model A slowed to a halt.

"Who put the rag in the oil filler?" asked the mechanic from the Ford agency at Sterling, Colorado, who came with a wrecker to fetch the car. He then gave them a lecture on how important it is for the crankcase to breathe, otherwise pressure builds up and forces all the oil out of the rear main bearing. Forrest's Model A's engine had self-destructed. They did not make it to the meeting.[58]

The holding of annual conventions by amateurs was just one indication of the energy and enthusiasm enjoyed by the hobby. More evidence was the number of clubs that had been formed by hams. There was the WAC Club—Worked All Continents Club. Members had to prove that they had established two-way communication with at least one ham on each of the six continents. To qualify, a letter or card verifying the contacts was sent to ARRL headquarters. An official WAC certificate would then be mailed to the applicant, signed by the Grand High Wacker himself. Those receiving this highly-prized certificate would be listed in the IARU section of QST every month.[59]

Then there was the RCC, or Rag Chewer's Club. It was initiated by hams who remembered the "good old days" when a ham worked another ham and chatted for hours with him (or her), thus getting to know each other personally. With the advent of the tube and the regenerative receiver and the challenge of DX, such lengthy, friendly conversations had become passé. This, some hams believed, was too bad. To encourage more lengthy, friendly conversing the RCC was created. Anyone who worked another ham and "talked" with him or her for a half hour, followed by both of them writing to a specified employee at ARRL headquarters, would be sent a card verifying their membership. To "get out" of membership all one had to do was call a fellow ham and then say something like "nil hr OM cul 73" ( "Nothing here OM see you later 73").[60]

At the second national ARRL convention, held in October 1923, again with headquarters at the Edgewater Beach Hotel in Chicago, was held the initiation ceremony of the Royal Order of the Wouff Hong. Hiram Maxim, who loved to create new words, had produced the Wouff Hong, a non-existent instrument that was vaguely pictured as a rough piece of wood shaped slightly like an axe head. The Wouff Hong was used to punish recalcitrant hams, especially the "young squirts" who violated federal regulations and caused interference that bothered the real hams, the ones with expensive rigs like Maxim's.

The ROWH was described as "a fraternity, a lodge" of one degree, strictly of and for the good ham, be he an ARRL member or not. The organization originated with the Flint, Michigan, amateurs, or gang, as they liked to call themselves. With the 1922 Michigan convention just days off (and this was prior to the 1923 ARRL convention), it was suggested that something light-hearted should end the meeting. Someone came up with the idea of an initiation of amateurs into some kind of club. One of the hams said that he had in his desk the ritual for initiation into "the Ancient Order of Mop Handles," whatever and whoever they were. They found that order's initiation just too crazy, too foolish; even so, it gave them something to work with. They put together a new, somewhat less ridiculous ritual. Now what should they name the organization? Someone remembered Hiram Maxim's Wouff Hong, and that did it: the Royal Order of the Wouff Hong was born.[61]

The first initiation must have been a farce. No Bible was available for taking the oath, so a Sear Roebuck catalogue had to do. When the candidates appeared before the Power Amplifier it was discovered that each of the characters conducting the ritual had been given the wrong part. Ad-libbing set in. In spite of the mishap the characters did well, possibly because the audience had no idea of what to expect. Representatives from the ARRL attended and were impressed; they approached the Flint gang and suggested that the RORH be a part of ARRL activities. If the Flint fellows would get to work on the ritual, polish it up, make it more solemn and dignified, then the ARRL would approve the ROWH as a part of amateurs' conventions. How about having a ceremony ready for the Second ARRL National Convention?[62]

The Flint gang worked three nights a week, twice asked for suggestions from the Hartford office, and were prepared to present a solemn and dignified Wouff Hong initiation at the National Convention. When the time came, their $1200 worth of robes arrived just in time for them to don them, apply makeup and wigs, and put on a good show. Everyone was impressed. The ARRL accepted the offer of the Supreme Council of the ROWH (all Flint amateurs) to make the order a part of the league connections with ham conventions.[63]

As the Twenties decade progressed, Americans increasingly recognized amateurs as useful hobbyists. In years that witnessed severe hurricanes, floods, and other emergencies; when there was great interest in explorations of the polar regions and of other places not yet thoroughly known; when readers avidly read newspapers about the accomplishments of airmen; mention of amateurs and their part in all of these activities registered upon the American psyche. As was written in an editorial:

> [The amateur] has perfected a wonderful communication system that is available for public service in time of community peril or national emergency.... Governments possess in their amateurs a self-supported reserve of trained specialists.... What other hobby can boast as much value to the government...? The United States must ask of what use twenty thousand golfers, pinochle players, stamp collectors or pool players would be to this nation some nice fine morning when the sky goes black with enemy aircraft.[64]

# 7

## Amateurs as Experimenters and Adventurers

Most true amateurs are experimenters. In the two decades 1920–1940 radio was still so new an art that almost everywhere an amateur looked, from antenna to ground and all the parts to his transmitter and receiver in between, he or she saw reason to experiment. Some of their early experiments are amusing, yet they reveal serious thought, an urge to improve, and delight should the experiment work. More adventurous hams accompanied expeditions into exotic places. Having radio along was in itself an experiment, and the amateurs operating it experimented constantly. Who could they contact and how far away? Where to place the antenna? How to travel under primitive conditions and keep the rig functioning?

They were hobbyists and most were young, so some of their experiments were amusing. To advertise their meetings at a university, hams mounted a receiver and Magnavox loudspeaker in a baby carriage with which they serenaded sorority girls in the evening. The transmitter was mounted on an automobile that followed several blocks behind the perambulator.

An example of the constant ventures going on was the report of a "Radio Lizz," a Model T Ford coupe with a radio in it. (Model Ts were so often called lizzies, tin lizzies, and flivvers that any of the three nicknames meant Model T and nothing else.) The antenna began in the middle of, and at the lower part of, the radiator and with insulators went up over the top and down to the spare tire in the rear. The receiver was just inside the rear window behind the driver. The "long, slender, loud-speaker" was attached to the top of the interior, crosswise so that the opening was right above the driver's head. It was a small set for the time, just five and a half inches high, eighteen inches long, and four inches deep. From left to right of the control board were the dials: a primary tuning condenser dial, tickler dial, inductance switch, detector amplifier rheostats, and an amplification switch. The experimenter admitted that QRM (interference) was pretty bad when the motor was running but when just parked, the receiver had brought in amateur stations up to a thousand miles away. Public interest ran high: the aerial just above the radiator cap had lost its insulation from passers-by feeling it. The year was 1922.[1]

The "Radio Lizz" was just equipped with a receiver. It would not be long before some experimenter tried to add a transmitter to his flivver and thus have a complete rig. The first attempt to be noted of a complete ham set in a tin lizzie took place in 1925. Oliver Wright, 6BKA, was the young ham's name. He described himself as an "inmate" of the University of Arizona. With the help of one of his professors and amateurs at the college radio station 6YB, they built a radio transmitter-receiver designed to be installed on the right side of the seat in a Model T Ford roadster, a true "radio flivver." On the rear tire cover Mr. Wright painted his call letters, his initials, and "73." He inserted a telegraph key on the steering wheel brace and connected it to the horn. This, powered from the magneto, gave an excellent imitation of a rotary gap. Then he mounted a defunct 50-watter (a 50-watt transmitting tube that had outlived its usefulness) on the radiator cap, which glistened in the sunlight "and caused more near-accidents than a pair of rolled top socks." The loop aerial was mounted on the right side of the car, fore and aft. One unit was wired to act as both transmitter and receiver. It was placed on blocks of sponge rubber in the extreme rear corner of the seat. "The set itself," Oliver reported, "took up such a small space that there was plenty of room left over for two operators without crowding." At 9½ × 11 × 15 inches, the set could be picked up by one person and taken elsewhere.[2]

Signals from the car could be copied letter perfect on a loudspeaker at 6YB while the Model T was traveling thirty-seven miles an hour. The receiver did not work as well because of the ignition system and loose leads. To check this difficulty the ignition was shielded by winding heavy rubber-covered wire around the various leads and grounding these to the car's frame. (The ignition was a Bosh timer, not the standard Ford equipment which would have posed a much more serious problem.) The car battery was used to light the filament (tube) and a dry cell was placed in series to boost the voltage. A notice was stuck on the windshield giving their call letters, the purpose of the trip, "and a fervent request that no fool questions be asked."

The trip, meant to give the radio a good try, was thirty-two miles over the mountains north-northeast from Tucson to Oracle, Arizona. The young hams had good luck with both transmitting and receiving, even when the flivver was running along at a moderate speed. The only real problem was the near wrecks caused by curious motorists.[3]

Forrest Bartlett, W9FYK, in Boulder, Colorado, was likewise an experimenter. He had a 1929 Model A Ford cabriolet. Why not, he thought, string an aerial back and forth in the underside of the canvas top? He tried it and it worked very well—until it rained.

The desire to become mobile, to be able to communicate from a car, was widespread. A ham named F. Johnson Elser, who lived for years in the Philippine Islands, installed a transmitter and receiver in his 1926 Ford Model T coupe. He described how his car was equipped with a pole antenna mounted to the back bumper with his rig installed behind his seat, how he powered it, and how he eliminated static. He returned to the States via the Indian Ocean and the Suez Canal to Italy, where his car was unloaded. From there he made his way through Switzerland, France, and England before embarking for the States. He enjoyed visits with a number of hams in the nations he passed through, and although he sometimes had trouble with the authorities, he claimed to have had a wonderful time.[4]

For thousands of hams, mostly young, amateur radio was indeed a joyous hobby, and nothing more. Most of their significant experimentation was with schematics and the various uses of the parts inside the rig, or possibly with the antenna, rather than with automobiles and other new contrivances of the young twentieth century. Yet some advanced, as did radio, to far more serious efforts. Some joined expeditions into remote regions.

The years between the two World Wars were also the last great period of on-earth exploration. One thinks of the expeditions of naturalist-explorer Roy Chapman Andrews into Central Asia or of Lowell Thomas, later a radio news commentator, traveling through the Khyber Pass into Afghanistan, or of Osa and Martin Johnson in Africa. All of them understood good public relations. However, there were many other expeditions in the two decades 1920–1940, forgotten today but at the time making significant progress towards knowledge of the remote parts of the earth. In this period QST ran a department variously titled "Contact with the Expeditions" or "The Month with the Expeditions." All of them, or certainly the great majority of them, had amateur radio operators along. Radio was the only contact these groups had with the rest of the world. Hams in "civilized countries" were urged to contact them and carry out their requests.[5]

Reports of these expeditions are to be found gathering dust and largely forgotten among their sponsors' archives. Some of the reports undoubtedly include mention of the amateurs along, their hardships and contributions. Fortunately an occasional ham saw fit to write of his experiences and submit them for publication. One such person was Harry Wells, W3ZD. He was employed by Westinghouse in Pittsburgh but had traveled to Washington, D.C., to attend a Maryland-Virginia football game. An item in a Washington paper caught his eye, an announcement of the All-American Lyric Malaysian Expedition. It was to consist of a small group of scientists bound for Borneo "to study the primitive natives, to obtain geographical data, and ... to make observations on tropical and equatorial radio conditions." Wells was intrigued. He contacted the leader, Theodore Seelmann of Chicago, described his ham activities, and was hired.[6]

The island of Borneo is tropical, warm and humid, with swamps along the south and southwest lowlands while mountains in the interior rise to more than 13,000 feet. Under the towering trees of the rain forest are at least 15,000 species of flowering plants (nine percent of the world's total) and 185,000 animal species (sixteen percent of the world's total). Tigers, elephants, and rhinoceroses are among the fauna that occupy, or did inhabit, the island. When Wells accompanied the expedition there more than seventy years ago, some of the island was still unexplored. Cannibalism and warfare among the native peoples still existed. In 1929, modernity was hardly affecting Borneo.[7]

Wells was well equipped. His three transmitting and receiving sets were purchased with the advice of experts at the U.S. Bureau of Standards and the U.S. Naval Research Laboratories. His primary unit, described as "semi-portable," had a "fairly long range"; the second unit was "an emergency transmitter to be used in case of any serious breakdown"; and the third was a portable set to be used by advance parties to contact base headquarters. Heintz and Kaufman, the shortwave specialists of San Francisco, furnished the first unit, the same as used by the Trader Horn radioman (see below). The manufacturers of the other two units were not stated. Wells had little to say about the second unit, but he described the portable rig as a Burgess aircraft type with power supplied entirely by batteries. Including a little gas engine generator, the equipment—all three units—altogether weighed less than two hundred pounds.

In March 1929, four Americans—Wells, Theodore Seelmann, and Mr. and Mrs. John H. Province of the University of Chicago—embarked from Seattle; Mrs. Province would accompany her husband only as far as Java. John Province was a graduate student in anthropology at the University of Chicago. Seelmann was employed by the American Mohawk Corporation, described as a radio concern. By maintaining constant supervision over their supplies, the expedition arrived in southern Borneo with the radio gear in good condition. Only once, when the carton containing battery acid was placed upside-down, was there a crisis, but the corks held and the acid remained capped in the containers. Being aware of the excess heat and humidity they would encounter, efforts had been made to seal or impregnate the apparatus as protection from the tropical moisture.[8]

Their first glimpse of Borneo was not encouraging. "The heat," Wells reminisced, "seemed to come rolling out to meet our small coastwise steamer. The shoreline was indefinite and appeared as a rather depressing maze of swamp and jungle." They landed on the extreme south end of the island at the Dutch settlement of Bandjarmasin, population forty thousand, of whom just a few hundred were whites. From there the expedition started up the Barito River aboard a little Dutch river boat, the *Niagara*. Sometimes progress was almost blocked by the hyacinths growing from bank to bank. Wells was fascinated by the flora and fauna:

> The strange jungle odors, the bright-hued tropical birds flying overhead, the herds of chattering monkeys playing along the banks, the occasional wild boar or deer seen cautiously quenching its thirst, the crocodiles or snakes gliding through the muddy, sluggish water, all seemed to be saying, "this is the road to adventure and the real things of life."

After a week the *Niagara* reached the village of Poerock Tjahoe, the last Dutch military outpost on the Barito, two hundred fifty miles from the coast and dead on the equator. The white
population consisted of the post commanders, two young lieutenants, and a doctor. Poerock Tjahoe would be the expedition's headquarters. Here they assembled the radio while natives emerged
from the jungle and squatted around, curious about the strange contraption with the little gas
engine generator put-putting away. Wells put on his earphones, grasped the tuning dial and
"through the terrible QRN from the gas engine's ignition, heard 'CQ CQ CQ de W6BYY
W6BYYW6BYY ar...' Ted gave the gas engine a twist and off she roared. I answered with a long
appealing call—the first time on the air for PMZ.... By all that's holy in ham language, W6BYY
[Robert C. Walton of Mountain View, California] came right back at us. Those thousands of
miles which separated our little group in the heart of wildest Borneo had vanished into the ether.
Gentleman, that *was* a thrill."[9]

Radio had its problems. The humidity affected the gas engine carburetor so badly that when
Wells turned off the engine to listen, the carburetor flooded or choked so that he had trouble
starting it. He had to pull the spark plug, wipe it off, replace it and restart the engine, by which
time the QSO (radio contact with another station) was lost. He soon learned to shield the ignition to prevent interference (QRM) while the engine ran at low speed with the clutch disengaged.
In a short time Wells had established schedules with hams in the Philippines and California;
they, in turn, relayed messages, including a daily report, to the expedition's Chicago office.

Once they were acclimated, more or less, to the heat and direct rays of the equatorial sun,
plans were made for an exploratory trip to the headquarters of the "treacherous Murun River—
territory never before seen by a white man—while at the same time making a search for the most
primitive of natives, the nomadic Punan Dyaks." The portable rig along with batteries and spare
parts was packed in a watertight metal box; it weighed about sixty pounds, "a good load for a
coolie." Captain J.C. DeQuant, the post commander and *controleur* of a portion of central Borneo larger than all of Holland, was in charge. Five convicts serving time for murder, as well as
a few other natives, were to assist in paddling the boats and carrying the supplies. Perishables
were packed in five gallon gasoline tins, indispensable and ubiquitous in the tropics.

The first two days out the heavily laden boats paddled through muddy, sluggish water. On
the third day the banks became steeper and the water more rapid. By evening they were at the
edge of the Kiham Hatas, the longest single rapid in Borneo, "600 yards [3+ miles] of water fury."
Wells added:

> The ensuing month was one continual story of man's battle with the elements. There were days
> of hard paddling—days of roasting in the intense heat; sudden showers would soak us through;
> then the slightest breeze would chill us to the bone; swarms of insects gave one little rest. Rapids,
> waterfalls, narrows and whirlpools had to be encountered, where the slightest error in judgment
> might spell destruction for all.
> We were too busy and tired to heed any rumors of unfriendly natives.

At last they arrived at Toembang Topus, the last village on the Murung and close to the
headwaters. They had to continue their exploration without Captain DeQuant, who had business at another isolated village. Every few days, or when conditions were favorable, Wells set up
the portable radio. KA1CY at Manila was on the air every evening at 6:30 "looking for my war-
ble," Wells reported. With a one-hundred-foot antenna he was almost always able to work the
Philippines. When not attempting to contact other hams he sometimes tuned the receiver to
stations broadcasting music. The natives enjoyed hearing it but were amazed when Wells
unhooked the aerial and the music stopped. Kindly spirits were at work, of course.

Making sure that the camp with the radio was well guarded, Seelmann, Wells, and two Dyaks
made the dash to the headwaters of the Murung. The river became so shallow that they soon
were wading, while logs and overhanging creepers impeded their progress. By afternoon they

had reached uncharted waters. They cut a small clearing in the jungle "and erected a shrine to the Goddess of Fate who had guided us so far," wrote Wells. A signed statement was sealed inside a gourd, their homemade PMZ flag bearing the ARRL emblem was raised, and an old battery and radio tube were left on a platform. With their rifles they fired several salutes in the air. "The Dyaks," Wells recalled, "seemed deeply impressed by our solemn ritual."

By evening they were back at the advance camp. Unfortunately the batteries powering the portable radio had lost so much power that messages could not be transmitted. Manila at the time was experiencing a typhoon. Philippine papers theorized that the little expedition was lost, strayed, or even eaten; when such reports reached America there was genuine concern for the safety of the All-American Lyric Malaysian Expedition. Two days later Wells and Seelmann reached base camp, where the number one transmitter was located. They informed the world that all was well. In days to follow Wells worked Australia, New Zealand, and Singapore as well as the Philippines and the United States.[10]

Their adventures were not ended. On one excursion they arrived at a village in which the women and children were cowering in their huts while the men prepared to defend themselves against two hundred enemy warriors suspected of hiding in the jungle, planning to attack them. Nothing happened. On Dutch Queen Wilhelmina's birthday the resident governor of Borneo and the commanders of the Dutch forces made a special trip to Poerock Tjahoe, where they enjoyed "chow" at the expedition's camp. The Dutch officials were impressed by the power of the shortwave radio, and especially appreciated the messages of greeting from American hams on behalf of the Dutch queen's birthday.

Weeks later, but while the expedition was still in Borneo, Captain DeQuant was brutally murdered less than an hour's hike from the camp. Without radio the quickest the Dutch colonial government could be informed and take action was by boat to the coast and return, a matter of two weeks. Residents of the entire area in which the expedition was camped could have been massacred by then. With the expedition's radio, however, notification went to ham stations KA1AF at Fort Mills and KA1CY in Manila and from there to Bandjarmasin where, on the very night of the disaster, the Dutch received the message. Wells's radio, by way of these relays, continued to aid the Dutch and helped prevent a general uprising.

Theodore Seelemann wrote a letter to *QST* thanking the ARRL for its cooperation with Wells in getting information to America. He also mentioned the aid given by radio to the Dutch in putting down the rebellion. "There is a feeling here," he wrote, "that the Dutch colonial government will elevate the status of amateur radio operators in these great islands and license them just as our own government does.... If this comes about we shall feel mighty proud, and I personally shall bring the matter to the attention of the Governor-General at the first opportunity."[11]

Shortly thereafter the little gas engine was shut down for the last time. Central Borneo was left for good. As for radio, Wells emphasized "what a wonderful and staunch machine it is, carrying us all forward into the realms of science and adventure." Certainly the expedition was a great adventure for the young man. Radio was demonstrated not only as a practical way of keeping in touch with the outside world but was also shown to be of great benefit to governments. It really had helped put down an incipient rebellion.

Just before the depression of the 1930s set in—remember that it took a couple years after Black Friday in October 1929 for pessimism to prevail—a number of ambitious plans were fostered in the realm of transportation and communication. One of them involved Mexico and Central and South America. Before the plan ended automobiles and shortwave radio, to say nothing of the intrepid pathfinders involved with them, were tested to their ultimate capacities. In 1925 the first transmitter and receiver in a Model T were described; by 1930 building a portable rig for a Model A Ford was considered a simple task.

The challenge was a highway from Fairbanks, Alaska, to Buenos Aires, Argentina. To make this possible the Pan-American Highway was suggested in 1923 at the Fifth International Conference of American States. The most-used section of the proposed route, known as the Inter-American Highway, runs today from the U.S.–Mexican border to Panama. Interest in such a highway increased during the 1920s and early '30s. The states of Mexico, once the revolution in that country was over, established committees which met, raised funds through their legislatures or from the central government, and in at least some instances worked hard to build their sections of the road. Central American states through which the proposed highway would pass at least showed interest.[12]

The routes of choice into Mexico were from Laredo and El Paso, Texas. This troubled Americans on the western borders, especially Californians. Concern promoted action. The Automobile Club of Southern California proposed construction of a western Pacific Highway. Dubbed the International Pacific Highway Project, the grand scheme called for a route bordering the Pacific Ocean from Fairbanks, Alaska, all the way to Chile and then east to Buenos Aires.

The immediate goal was Panama, but first a route had to be forged to Mexico City. So determined were club members that they raised funds for a pathfinder expedition. Its orders were to trace a route as close by the Gulf of California and the Pacific as possible. The first expedition set out on March 15, 1930; the second began in 1931 where the first left off, making its way through the jungles and mountains to El Salvador. There was no third expedition.

The 1930 caravan consisted of nine men and five vehicles, three trucks with heavy special bodies, and two passenger cars, apparently all Model A Fords. One of the cars was designated the radio car. Their radioman, amateur Bertram Sandham, W6EQF, was given just ten days to assemble his rig. It was to be a portable high frequency station with call letters IPH. For the transmitter and receiver he designed a single carrying case, assembled spare parts, built a portable mast, and hung an extra storage battery beneath the rear floorboards with switches "for throwing both batteries on charge or in series for [the] transmitter." He took along two antennas wound on a reel, one for each of two bands. Between the front and rear seats a large waterproof box was positioned to contain fourteen 45-volt "B" batteries, two for the receiver and the rest for the transmitter. If the rig was set up in a tent or a town square, a seven-wire cable from the power source in the car led to it. A portable table and camping chair completed his equipment. Two, possibly three, Los Angeles ham stations were scheduled to be in contact with him.[13]

"Sturdiness," wrote Sandham, "was the paramount consideration in the construction of the unit." All parts were carefully anchored and heavily soldered. "How wise the extra effort in this regard," he recalled, "for the radio equipment took a terrific thrashing and was almost daily subject for discussion.... Despite the well-soldered joints, they were broken loose occasionally."[14]

Interest in the project was high. The *Los Angeles Times* would follow the expedition's progress by means of Sandham's shortwave radio. "A portable sending apparatus will give the story of the day's adventure," the *Times* reported in the first of many articles on the journey. "It will be possible for amateur radio operators to tune in every night." One of the nine men was an *L.A. Times* journalist, Harry Carr; he may also have represented the North American Newspaper Alliance. Many of his articles, sent out through Sandham's transmitter, ran on page one while extensive feature stories also appeared in several Sunday issues.[15]

The careful preparations for the radio were equaled for other necessities. Water and fireproof compartments behind the drivers' seats were filled with motion picture and still camera equipment. Delicate engineering instruments and ammunition for rifles and automatics were all carefully packed. One of the cars was equipped with a compact kitchen with compartments for food and camping equipment. The two passenger cars, one of which carried the radio, had been equipped with truck springs to bear the heavy load. Sandham's car was loaded with non-radio equipment including a rifle mounted behind the driver; the other car had two compasses mounted on it, an altimeter, a mileage computer, and other devices for the survey work.

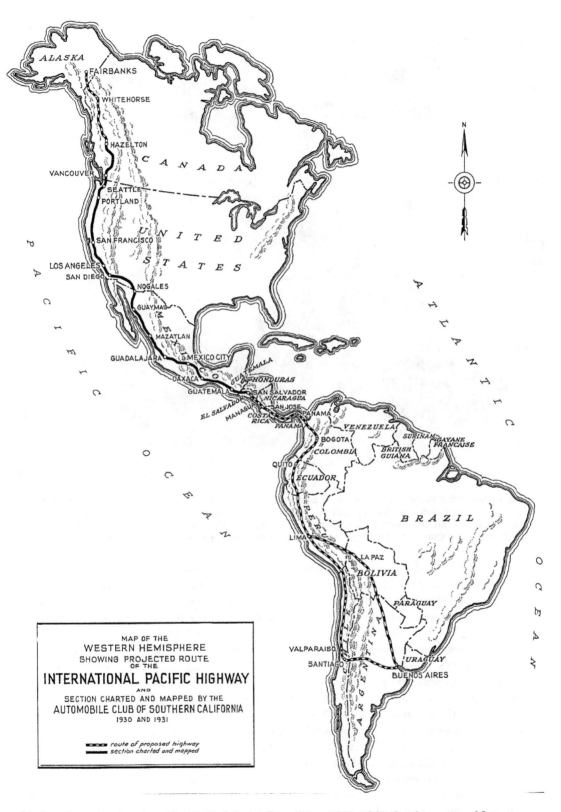

Section charted and mapped by the Pathfinder Expedition, 1930–1931, for the proposed International Pacific Highway (courtesy Automobile Club of Southern California Archives).

**Members of the Pathfinder Expedition in front of their cars. Standing, left to right: J.R. Deason, Ernest E. East (leader of the expedition), unidentified woman (not part of the expedition), Douglas C. Rhodes, and Kenneth Keefe. Seated, left to right: Phil T. Hanna, Bertram E. Sandham (radio operator), and Carl E. McStay (courtesy Automobile Club of Southern California).**

Front page headlines announced their scheduled departure on March 15th. By March 17th the caravan had reached the Mexican and American cities of Nogales. Harry Carr, the *Times* reporter, used some purple prose to depict the contrasts at the border. "On one side of a chicken-wire fence is a little American border city with Rotary clubs, fan magazines and real estate subdivisions. On the other side ... is mysterious, inscrutable, changeless Mexico, looking out from under its rebozos [long scarfs worn by Spanish women around head and shoulders] with melancholy eyes."[16]

Sandham's unusual adventures began at the Mexican Nogales where he set up his rig for the first time. Immediately the mayor, the representative of the governor of the State of Sonora, and other officials filed messages. "Had it not started to rain," he wrote, "I believe I should still be sitting in the center of a street in Nogales pounding out the messages." His real assignment, to make contact with hams in Los Angeles, did not go so well. He was unable to contact them at first although he was heard by a boy amateur in Cambridge, Ohio; when he finally did contact Los Angeles, the signal faded out. Fortunately he was heard by a military operator at Schofield Barracks in Honolulu, who relayed Sandham's message back to Los Angeles."[17]

After several days at Nogales waiting for the sun to come out and the road south to dry, Sandham wrote, "we started into Mexico in earnest and the battle was on." At villages and cities the radio rig was always surrounded by jabbering natives who had never seen a radio, while some

Bert Sandham, the radio operator of the Pathfinder Expedition, working his rig that is installed in the car (courtesy Automobile Club of Southern California Archives).

of the children had never seen an automobile. *Times* correspondent Harry Carr's dispatch, dated March 23, was from Hermosillo. "Pathfinders Ride into Burro Era," headlined his article, adding, "Auto Club Party Discovers Welcomers All Abed on Reaching Hermosillo." The lengthy article, mostly of the travelog variety, was sent in entirety from Sandham's portable transmitter and received in Los Angeles by amateur Charles A. Hill, W6BRO, who relayed it to the *Times*.

One of Sandham's constant problems was with raising his antenna. "At one time or another," he recalled, I used "ice plants, breweries, flour mills, penitentiaries, cathedrals, city halls, weather observatories, governor's palaces and state buildings." Sometimes the most logical place proved impossible because a nearby telephone pole supported a bank of transformers. "In some cities," he added, "the electric wires hung so low that women hung their washing on them!" At other sites the wires of telegraph and telephone lines were supported by old tree trunks; in one place barbed wire had replaced the copper wire, which had been stolen by farmers.[18]

Humanity, he pointed out, was not the only problem. "The infuriated dogs—and there are millions of them in Mexico, all the nondescript breeds—would occasionally snap at my heels as I sat at the transmitter…. At Navajoa a burro lazily made his way up to my side and, flipping his upper lip several times, started to eat the power cable running to the car. At Guamachil I felt a warm draft of air on my neck and turned in my chair to face an immense long-horned bull with his horns encircling my head like a halo…. At Santiago several soldiers endeavored to detour a flock of pigs around my QRA, but managed to get them in a frame of mind that the only way out was beneath my table and chair."[19]

As they traversed Yaqui country, so named for the warlike tribe inhabiting the area, two truckloads of Mexican soldiers, one in the front and one in the rear, and in the middle of the caravan a truck half filled with friendly Yaquis as hostages, gave protection. *Times* correspondent Harry Carr, again sending his story via Sandham and Los Angeles ham W6BRO, described the country "as the Rhine frontier must have looked under Caesar—miles of mud forts within sight of each other—sentries standing on the battlements scanning the horizon." Southward advanced the expedition, from Sonora into Sinaloa.[20]

The caravan continued down the western edge of Mexico. Sandham's tasks included driving the radio car. It was a tough job, and often a scary one. Some of the roads were little more than ox cart trails. When they arrived in cities, sometimes between 8 p.m. and 4 a.m., they found the city fathers waiting for them. Dead tired, the pathfinders had to attend the festivities. Their most difficult progress was in the Mexican state of Jalisco. By now they were heading southeast toward Mexico City. The barrancas (gorges) of that state have been compared to "several Grand Canyons scrambled in all directions." Just two of the five vehicles remained to traverse the region; the others, including the radio car, were placed on Southern Pacific railroad cars and thus transported to Mexico City. Sandham remained behind with the others, all of whom were determined to get the two cars through. Three days were spent surveying on foot the area to be traversed. A road did exist—a five-hundred-year-old trace used to carry Spanish treasure across

A view of one of the smoother roads over which the Pathfinder Expedition advanced (courtesy Automobile Club of Southern California Archives).

to the Atlantic. Long lines of oxen had pulled the primitive, squeaking high-wheeled vehicles. How would automobiles fare? On the first day, from sunup to sundown, three miles were covered. The grade out of the first canyon was 43 percent "and block and tackle were used for 400 yards, with oxen holding the car on the road so that it would not disappear with its driver into the oblivion in the abyss beneath." They progressed sixteen miles in three days. The party ran out of water and food and when Magdalena (a small town west-northwest of Guadalajara) was reached, their mouths were so parched they were unable to talk. In other areas they confronted for many miles solid masses of boulders The cars took terrible abuse: frames were bent, a tie rod would bend double and the front wheels turn out in opposite directions, and many a shock absorber was removed. When they finally arrived at Mexico City, they estimated their average speed for the 2,280 miles to Mexico City at 11.4 miles per hour.[21]

In Mexico City, Sandham erected his antenna on an oil derrick five miles from their hotel. Then the rains came. The decision was made to cease the southward trek for the year and head for home. Now the expedition advanced northward along the east coast of Mexico. IPH went off the air and the sole aim of the group was to return to the States via Laredo, Texas, then westward to California across New Mexico and Arizona. The roads were not a challenge, at least not in comparison with what they had experienced down the west coast of Mexico, and they returned safely.[22]

Not to be deterred, the next year (1931) the Southern California Automobile Club financed a second expedition. This time they would travel in just three strengthened Ford trucks which were shipped from El Paso to Mexico City by railroad. From there they would start south through southern Mexico, Guatemala, Salvador, Honduras, Nicaragua, Costa Rica, and finally to Panama—or as far toward Panama as they could go before the rains set in. Much of the terrain, they knew, had never been seen by a white man. Serious obstacles lay along their route, but, said E.E. East, the engineer in charge, nothing short of complete disaster would stop them.[23]

On January 25, 1931, the crew set out. The party again included Bertram Sandham, W6EQF, as radioman, and five others including a journalist for the Los Angeles *Times* named Clarence Martin. As for the radio, improvements had been made based upon the past year's experience. Instead of batteries a small, two-cycle single cylinder gas engine powered a generator. Weighing just thirty-one pounds, it ran as high as four thousand revolutions per minute and gave promise of excellent service. Again, the radio was expected to be in communication with the automobile club headquarters and with the *Times*.

The first eighty-two miles, to Puebla, Mexico, were accomplished easily over a passable road. From then on the expedition was on its own, making its way through jungles, mountains, and across rivers. Sandham wrote that the previous year's expedition, over ox-cart trails and rocky terrain, seemed like a pleasant vacation trip compared with the problems they encountered as they made their way slowly—very slowly—into Central America. Even so, the "biggest bugaboo" of the trip, he wrote, was the curiosity of the natives. The racket of the little one-cylinder gas engine brought them to the source of the sound, and he found himself soon surrounded by curious onlookers.[24]

When they traveled down old ox-cart roads, the width of the ox-carts being six inches less than the width of the car wheels, one side of the car ran along the ruts, which shaved the sidewalls "like an emery wheel." Once the pathfinders were held up by brigands. Fortunately the interpreter talked the bandits into just taking sardines, ammunition, cigarettes and matches. At places "boulders half the size of the cars" blocked their way. In the mountains as many as forty Indians were employed in helping the struggling engines pull the trucks up the hills, while they held ropes to prevent the cars from breaking away on the way down. It took 50 days to travel 70 miles and the Tehuanapec River was crossed 88 times in 17 miles, "about every third crossing requiring a block and tackle." By the time the expedition reached the Salvadoran town of La

Liberdad, it had traveled 1435 grueling miles. The rains had come by then, so the trucks were lightered out to a vessel and they and the expedition members sailed for home.[25]

And what of the radio? As *Popular Mechanics* reported, "'CQ CQ CQ' sang the tiny set.... 'IPH calling CQ CQ CQ.'" Eventually contact with a ham was made. However, the radio did not come through as well as it had the previous year. The problem started with erecting the aerial. The best tree was always on the other side of a river, necessitating hip boots. Or mineralized cliffs of deep canyons reduced signals "to a whisper." The small villages had just one story buildings. Moreover, noise from the engine exhaust made copying signals difficult even though the engine was running at the end of a cable twenty-five feet away. Skipping and fading were constant. Mosquito netting was a must while working the rig. At high altitudes the little gasoline motor did not produce the power needed to operate the transmitter.[26]

The result was that reporter Clarence W. Martin was unable to send most of his dispatches via shortwave radio. Several of his articles say "exclusive," meaning that the reporter telegraphed or even mailed his story. From Oaxaca, Martin wrote that the high altitude was believed to make contact impossible with Los Angeles. A dispatch from Chiapas, Mexico, was relayed to the *Times* by ham station W9EUU (name not listed) in Sibley, Illinois. Martin's story from "One Hundred and Thirty-four Miles North of Tehauntepec" was relayed from amateurs W.A. Syles in Vermillion, Ohio (call letters not listed), and from A.L. McIntosh, W6CYR, of Los Angeles. It told of men slashing at the undergrowth with their machetes and camping in a jungle, with its "incessant cackles, squeals, grunts and the tread of padded footsteps attracted by our campfire." They were averaging five miles a day, three-quarters of a mile an hour.[27]

Now the great depression was upon the United States in earnest. For a few years, interest in the International Pacific Highway waned. But during the two years in which they were active the *Los Angeles Times* had given amateur radio considerable favorable publicity. This was especially true when its articles bore in small print, just below the headlines, as in the March 14, 1930 issue:

> This story by Harry Carr, *Times* staff representative with the Southern California Auto Club pathfinding expedition in Mexico, was transmitted directly from the expedition's portable radio station, IPH, and was received locally by Clarence Hill, owner and operator of amateur station W6BRO.

Another expedition enlisting automobiles and amateur radio illustrates the growing link in the 1920s and '30s between innovation and commercial use. In 1929 Metro–Goldwyn–Mayer, the Hollywood motion picture corporation, sent a troupe of thirty men and women (or at least one woman, the heroine, a blond beauty named Edwina Booth) to East Africa to film *Trader Horn*. This epic motion picture was loosely based upon a best-selling book by Alfred Aloysius Horn bearing the same title. A browse of the book makes clear why Hollywood made it into a movie. Trader Horn rescues a young white beauty who had become the goddess of a tribe of fierce, even cannibalistic, natives! She had been taken from her missionary parents when an infant.[28]

Filming was under the direction of noted director W.S. Van Dyke at a cost of $5,000 a day (excessive when one considers the value of a dollar in 1929—one suspects Hollywood hype). The expedition headquartered at the city of Nairobi in Uganda and spread out to many locations.[29]

How to keep in touch with the base was a primary challenge. How to keep units in the field with supplies, how to give them orders, and how to make arrangements for the sending of film to Nairobi where it would be processed, all demanded a dependable, rapid means of communication. The answer, of course, was by shortwave radio. Ralph Heintz, the well-known San Francisco radio manufacturer, was brought to Hollywood with one of his state-of-the-art portable transmitters; another well-known amateur, Fred Roebuck, accompanied him. The two radiomen, experienced experts, demonstrated the rig at the studio one afternoon where it was attended by many employees of the sound department, a number of whom were amateurs. The set was considered a success and six units were purchased.[30]

The Trader Horn Expedition's radioman, Clyde De Vinna, W6OJ, was also the project's chief cinematographer. With these two assignments De Vinna certainly earned his keep. He considered the transmitter "the berries" and noted that more than 700 messages were exchanged between his rig in the bush and the base station at Nairobi, "working from 50 to 1500 miles and under nearly every conceivable condition."

Their assignment carried with it the usual problems and inconveniences experienced by radio operators in remote areas. First, officials at Nairobi were reluctant to grant the station a license. That problem was solved when a British resident, an amateur named Sidney Pegrume, was given a special license, FK5CR, and assigned the job of handling traffic at the Nairobi headquarters. He did his work well. The next task for the expedition was a preliminary scouting trip to determine where to do the filming. This also served as a trial for the radio. Problems were ironed out so that when production got under way, the radio situation was well under control. A specially equipped Hudson car carried into the bush a radio as well as vital motion picture equipment. A bit of serendipity helped De Vinna: his radio tent usually occupied the best site on the camp grounds because the radio worked best there.

The natives, as always, were fascinated by the radio. De Vinna was dubbed "the man who talks with the winds." At first, when breaking camp, natives tried to dig up wires which they thought were buried out of sight, for they believed the radio must be a telegraph. They refused to touch wires or tubes because the safari superintendent warned them that if they touched any of the equipment fire would flash out and consume them. They believed him. De Vinna had no more trouble.

Amateur radio made progress more efficient in many ways. At one time director Van Dyke divided the outfit, sending one segment into the Belgian Congo for three or four weeks and another party to Kampala on the northwest shores of Lake Victoria. After a few days dominated by bad weather Van Dyke called back the two parties. With runners it would have taken ten days to two weeks, wrote De Vinna, but "thanks to the radio we were all back together and ready to resume work in four days." Another time the safari surgeon, Dr. Clark, was missing "some of his pet serums." A message to Nairobi started the medicine on the way that very night. Instructions were sent from the field for a car to photograph animals. When the party returned to Nairobi the car, the necessary alterations made, was waiting for them.

De Vinna still found time to contact hams throughout the world and especially in the United States. One night he contacted an American operator who protested the static he was getting from passing automobiles. He was told that in Africa the problem was hyenas, which were howling up a storm around camp. To those hams he had worked but had not yet sent QSL cards, De Vinna urged patience; it would take time, but they would receive the cards.

As for the picture, *Trader Horn* was a great success.[31]

Clinton B. De Soto, in his book *Calling CQ: Adventures of Short-Wave Radio Operators*, relates how De Vinna's life in Alaska was saved years later by amateur radio. His rig was in an isolated shack. He stoked up the little stove, made sure the door and windows were tight, and got on the air to a ham in New Zealand. De Vinna's tapping became more and more eccentric, finally stopping. The New Zealand ham, suspecting trouble, notified a ham in Hawaii who relayed the message to Nome where the ham contacted by telephone De Vinna's camp at Teller. Hastily the local doctor and others broke down the door and found De Vinna unconscious from carbon monoxide poisoning. Ten minutes more and it would have been too late.[32]

Other expeditions involving automobiles and shortwave radios were active in the late 1920s and early '30s. Unless their radioman was a writer, it is unlikely that much is said about radio beyond the statement that the expedition was so equipped. This does not mean that radio was not important. Equipment, however, was still fragile and heavy. These drawbacks may have persuaded some explorers to do without wireless. Moreover, a dependable amateur had to be recruited, fed, and paid. Most expeditions operated on a tight budget. Some considered a short-wave radio and its operator as dispensable.

One of the better funded, more ambitious excursions with radiomen along was the Citroen-Haardt Trans-Asian Expedition of 1931–1932, call letters FPCF. Georges-Marie Haardt was an adventurous Frenchman who, among other achievements, had established a first by driving automobiles across the Sahara Desert successfully in 1922–1923. Two automobile caravans were to accomplish his new mission, one working east from the Levant, the other west from China.[33]

Among the thirty to forty members of the east-bound contingent was Maynard Owen Williams, chief of the foreign staff of the National Geographic Society and a radio ham. His rig was never described except for the facts that it was housed in one of the cars and had a sixty-seven-foot telescoping antenna. It was said that it could be set up in a half hour and had sufficient power to send messages five thousand miles. Plans were for him and the radio car to stop somewhere before the mountains were reached and relay messages from the west-bound contingent, which had a similar rig. Williams was also to keep in contact with the society in Washington, D.C., in the process asking hams whom he worked to relay messages to that destination. He also wrote a lengthy article on the expedition for *National Geographic*.[34]

Williams describes the half-track vehicles and expedition personnel. Through the *National Geographic* articles one can follow the expedition through regions that today are in flux (at this writing, the United States occupies Iraq) and meet "Syrians, Mesopotamians, Jews, Arabs, Persians, Afghans, Kashmiris, Tibetans, Ladakhis, Turks, Kirghis, Pamiris, Mongols, Chinese, Annese, Cambodians, Siamese, Burmese, Rajputs, Baluchi, and numerous tribal variations of all of them." Along with the photographs, the narrations can occupy a delightful hour. Yet, with the exception of a paragraph or two in the first essay, the gist of which is given above, nothing more is said about radio.[35]

Of the many forays into remote areas in the 1920s and '30s, the Terry-Holden British Guiana Expedition may be cited as an example of radio's progress. In 1937 the American Museum of Natural History sponsored this group of scientists. Accompanying them was O.W. Hungerford, VP3THE, equipped with a state-of-the-art RCA transmitter and receiver. His experience is of particular interest because he was in contact via shortwave with NBC, whose Blue and Red networks patched in his reports and broadcast them nationwide.[36]

Guyana is a narrow strip of land consisting of about 83,000 square miles in northeast South America. It has a coastal plain where sugar and rice are grown, savannahs inland where cattle are raised, and timber, gold and bauxite deposits are exploited. The inhabitants—724,000 in 1994 but considerably fewer in 1937—consist of East Indians, blacks, native Indians, and whites. The Essequibo River flows down Guyana's center and mountains rise inland. It is, of course, a hot, humid land.

The expedition was headed far up the Essequibo, then up a tributary, the Rupununi, to a headquarters more than 300 miles inland, called Isherton. The long boats that carried the expedition were motor driven. One of the problems was keeping the crews from racing each other. Eventually the river narrowed and dangerous rapids appeared, but they were traversed satisfactorily. Getting the 30,000 pounds of equipment to the base camp was indeed quite a task. By now some of the travel was overland by bullock carts. Nights were usually spent at traders' camps or remote ranchers' dwellings, the ranch owners often being Englishmen. At Isherton they were into the territory of the primitive Mascusie tribe, noted for its use of the blowpipe and the poison curare.

Here they dug a well and erected a "little grass shack" of pole framework with a foot thick thatched roof of palm leaves; within this enclosure, though it was open on all four sides, VP3THE set up his rig. Indians found tall trees for the receiving antenna, 100 feet high, and three 65-foot poles which were placed in a "V" for the transmitter. Thanks to good RCA construction, the equipment had weathered successfully the trip upriver, over rapids, and overland in ox carts.

Hungerford's rig worked well; on some days he heard "nigh on 200 reports." He made contact with a New York City ham and "through his splendid cooperation" was able to arrange contact with RCA's commercial station and maintain continuous schedules. VP3THE had plenty to talk about as his voice appeared on programs of the day with names such as "The Magic Key," the "Variety Hour," and even a special Christmas broadcast. The latter included news via a mobile transmitter from an expedition party 100 miles away in the jungle.

What of interest did the Terry-Holden members have to tell their United States listeners? According to Hungerford, they described the inch-long lightning bugs whose illumination was strong enough to read by; a "navigation worm with red lights on one side and green on the other;" the antics of Icky, their Marmoset monkey; and descriptions of their other animals: "Plato, Pluto, and Oscillator, their three jabbering parrots; a live coral snake, and Oswald the Owl." They also complained of the mosquitoes and bugs, universally present but "uninvited guests."

Gasoline ran low after Christmas so Hungerford had to reduce the use of his rig. When he went on the air he warmed up the little engine with gasoline and then switched to less precious kerosene, hoping the heated engine would accept the cruder fuel and keep running, and usually it did. On one occasion VP3THE was prepared to broadcast when a tropical storm came up, soaking everything including the engines (another one being used for lighting). Wearing just their pajamas (their usual night wear) Hungerford and a fellow member had to run out in the rain, dry the engines, and get them restarted. This they accomplished just in time for their scheduled broadcast. Listeners may have been surprised that Hungerford, soaked to the skin, reported his body shivering and teeth chattering while living nearly on the equator.

On the morning of January 15, 1938, the expedition left the Isherton base, returned to Georgetown and eventually to the United States. Broadcasting from a thatched hut on the equator to New York and from there over the air waves of NBC's Red and Blue networks speaks well of the improvements made in radio by the year 1937.

Not all adventurous amateurs were involved in honest enterprises. Bootlegging was active in Wyoming in the early 1920s. The Laramie *Boomerang* for November 15, 1922, ran these front page headlines: "RADIO FAILS TO AID WHITE MULE MAKERS IN R[OCK] SPRINGS RAID. "Distilling white mule with the protection of a complete radio receiver and sending apparatus was put to an end today at a place 40 miles north of Rock Springs.... The still is one of the largest yet to be found by dry workers in this state."

# 8

## The Olympics, World's Fairs, Trans Radio Press Service, and International Radio Conferences

The Great Depression worsened from 1929 until it hit rock bottom in 1932, then drifted along with hopes and experiments under the New Deal but did not end until the coming of the European war in 1939. Amateur radio does not appear to have suffered much from the economic disaster. The number of licensed hams rose steadily through those years. In 1934 there were 41,555 licensed amateurs; in 1939 there were 51,000 and by the beginning of the '40s decade, 54,000. The average ham was just twenty-two years old in 1926 but by 1938 he or she was an even thirty. Radio had become more complex. In a rather pessimistic editorial it was stated that the day of the "attic experimenters" and "basement laboratorians" was over. The high school kid was no longer graduating from his Erector set or Chemicraft kit to radio. Youngsters showing interest, it was said, now purchased fully assembled transmitters, readily obtainable and less expensive than a decade before. Were it not for the necessity to know basic theories and technologies of radio in order to pass the federal examination, many a ham would never know them.[1]

The editorial writer forgot something about that curious kid. True, more hams than ever mastered, just minimally, the code or the technicalities to pass the FCC examinations; they never looked inside their rigs. But always some of them were—*curious*. They hoped their transmitter would "go on the fritz" so they could attack it with screwdriver and pliers and take a look at the components. With the schematic they would figure out why the rig failed to function. Some found short-cuts or new ways to wire their transmitters; some became inventors.

What of the personalities of these people? In 1939 a psychology professor, E. Lowell Kelly, set up an experiment to test amateurs' personalities. Are amateurs different from any other group of the population? Kelly found the results "quite flattering." Hams were rated as superior on twenty of the thirty-six traits surveyed when compared to a group of college men, and were inferior in only six. They were "better adapted to social situations, more friendly, more punctual, more cooperative, more persistent, more honest, more optimistic, and more dependable than the college men." What is perhaps more amusing, and for those of us who know hams appears incredibly valid, were their negative traits. Hams were "somewhat less handsome, considerably less religious, somewhat less courteous, somewhat poorer dressers," had narrower interests, and were generally less conventional than the average college man.[2]

Whether hams are shy or social, reticent or loquacious, it is a certainty that they are proud of their hobby and delight in seeing it publicized. They were reminded, however, that most

116

events that are news to hams are not news to the man on the street. But if an amateur worked a celebrity or if a Very Important Person visited the ham and his rig, that was news. Being in contact with a place of disaster was news, and certainly participating in aid during a disaster warranted a newspaper story. If the ham struck up a romance with a YL and it materialized into a marriage, via dots and dashes or 'phone, *that* was news.[3]

Individual amateurs may or may not seek publicity, but it is a certainty that as an international hobby amateur radio seeks notoriety. The art does perform services that will attract the populace. Clubs try to inform the citizenry of their organizations' preparations for emergencies, and are pleased when the press gives them credit for what they have accomplished, something the press, including radio and television, often fail to do. From county fairs and sidewalk arts and crafts shows up to the national level of world's fairs and international sports events, amateur organizations can be found exhibiting their rigs and giving demonstrations. The public can ask questions, be informed of the emergency network maintained by amateurs, and be shown how one can become a ham in his or her own right. Hams will be on duty to send messages to observers' friends or relatives in other places—messages that should be delivered by the receiving ham.

One of the high points in amateur history is the part played by W6USA at the summer Olympics held in Los Angeles in 1932. It was a dismal time worldwide. Six months before the games were to begin on July 30, not a single nation had responded to the invitations, nor had many spectator tickets been sold. Favorable publicity, including Hollywood stars volunteering to entertain spectators, resulted in more sales. Meanwhile the Olympic Committee had erected the first Olympic Village. It was located on the eastern end of the Baldwin Hills in southwest Los Angeles. Within its fenced 321 acres were erected 550 two-bedroom white stucco bungalows for the male athletes, plus a hospital, post office, library, administration building, recreation buildings, a fire department, and several dining halls. (Women athletes were housed in a Los Angeles hotel).[4]

One small building could be spotted by its proximity to two fifty-foot telephone poles with a wire strung from the top of one to the other. It was the ham station, W6USA. Early in May the Olympic Game Committee had approached a leading local ham about amateurs manning a station at the village. Of course, the hams accepted the invitation. A telephone and necessary carpentry work was done in the cottage assigned, and two Los Angeles amateur radio retailers supplied the equipment. California's U.S. senators prevailed upon the bureaucracy to grant to the station a three month's license with call letters W6USA. By June the station was on the air although the games would not begin until July 30.

The lion's share of operations was accomplished by two amateurs who lived at the cottage and worked a fifty hour week at the rig. These were Don Shugg,W6ETU, and "Chuck" Perrine, W6CUH; four other hams, including W.A. Lippman, W6SN, who reported on the activities, split up the remaining hours. In the weeks prior to the games these men "worked traffic." They contacted amateurs who volunteered to keep schedules come static, fading, breakdowns, and other calamities. By the time the games began, five continents and eighteen countries were represented on W6USA's weekly chart of schedules. Even South Africa came through.[5]

Ham radio at the Olympics proved a godsend. It helped maintain morale among the many athletes from other countries residing in the village. Because the athletes were far from home, most of them with too little money to make long distance calls, the totally free amateur service that made possible contact with their loved ones at home buoyed their spirits. They handed their message to the operator on duty at the ham shack. It was promptly relayed as closely as possible to their home, with the final ham on the relay promising to have the message delivered within two days. Each schedule lasted about one hour, and a 23-hour day for the hams on duty was not unusual. The grand total of messages sent during the 60 days on the air was 5,682. All

**Amateurs aided morale at the 1932 Olympics held in Los Angeles by having their own radio shack at the Olympic Village and sending messages from athletes to their homes in dozens of countries (courtesy Amateur Athletic Foundation of Los Angeles).**

6 continents were contacted, embracing 24 countries. The cooperation of totally volunteer amateurs was reported to have been almost unbelievable.[6]

Another of hamdom's greatest publicity successes was the amateur display at the Century of Progress Exhibition, also known as the Chicago World's Fair of 1933–1934. Of the several world's fairs held in the twentieth century, including the New York World's Fair of 1939 and the Golden Gate International Exposition at San Francisco of the same year (sometimes called the Treasure Island Exposition), Chicago's was the most successful. The fair served as a beacon of hope for a nation caught in the most serious depression of the century. It is a commentary on the human psyche that both the World's Columbian Exposition of 1893, also held in Chicago, and the 1933 exposition are most likely to be remembered for the bold dancing of "Little Egypt" in 1893 and Sally Rand and her Fan Dance in 1933. (When Miss Rand had to defend herself against a lawsuit claiming obscenity, Judge Joseph B. David threw out the case out, commenting that "some people would want to put pants on horses.")

There was much more to be seen by the thirty-nine million visitors on those forty-seven plus acres besides Miss Rand's lovely body, or what one saw of it between the fans. The Fair was claimed to have ninety miles of displays. Nations, American states, corporations, industries and various organizations were represented. In spite of the depression, many of them had expended substantial sums to place their particular interest in the best possible light. One of the outstanding exhibits was that of amateur radio.

For a solid year the World's Fair Radio Amateur Council had worked on the exhibit, which was located on the second floor of the Travel and Transportation Building. Visitors could not miss the big red letters AMATEUR RADIO that signified its location. Here, along the walls of the 2,000 feet allotted, were 25 manufacturer's exhibits and apparatus of historical interest. In the center of the huge room was the transmitter cage, glassed in. It was estimated that in the first year (1933) 4,000 hams and 400,000 of the general public visited the display. They could purchase souvenirs and ARRL handbooks. Amateurs supplied by the Chicago radio clubs were on hand to answer questions. Many visitors had messages sent to friends and relatives from stations

W9USA and W9USB. Most were "wish you were here's," or informed those at home that they had arrived safely at the fair and were enjoying it.[7]

The high point of the amateur experience at the Century of Progress in 1933 was the Central Division World's Fair Convention held August 3rd, 4th, and 5th. Aided and encouraged by the World's Fair Radio Amateur Council and the affiliated Radio Clubs of Chicago, the meeting had the trimmings of a national rather than a regional meeting. Headquarters were at the Medina Michigan Avenue Club where rooms were $3.00 single and $2.00 each for groups of four per room. Special travel rates, it was reported, had been arranged from "all parts of the world."[8]

Distinguished speakers included Hiram Maxim, president of the ARRL, Clinton B. DeSoto of the league; a former president of the Chicago Stock Exchange, the FCC inspector in charge of the Chicago District; and other men distinguished at the time, or experts in areas of amateur technology. On August 4th and 5th were held the World's Champion Code Speed Contests. At 57.3 words per minute Joseph W. Chaplin of Little Neck, Long Island, won in the Class A competition. Winner at Class E, twenty words per minutes, was Miss Jean Hudson. A reporter told how she might have done even better had her Underwood typewriter not jammed, a result of her "too-enthusiastic touch.... The tears stream down her cheeks as she frees the bars—but little 'brick' that she is, she picks herself up to win in that class with only one word missing." Interviewed, she said that she preferred doing code to arithmetic but really preferred playing with dolls. This young lady was all of nine years old. Whether she attended the final banquet is not known: dinner, speeches, awards, and the drawing of 250 prizes took time. It broke up at 3:00 a.m. Sunday.[9]

In the attempt to pay off the fair's indebtedness it was decided to continue the Century of Progress through the spring, summer and fall of 1934. The entire amateur exhibit was remodeled and redecorated. The most intriguing attraction for non-amateur visitors was an old-time spark station at one end of the room. Here the visitor could send a message by radio using an oscillator instead of an antenna. The message was received at a state-of-the-art rig at the other end of the room where it was typed on a special "Souvenir Radiogram" blank and presented to the visitor. On the flip side of the radiogram was a brief history of amateur radio.[10]

Amateur radio's Worlds Fair station WSUSA was to be the stellar attraction for hams. With licenses they were allowed to operate the 750-watt 7030 kc transmitter. (In 1934 a special transmitting exhibit was installed in the Hall of Science Court also.)[11]

Forrest Bartlett, W9FYK, was one of the many hams who took advantage of the privilege of transmitting from WSUSA. Having experienced the sight of a lot of fan but little of Sally, Forrest and Curt Roche, W90SF, both enrolled at the RCA Radio Institute in Chicago, headed for the ARRL amateur exhibit. After having his ham license verified, Forrest took the key at W9USA and called CQ. The log he kept reads, "w W3CQQ , Pottsdown, Pa 500p" and at 5:15 "w W4CBY, Atlanta." (The "w" preceding the call means "worked," ham talk for "contacted.")[12]

Five years later there were two more world's fairs. At Treasure Island in San Francisco Bay was held the Golden Gate Exposition. It had an amateur exhibit complete with a station, W6USA. It displayed six rigs ranging from 5 meters to 160 including 'phone and CW. A committee made up of members of amateur clubs in the Bay Area manned the exhibit. An official ARRL joint convention of the Pacific and Southwestern divisions was held there in early September 1939.[13]

At the other exposition, the New York World's Fair, the operating station was W2USA. It consisted of a number of transmitters, making possible simultaneous operation on several bands. It was operated by the World's Fair Radio Club. In the Health Building was a rig specially constructed for use of the blind and operated by blind amateurs. Of greatest interest was the display in the Westinghouse Building. It was an animated diagram of a 'phone transmitter and receiver. Using 7,000 lights flashing on and off as an electric sign is programmed, it traced the circuitry of both a transmitter and a receiver, the former from the first voice from the 'phone through to the antenna, the latter from reception through the antenna to the loud speaker. Forty feet long and three feet high, the exhibit, with the lights flashing through tubes, coils, condensers, and the other components of the rigs, carried the spectator through a radioman's schematic. The display had been designed at league headquarters and was built by technicians of Westinghouse, the American Institute of Science, and RCA.[14]

Amateurs kept their art in the public mind in other ways. They were of considerable aid to sports judges where distances were involved, such as ski competitions. They informed officials when a contestant left the starting post and when he or she reached the finish line. Outboard motorboat races were likewise timed with the aid of amateurs, as at Lake Wingra at Madison, Wisconsin, and at Lake Merritt in Oakland, California. Every New Year's Eve in Colorado the Adaman Club climbs Pike's Peak and sets off fireworks. Experimental work with a five meter rig from the summit, with its broadcasts picked up by Denver's radio station KOA, were achievements of Colorado's hams as the year 1937 came in. Rutgers University's football games at visiting fields were shortwaved and placed on a score board at the Rutgers' campus. Washington State University at Pullman received play by play from Seattle, where its team was playing the University of Washington. Ham radio was used to convey scores of the New York Stock Exchange rifle team with their opponents in other parts of the country. Such activities kept ham radio in the public's eye.[15]

Amateur radio entered into the minds of many ordinary people by way of the Civilian Conservation Corps, the CCC. Amateur radio within the organization was widespread. Over 300,000 men enlisted in the corps. They were distributed at 1,430 camps in the army's nine corps areas. Funds were made available for rigs and some CCC amateurs brought along their own sets. The first heavy activity of amateurs was in the Ninth Corps area, which embraced most of the far western states. Here 459 camps were located, many of them situated miles from populated places. For purposes of morale as well as for the handling of corps business, the ham radio net was of vital importance. In the Third Corps area, which embraced Maryland, Pennsylvania, Virginia, and the District of Columbia, all thirty-five CCC amateur stations were permitted to send messages at any time for the thirty thousand men in the camps. An active campaign was waged to commit civilian amateurs to deliver the messages to enrollees' families. "The U.S. Army has gone to bat for the amateur at many radio conferences," editorialized QST. "Here's a chance for the amateur to go to bat for the Army."[16]

In the Sixth Corps area, consisting of Illinois, Wisconsin, and Michigan, army station WEUL maintained regular schedules with CCC camps throughout the region. So successful was the system that a unique radio university of the CCC was established. It brought along selected students to the minimum speed of copying thirteen words a minute as well as teaching them the basic theory of radio. It is safe to say that many continued as amateurs after their tour of duty with the CCC was ended.[17]

In the 1930s hams continued many of the activities that had begun as early as the 1920s. Membership in the Royal Order of the Wouff Hong was the ambition of many a ham. To qualify for membership in the Brass Pounders League, the Worked All States Club (WAS), the Worked All Continents Club (WAC), the Worked British Empire Club (WBE), the DX Century Club, and possibly others, and display the certificates of achievement on the wall above the rig, along with QSL cards, was the ambition of many an amateur. A small town newspaper might report the local ham's attainments. (A few organizations failed to last. Among them was the Boiled Owl Society, made up of hams who could prove they had worked their rigs through the night.)

Membership in the ARRL and in state, regional, and local amateur clubs was widespread. These societies were very active; many gave awards for various aspects of ham expertise. In 1932

QST listed fifty-nine "affiliated clubs" in the forty-eight states. In 1937, a typical pre–World War II year, QST reported awards in thirteen separate categories, announced forty-six contests and tests or their results, and recorded thirteen conventions. Activities of the Army-Amateur Radio System and the Naval Communication Reserve included Armistice Day and Navy Day relays.[18]

Upon Franklin Roosevelt's first and second inaugurations (1933, 1937) amateurs held the usual Governor's-President Relay. (They had held them in 1921, 1925, and 1929 also.) In 1933 thirty-five states and three territories contributed messages received by the Washington, D.C., Radio Club, which then prepared them for delivery. President Roosevelt replied with a sincere thank you message. The 1937 relay was even more successful, with thirty-nine states and three territories relaying messages. Again they conveyed their best wishes to the president by way of amateurs contacting the hams of the Washington, D.C., Radio Club. This time members were invited as a group to deliver the messages in person. F.D.R. accepted them, whereupon the amateurs, considering the event at an end, started to follow the photographer out of the room. But Roosevelt stopped them. For twenty minutes they discussed amateur radio and the part it had played in serious floods. (There was no relay for Roosevelt's third inaugural in 1941, but amateurs gained some publicity by participating with the Red Cross in managing crises during the Inauguration.)[19]

In 1934 the ARRL staged a big twentieth anniversary celebration. Activities included invitations to relay to Maxim suggestions for forthcoming years. President Roosevelt again complimented hams, noting that "the future of radio depends to a large extent on amateurs, for it is their initiative, enthusiasm and ingenuity that overcomes radio barriers and leads to new frontiers." Herbert Hoover also noted that he had "felt over these many years since the association [the ARRL] started that the amateurs were making a positive contribution to the development of radio." Marconi also sent congratulations.[20]

Change and innovation are a part of the attraction of amateur radio. In the mid–1930s attention was being paid to experiments with 5 meter, 56 mc rigs. The sets were small and light and thus could be used as portables. They were being used in timing ski runs, motor boat races, and keeping track of gliders. Amateurs were climbing mountains from which to transmit on 5 meters, carrying out experiments to discover the distances that could be attained. Radiomen, all amateurs, employed by the U.S. Forest Service, were in the vanguard of the experimentation.[21]

In 1932 a new contest was held involving portables. Known as the ARRL Field Day, the idea became one of the most popular of the contests in which awards were given. The exercise, it was said, "combined the fun of real outdoor activity with the more serious problem of testing portable/emergency equipment." Rigs ranged from high and medium powered transmitters obtaining their energy from portable power plants to rigs energized by hand driven generators or "B" batteries. In 1937 sixty clubs involving 405 individuals participated, as did 177 participants who were alone or with small, unaffiliated groups. A total of 550 different portable stations were worked or logged, many contacts being from portable to portable.[22]

On August 15, 1935, two celebrities crashed to their deaths in a fog-shrouded lagoon about thirty-five miles south of Point Barrow, Alaska. The pilot was a one-eyed Oklahoman named Wiley Post. He was small of stature, his clothing never seemed to fit him well, and he was impatient and was known to take chances; although a brave, dedicated pilot, he lacked the charisma of a Lindbergh. His passenger was the well-known actor and newspaper columnist Will Rogers.

It was August 15, 1935. Forrest Bartlett, now out of the RCA Radio Institute, and licensed, was sitting at a typewriter under a shade tree at Radio Station KIUJ in Santa Fe, New Mexico. He was demonstrating a new press service known as Trans Radio. The receiver was perched on top of its packing crate because the manager had not yet had time to install it in the station. Would Forrest demonstrate how the shortwave delivery system worked? Of course Forrest complied. He knew Trans Radio's schedules and sat down at the makeshift "operating position." He was taking copy—he could type code at thirty-nine words per minute—when the first words came through: "FLASH WILL ROGERS AND WILEY POST KILLED IN ALASKAN PLANE CRASH."

"KIUJ's management was impressed ... the first day of their news service and they had a scoop with the big news of the day.... The station manager tried to hire me on the spot," Forrest recalls, "but I had already accepted a job in California."

How did news by code come about? Therein lies quite a story.[23]

That radio and television should broadcast or telecast news seems so logical today that the popular assumption is that it has always been so. News, after all, is about something that has just happened, and radio and television are the fastest methods of conveying it to the public—*providing* the great news associations will allow it. Initially, they did not. Were it not for amateurs, news dissemination today could be much different from the way it is delivered.

Certainly early radio operators thought the dissemination of news should become one of their services. As previously noted, on August 31, 1920, the very few people who owned radios in the Detroit area heard the following: "This is station 8MK, the *Detroit News* brings you the early returns from today's Michigan voting." According to Mitchell V. Charmley in *News by Radio*, "Regular broadcasting of news was born that night."[24]

Within two years more than a hundred newspapers started radio stations, primarily to advertise their papers. As commercial radio grew by leaps and bounds, regular programs were beginning to be broadcast, and advertising was entering in, the great press services of the time began to view the situation with alarm. The news was sold to newspapers, not to radio. If radio used news distributed by the Associated Press, International News Service, United Press International, or from other services, then it was stealing news meant only for newspapers. Most especially the powerful Associated Press rose up in protest. On February 20, 1922, it asked its 200 member newspapers to forbid the broadcasting of news that, according to AP's by-laws, was its property. Why? If radio disseminated news, it was argued, why subscribe to the local paper? And if radio accepted advertising, a principal source of newspaper income, then surely newspaper revenues from that source would fall also.[25]

An uneasy modus vivendi ensued throughout the 1920s. The other major news services—International News Service and United Press International—continued to allow broadcasting of their news. A few radio stations began hiring their own reporters for local coverage. Slowly, grudgingly, the Associated Press gave ground. It allowed the broadcasting of baseball scores and of "transcendent [extraordinary] news." Meanwhile transcontinental radio networks had come into existence. (The Federal Radio Commission, predecessor of the Federal Communications Commission, was created in 1927—it would become the FCC in 1934). In 1928 the three news services allowed the broadcasting of the Hoover-Smith election. But the news services and the newspapers were uneasy. If radio could broadcast both news, and advertising with the news, did newspapers have a future?[26]

Then came the Great Depression. As with every other industry—save *one*—newspapers suffered reduced profits. That *one* was the great radio networks, NBC and CBS, which were making money. Moreover, the era of news commentators led by Floyd Gibbons, David Lawrence and H.V. Kaltenborn further convinced newspaper owners that radio was responsible for their hard times. So, early in 1931, "the papers said, in effect, that they would be good and damned if they were going to give free advertising to commercial concerns who deserted them for radio, and henceforth all commercial names were to be omitted from program listings." That explains why the Atwater-Kent program became "The Concert Hour" and the Cliquot Club Eskimos became "Eskimos Orchestra." Convinced that radio news was detrimental to the future of newspapers, in 1933 executives of the Associated Press, International News Service and United Press set out to crush radio news.[27]

Radio did not want a quarrel with the all-powerful newspapers. After all, radio was and still is subject to the regulations of a federal commission; at that time any station could have its license revoked within six months. Criticism from newspaper publishers could conceivably result in a station's losing its license and thus being thrown off the air. But not only the depression and loss of revenue to newspapers prevented a permanent modus vivendi; news of increasing interest to the public entered into the story. The Lindbergh kidnapping, the bonus marchers, and the two political conventions of 1932 and the subsequent campaigns, were all of great interest to the citizenry. Radio, intentionally or not, was presenting more and more news, often ahead of the newspapers.

In the spring of 1933 the National Association of Newspaper Publishers flatly decreed that they had property rights to the news and stated unequivocally that no news could be broadcast unless it was gathered directly by the networks or individual stations. This could have thrown out the programs of the news commentators, now enlarged to include Lowell Thomas, Boake Carter, Edwin C. Hill, and others. They continued on the air, basing their talks primarily on handouts from police, government agencies, and the like. Nevertheless it was a crippling ukase. Radio executives vowed this time to fight.[28]

At this stage the story becomes complex. The reason is that the networks faced the problems differently (CBS set up its own news organization but NBC did not); the great press associations, AP, UPI, and INS, did not always work together, and individual radio stations reacted differently, some of them, especially those owned by local newspapers, employing journalists to cover local news. Some used INS or UPI sources in defiance of the ruling. Thus the attempt of the newspapers and press services to put up a solid front at first met with some success but ultimately failed.[29]

In December 1933, meeting in "smoke and hate-filled rooms in the Hotel Biltmore in New York City," radio, newspaper, and news services came to terms. The agreement specified that the two major networks should cease gathering their own news (actually NBC had not entered the field); their commentators must restrict themselves to interpretation and comment; and actual broadcasting of hard news should be restricted to five minutes, once in the morning and once in the evening. The broadcasts were to be five to eight hours after newspapers had delivered the same news. News announcements could not be sold commercially, should be limited to thirty words per item, with special news issued only of "transcendental" importance. It was to be supplied by the Press Radio Bureau, which was to collate the news from all three services. One office was to be on each coast, and they were to be funded by the radio stations. The bureau became operational on March 1, 1934.[30]

Of the 600 radio stations in operation at the time, just 200 were affiliated with NBC or CBS and most of these were not entirely owned by the networks. The other 400 stations, most entirely independent but a few affiliated with small networks such as the Yankee Network in the northeast and the Northwestern group in and around Seattle, rebelled. To them this was a draconian decree, and they vowed to fight it. They were especially incensed at the ban on advertising, for news had become a prime "spot." The veiled threats to bring on a congressional investigation that could cut off a station's license increased their wrath. Another threat, occasionally carried out, was the elimination of radio station listings in the local newspapers; CBS and NBC affiliates were especially coerced by this intimidation. Some stations set up their own press bureaus for local news and vowed to obtain national and international news as best they could. Press Radio's unpopularity is indicated by the fact that at its peak it serviced just 220 stations.[31]

Into the story comes a former CBS newsman, a young Southern journalist named Herbert L. Moore. It has been said that Moore's incentive was that he was incensed at what he considered the totally unreasonable regulations. Supported by former CBS newsmen and an unknown venture capitalist, he proceeded to found Trans Radio Press Service. It would have correspondents delivering news which would be sent via shortwave to subscribing stations. Moore clearly had financing, and within a remarkably short time Trans Radio was sending out news to an ever-increasing number of stations. Some still paid for their Press Radio service while subscribing to Trans Radio, which was operating in competition with it. And apparently a few other smaller and less successful press services for radio sprang up. Trans Radio was by far the most successful.[32]

As early as May 1935, twenty-seven of a hundred and fourteen newspaper-owned stations were buying news from Trans Radio. After all, under Trans Radio's contract a station could sell sponsorship (advertising) with no strings attached. The service was sending out as many as 30,000 words a day with the lowest number for regular service at 5,000. Moreover, Trans Radio news was sent in a form designed for immediate broadcast whereas much newspaper material had to be edited into a broadcast format.[33]

On May 21, 1935, Trans Radio filed suit for $1,170,000 against Press Radio charging conspiracy in restraint of trade in violation of the Sherman Act, the Clayton Act, and the Federal Communications Commission Act. It claimed that the defendants had indulged in "a campaign of vilification, abuse, and denunciation ... and had by false witness, sarcasm, irony, and defamation" caused harm to the dissemination of fresh news by Trans Radio." Several examples of such efforts were presented, along with a case in which the Continental Baking Company, which had contracted with Trans Radio, was not allowed to broadcast over certain CBS stations. (At the time, Trans Radio accepted sponsorship from commercial advertisers.) By July of 1935 cracks had appeared in the news services' wall. UP had weaned away eighteen of Trans Radio's clients by selling news to individual stations below cost; a year later UP inaugurated its own radio-service teletype circuit. INS merely began selling its news to broadcasters "for peanuts" in an attempt to crush Trans Radio.[34]

The wrath of the Newspaper Publisher's Association knew no bounds. They were determined to win out. In 1937 they urged Congress to enact a law that would require the FCC to designate "an appropriate time each day for radio stations to broadcast newspaper and press association news reports as furnished by the Press Radio Bureau without exploitation by advertisers." The request got nowhere but was a permanent stain upon newspaper publishers, defenders of the First Amendment.[35]

The complexities of the situation simply got worse. After sixteen months the Press Radio Bureau in Los Angeles closed. It was down to just 45 subscribers from a peak of 65. Deals with radio stations were being made by the news services. Some of the most restrictive rules set up in the Press Radio agreement were rescinded. The *Christian Science Monitor* began sending "The Monitor's View of the News" to 187 stations. All the while, Moore had expanded rapidly. Within a few months he had his own correspondents and staff in European capitals and in American cities. By March 1936, Trans Radio was servicing 250 stations. To transmit news it used telegraph, shortwave, and a spin-off of Trans Radio dubbed the Radio News Association. Moore's concept was a great success. He "makes free use of the technique of direct telephone to the scene of news; maintains the only twenty-four-hour open-wire service in the world; and above all keeps its channels always open for flash, headline news, without clogging the system with the endless columns of background, supplementary, feature and 'dope' material." Trans Radio had "scored some notable 'beats,' such as the Dillinger escape, the correct Hauptmann verdict and the Italian invasion of Ethiopia, and its fast, skeletonized service points the way to news coverage of the future—radio," wrote a *New Republic* correspondent.[36]

The lawsuit was settled out of court and by mutual arrangement the details were not made public. With a grudge, the great news services slowly capitulated and before 1935 was out INS and UPI were openly selling news to radio—although below cost and with the single desire of driving Moore's Trans Radio out of business. Radio stations continued to be purchased by newspapers, which brought on an obvious sharing of the news. Gradually the great news services restored their dominance although Trans Radio continued in operation until December 1, 1951. By then it had just fifty clients. Even then, the wrath of the newspapers and press associations was apparent. A radio commentator asked permission to tell the story of Trans Radio's fight for freedom of news over radio station WLS in Chicago, but the station manager lacked the courage to let him do it.[37]

As for Forrest Bartlett, he worked Trans Radio at a station in Marshfield (now Coos Bay), Oregon, then went to California, where he obtained work with Hollywood station KNX. That station was sold to CBS, which cancelled Trans Radio. Still active, Trans Radio was in the process of installing a full-fledged bureau in Los Angeles. Forrest was hired to install equipment and do intercept work. As Trans Radio faded, its shortwave transmissions were handled by Press Wireless. Forrest made connections and eventually took employment with Press Wireless, being sent by them to the Philippines during World War II.[38]

A byproduct of Trans Radio's great campaign for news freedom was the several hundred amateurs who obtained employment doing what they most enjoyed—receiving and transmitting Morse code radio. In the depths of the Depression it was an unusually open job market. Many of the interceptors, as they were called, went on to important positions in radio. For the rest of their lives they all shared a pride in the contribution they made in destroying the newspaper-press association's monopoly over the news. To a substantial degree, we can thank Herb Moore's gallant battle for the freedom of the air we enjoy today.

In 1938 the number of licensed ham stations in the United States reached 50,000. A league national convention held in Chicago in September was attended by nearly two thousand registered hams. Forty companies displayed a hundred thousand dollars' worth of equipment and both CBS and NBC carried broadcasts from the convention hotel.[39]

By the 1930s there were so many new and improved parts for transmitters and receivers as to constitute an embarrassment of riches. Tubes, antennas, receivers with greater selectivity and manufactured transmitters at reasonable cost were available in increasing numbers. DX-ing was widespread, and amateurs were forging ahead with ultra-high frequency experiments. Nor should we forget television. From the early 1920s on, television was mentioned in ham radio journals. As the '30s drew to a close, more and more articles involving television were appearing. Amateurs knew it was a coming form of communication long before the general public was aware of it.[40]

They were part of an incredible generation, those hams of the years from 1900 into the 1930s, but time takes its toll. As the decade advanced, pioneers of amateur radio began to be listed inside black borders as "silent keys." On February 17, 1936, amateur radio's greatest pioneer, Hiram Percy Maxim, died at age sixty-six. The obituary in QST is perhaps the finest writing that has ever appeared in the journal. Try reading it without shedding a tear. On February 12, 1936, another less known but very important defender of the art died also. This was Charles H. Stewart. At the time of their deaths Maxim was president and Stewart vice president of the ARRL. Stewart had made a life-long study of radio legislation and was of great importance when hearings or conferences were held involving radio.[41]

Two years later, in September 1938, the league lost another of its stalwart leaders. This was a brilliant amateur, Ross A. Hull. He had come to America from Melbourne, Australia, where he had been a licensed amateur since 1922. He gained employment with the league and was soon head of a technical development program at headquarters. To him goes the credit for helping hams meet the restrictions set up by the Washington Radio Conference of 1927. He made several other advances in radio design. His abilities resulted in his being appointed editor of QST and of The Radio Amateur's Handbook. He was electrocuted while working on a television rig at his home. His obituary reveals him to be, as was Maxim, a renaissance man with many interests, and good at all of them.[42]

Fortunately other capable men came to the fore and continued the defense. During the '30s they were needed. The need for international agreements on radio did not stop with the International Radiotelegraph Conference of 1927. Meetings were held down to the outbreak of the Second World War and after it ended. In these conferences amateurs were always on the defensive.

The background of amateur participation is this: the 1912 law which specified 200 meters and down for amateurs did not restrict the allocation exclusively to hams. In case some group in radio wanted to operate somewhere within that 200 meter range, nothing official existed to stop them. Between 1912 and 1930 amateurs had proven the usefulness of shortwave. By the 1930s they were working ultra-high frequencies, down to just five meters or even two and a half. But their range below 200 meters was threatened.

Inroads were steadily being made on those 200 meters, restricting amateurs to specified areas—20, 40, 80 meters, for example. Airlines, broadcasting companies aware of the coming of television, and nations threatened with war were all requesting meters below the 200 level. Hams rightly felt under siege. It was not beyond possibility that amateur radio could be destroyed. Matters got steadily worse as the decade of the '30s advanced.[43]

Increasing tensions, armament, revolts and military aggressions characterized the decade of the '30s. That the hope for continuing peace was retained throughout the decade is demonstrated by the international conferences that continued to be held. This is especially true of meetings to regulate radio throughout the world. The participants met on terms of friendship and passed resolutions and regulations, as if peace was permanent. Defenders of American amateur radio attended them all, fought the good fight and experienced remarkable success.

A portent of the deteriorating international situation was brought home to ARRL officials one evening late in 1932. While attending an international radio conference in Madrid, league secretary K.B. Warner, Paul Segal, the league's general counsel, and a couple dozen other leading radio enthusiasts were being entertained by Spanish amateur friends in a small apartment. A famous guitarist supplied "gorgeous music" and there was lots of talk, as there always is when two dozen hams get together. As the party broke up, reported Warner, "we found the Director of Public Safety throwing a ring of gendarmes around the building, with motorcycles and armored sidecars *en route*. Somebody had reported to the police that another monarchist uprising was being hatched! It was a near-pinch and our friends had to do some tall talking to keep Paul and me out of the *calabozo* that night!"[44]

The Madrid conference was held from September 3 until December 10, 1932, with Columbus Day the only holiday. The American delegation consisted of 53 souls, including wives and children, plus representatives of American interests, bringing the total number of Americans to about 100. In all, 77 governments were represented plus nearly 100 international associations, bringing total attendance to nearly 600. The importance of the conference may be estimated by the positions held by officials of the American delegation: the head of the Federal Radio Commission, its chief engineer, an official from the State Department, and a major representing the War Department. The American delegation occupied sixteen rooms in the Palace Hotel. Warner and Segal unpacked their twenty pieces of luggage at rooms assigned them in the same hotel and set to work with their portable typewriters, charts, and files.

The Americans had come to the meeting well prepared. Their instructions were to fight to the limit for the bandwidths amateurs had allotted to them by the Federal Radio Commission. They soon discovered, wrote Warner, that amateurs had supporting them "the most friendly administration in the world, and where in any one of a score of scraps at Madrid we would have lost the fillings of our back teeth but for U.S. backing, with that backing we were invincible." The ARRL representatives had become aware very quickly of the hostility of most European countries toward amateur radio. Yet the Madrid conference was a success. "Amateur frequency bands have been preserved unchanged," Warner began his report. "The communications world recognizes the amateur as an accepted part of the radio picture. The job is done and the ARRL did it!" They had the considerable help of the official American delegation.[45]

A few months later, from July 10th to August 9, 1933, a regional radio conference was held in Mexico City at Mexican government invitation. Again the question was one of band allocation. The Madrid conference had made 1715–2000 and 3500–4000 kc bands for amateurs "and other services as well"; now ham representatives wanted the regional conference to grant those bands to amateurs exclusively. Just about all North American and Central American countries agreed to the grant; only Canada objected because it meant a narrowing of her maritime mobile band.[46]

Another worldwide radio conference was scheduled for Cairo in 1938. Ultrahigh frequency was coming into its own, airlines needed radio as never before, television loomed on the horizon, and so did war. The anticipated importance of the Cairo conference resulted in several prior regional meetings, held over a two year period, that laid the groundwork for the campaigns for or against changes that would be discussed at Cairo. At all of these meetings the league was represented.

From June 15th to 26th, 1936, the FCC, in preparation for the Cairo meeting, held an "informal engineering hearing" in Washington. All told, the radio industry made 103 appearances before the commission with most of the discussions about frequencies below 30 mc. The league made an excellent presentation replete with charts and statistical studies. It requested that its 4-mc band be expanded to 3500–4500 kc, and the 7 mc to 7000–7500 kc (80 and 40 meters). But when the Preparatory Committee on Allocations met in July, the request was denied. The league reacted by filing a minority statement in October.[47]

This statement was to accompany the majority report of proposals to be discussed at Cairo. "The amateur has been the 'Indian' of radio, his holdings periodically depleted for the benefit of other services," stressed the report. "Unlike the Indian, his population grows." The report stressed the great benefits that amateur radio had brought to the United States. It warned that if the federal government did not support its requests at the Cairo conference, the benefits of amateur radio to the United States would be lost.[48]

The request for an increase in amateur frequencies, rejected by the Allocations Committee, was then appealed to the Federal Communications Commission, which also rejected the request. The league then took a final appeal to the Department of State even as it was learned that a similar request in Canada by Canadian hams had likewise been rejected by their government. The Department of State turned down the request. In Great Britain the regulating agency likewise denied a similar request by England's hams. Months before the Cairo conference convened, it was clear to amateurs worldwide that they would not be granted greater frequencies. Implied in the peremptory rejections was the danger that their frequencies might well be even further reduced.[49]

About a year later, still prior to the Cairo meeting, a conference of importance to amateurs worldwide was held in Bucharest, Rumania. This was the fourth meeting of the CCIR (International Radio Consulting Committee—the initials are for the French name), May 21st to June 8th, 1937. Some 200 delegates representing 27 nations and 5 international organizations attended. The IARU (International Amateur Radio Union) was represented with funding from the league. For IARU representatives the principal purposes were to protect amateur rights and do as much proselytizing in behalf of amateurs as possible. A Canadian representing the organization was of particular help because he spoke French fluently. (In pre–World War II days French was still the official language at diplomatic gatherings.) The meeting was considered a success.[50]

Meanwhile the Book of Proposals for the impending Cairo Conference had arrived. It was 600 pages long and had to be translated from French into English. Studying it, league officials noted "that good old Europe hands us a stiff dose of its usual medicine." In other words, European nations, most if not all of whom had monopolies on communications and were increasingly paranoid as war clouds gathered, were out to restrict amateur operations in every way possible. The league very wisely studied the proposals meticulously and made an abstract of proposals affecting amateur radio. The American delegation to the conference, in consultation with league representatives, would have to deal with these matters and, hopefully, support the league position on them.[51]

Still another important conference took place prior to the Cairo meeting. This was the First Interamerican Radio Conference, held at Habana, Cuba (in 1937 Habana was the correct spelling). Representatives of seventeen Western Hemisphere nations convened there from November 1st to December 13th, 1937. The United States delegation (and only official delegates could participate and vote although others, such as K.B. Warner for the league, could attend and give advice outside of the formal meetings) wanted to ensure that all amateur bands should be confirmed as exclusively amateur in the American region. Another aim was to obtain from the many nations permission to convey third party messages "of a nature that would not normally go to any paid service."[52]

In his opening remarks the president of Cuba set the tone of the conference. "It is impossible to forget the extent to which radio amateurs can contribute toward the brotherhood of peoples," he said. "To them we also owe, from the beginning, the greater part of the progress made in the technique of radio-electric communications as well as the discovery of the unlimited field of shortwaves." The United States delegation won just about everything it wanted. It gained a whole-hearted pledge from members to support the present band allocations at the Cairo conference. The delegates agreed to the interchange of third party messages. "The contrast of philosophies between the New World and the old was never better exemplified!" Warner commented.[53]

So concerned were amateurs about the Cairo conference that QST ran two long articles on what it might mean for hams. For at least two years the countries of the world and vested radio interests had been preparing for the Cairo meeting. The ARRL alone drew up a 20,000 word, 13 chart "Presentation for the Amateur Service" which was widely distributed to Cairo delegates and was sold to amateurs for 50 cents a copy. It made the best case possible for amateurs: fair assignments in the uhf (ultrahigh frequency) category; more territory at 3.5 and 7 mc, the former going to 3500–4500 kc and the latter from 7000 to 7500.[54]

Finally the conference convened. The Habana conference paved the way for the American delegation's position: to continue the present amateur bands. The American nations appointed the United States delegation to represent them. League representatives were Warner and Segal, who also, with the British delegate, Arthur E. Watts, represented the IARU. Seventy-four national and colonial administrations, plus several operating companies and other organizations, finally assembled at Cairo on February 1st; they would not adjourn until ten weeks later, on April 1st. Antagonistic to amateurs were the delegates from Japan, the USSR, Italy, Switzerland, Germany, Rumania, "and most of the French and British representatives." It was clear, as Warner and Segel wrote, that the true friends of amateur radio were the Western Hemisphere nations "from Newfoundland to Magellan."[55]

In spite of the hostility towards amateurs, the Americans emerged from the conference pretty much as they had entered it. Amateur bands were unchanged throughout the Americas; shortwavers gained the non-exclusive rights to use 7200–7300 kc for broadcasting outside the Americas; proposals to reduce amateurs' power was defeated; and European powers hedged on subjects involving the 5-meter band in Europe. European amateurs did not fare so well.[56]

A year and a half later, in September 1939, the war came. In lamenting the course of events, it was noted that the conflict had affected amateur radio in about sixty percent of the IARU's membership. Outside of the United States about seventy percent of amateurs were now off the air. "Heck," wrote a DX-er, "we don't think of those fellows in other countries as 'foreigners,' but as our good friends, particularly when we've had some good rag-chews with them or when we've worked them year after year in the Contest [DX Contest].... It's really too bad that a mad dog [Hitler] can't get a ham license—this might be a better world."[57]

The world was about to change forever. What did it mean for amateurs?

We are not quite ready to discuss the role of amateurs in the Second World War. In the '20s and '30s some went to sea and some went into the air, some went into the polar regions and some participated in disasters. Their contributions are the subject of the next chapters.

# 9

## Adventurous Amateurs
## at Sea and in the Air

Adventurous amateurs went to sea and shared in the conquest of the air in the years between the two world wars. One league official with his shortwave rig went on maneuvers with the navy and convinced the service of the merits of shortwave. Others were to be found on sailing ships as well as on modern freighters and luxury liners. They did service on yachts of the wealthy as well as on research vessels funded by non-profit institutions. At least one ham participated in the great grain race run by sailing vessels between Australia and England. Another visited Pitcairn Island, where the descendants of the *Bounty* mutineers live. Two others tried to sail Chinese junks to each of the world's fairs of the late 1930s.[1]

At the end of the First World War the United states had the largest fleet of square riggers and fore-and-afters (schooners or ketches with fore and aft rigs) in the world. They died a slow death; even today there are a few stately sailing ships plying the seven seas. As commercial cargo vessels, however, the tall ships had pretty much come to an end by 1930 or 1931.[2]

The cargo most often carried by these sailing vessels was grain bound from Australia to England or European ports. With a dozen or more of them due to leave the "down under" ports for England, they indulged in what was called the Grain Race. It was of time: regardless of when the ship left port, how many days did it take it to reach the Thames estuary? In 1927, of nineteen sailing ships involved, seventeen raced from Australia to the English Channel. It took them between 88 and 167 days to reach port. Among those that did not make it to England under their own power (meaning the wind) was an American barquentine, the six-masted *E.R. Sterling*.[3]

A young ham named M.B. Anderson, whose Australian call letters were XOA5MA, wanted to meet some of the hams he had met overseas by shortwave radio. When the opportunity arose to ship out aboard the *E.R. Sterling*, which he described as "the largest of her kind in the world," he grabbed at the opportunity. On April 16, 1927, Anderson watched Australia disappear over the horizon. He was on his way to England. The ship had its own long wave radio at 600 meters but XOA5MA was allowed to take along his own shortwave rig. The trip lasted six months, five and a half of which were out of sight of land.[4]

Anderson was no Richard Henry Dana, whose *Two Years Before the Mast* is a classic. Nevertheless, by describing in plain, unvarnished English his experiences, the reader can envision what Anderson, a thorough landlubber, went through. As he sailed southward, the winds, he wrote, "were so cold that you seemed to be frozen stiff all over, especially in the finger tips. Due to this," he added, "the exhaust pipe of the gasoline-driven lighting plant interested me much more than did the radio installations. It was the only means of heating that was available." Because of this he failed to use his shortwave set every night, as he had planned on doing.

He described the seas as being "calm as a mill pond" as the ship rounded Cape Horn. Headed northward, after leaving the ice fields off the Falkland Islands, the E.R. Sterling was struck by a "squally fierce gale" that carried off the main mast. A half hour later, the mizzen mast also crashed to the deck. Both masts trailed over the starboard side but did not fall away due to the wire and rope rigging. Before crew members were able to cut them adrift the masts had struck the rudder with great force and damaged it.

Anderson was able to erect a new antenna. Weather conditions improved as they approached the tropics. He was able to contact hams about the disaster with his shortwave transmitter, while the ship's 600 meter radio simply failed to make contact with anyone. Then, on September 4th at about 2:00 a.m. he signed off his shortwave because the sea was getting too rough to work and so went to bed. About 4:00 a.m. he was awakened to discover that the ship was in the midst of

> a roaring hurricane. When I reached the deck there was wreckage everywhere and shreds of sail were whistling and flapping in the wind. At about 7:00 a.m. the large steel foremast with its six square yards of canvas crashed to the deck and the mountainous seas took charge of the ship. Ventilators, pumps and light woodwork were washed off flush with the deck. Leaving us a pitiful wreck, the hurricane subsided shortly after eight o'clock.

Worse, the injured chief mate was carried to the after cabin where he died shortly thereafter. With just the center mast still in commission, the E.R. Sterling drifted more than sailed until, six weeks later, it reached Sombrero Island, about 130 miles east of Puerto Rico. There it was rescued by the U.S. naval vessel Grebe, which towed it to St. Thomas. From there a tug towed it to England.

Anderson does not mention the part played by amateur radio in contacting help—assuming the 600 meter ship's radio did not function. Yet it had to be one of the two rigs and most likely his ham station that obtained help and brought the barquentine safely under tow to the protection of the harbor at St. Thomas. Nor does he mention the warm welcome he received by two hams who lived there. As for the E.R. Sterling, this was its last voyage.[5]

The twenties seem quiet and peaceful when compared with our world today, and although the thirties became increasingly violent, in America they were years of more concern over the economy than international relations. It was a period in which technology in the form of airplanes, automobiles, and radio, as well as in the myriad of smaller inventions and innovations, were making the world seemingly smaller, safer, and more enjoyable. The last remote areas of the world were being explored, mapped, and conquered. Wars were few and one could travel extensively with a minimum threat of violence. Besides listing expeditions of exploration QST followed the progress of ships, most of them yachts and schooners owned by wealthy men or research institutions, which were at sea for long periods, often sailing to remote parts of the world. Most had radiomen aboard.

The western novelist Zane Grey owned the Fisherman, which toured the South Pacific. The yacht Carnegie, sponsored by the Carnegie Institute of Washington, D.C., was at least two years at sea. Count Von Luckner sailed with fifty American boys aboard the Mopelia. He offered a cup to the amateur who "gave the best service in communicating with the yacht," whose call letters were DAIV. There were many other private vessels at sea, and most were radio equipped. One of the others was the schooner Yankee.[6]

We know some of the details of the Yankee's voyage because Alan R. Eurich, W8IGO, was aboard. It was in the spring of 1936 when he read an article by Captain Irving Johnson about an around-the-world cruise with a crew made up of college age young men. The captain wrote of sailing to out-of-the-way places, noting that on his last cruise he had discovered five new islands in the Pacific and charted them, had weathered a hurricane, and had discovered a waterfall in New Guinea five-and-a-half times as high as Niagara.[7]

Eurich, whose home was Youngstown, Ohio, must have had financial resources. He contacted the skipper, who informed him that the *Yankee* was to sail from Gloucester, Massachusetts, on November 1, 1936. Plans were for a year-and-a-half cruise in which the schooner would circumnavigate the globe. "After due preliminary negotiations, during which I looked over the boat and was looked over," reported Eurich, he was signed on.

Alan Eurich was the ideal amateur. Tall, slim, dark-haired and wearing glasses, this seventeen year old was possibly the youngest member of the *Yankee's* crew. Captain and Mrs. Irving Johnson, the owners, had assembled it with great care. The youngsters were to work as seamen and help pay expenses. Eighteen people were aboard: the skipper and his wife and their one-year-old son; the cook, two girls and twelve boys. Their average age was twenty-two. Among those on this, the *Yankee's* second cruise, was twenty-one-year-old Sterling Hayden, later a movie star, and David Donovan, whose mother and twenty year old sister Pat joined the crew at Bali. "Wild Bill" Donovan, head of the OSS (Office of Strategic Services, predecessor of the CIA) was the mother's husband and David's and Pat's father.[8]

Alan Eurich had informed the captain that the *Yankee* should be radio equipped; on the previous cruise there had been no radio. Johnson agreed, but the FCC refused to allow Eurich to operate on amateur bands. WCFT must operate on regular ship frequencies but was also permitted to work amateur contacts on those frequencies. Unfortunately, many hams could not be contacted because their rigs worked solely on amateur bands. Even so, WCFT talked with many a ham, some on regular schedules. By this means the crews' relatives were kept informed of the conditions and whereabouts of the *Yankee*. As would be expected in an article for *QST*, Eurich devoted about half of his essay to his radio work.

The *Yankee* was a seaworthy vessel. She was a ninety-two-foot, two-masted schooner about forty years old, built by the Dutch government to serve as a pilot ship. Her orders had been to stay at sea "until no other vessel could," and then weather the storm on her own. Eurich said that she was "as sound as the day she was commissioned." The skipper had refitted the living quarters to make them more comfortable, especially in the tropics, and had added a diesel motor for use in maneuvering at docks and through narrow straits. Another innovation installed by Eurich was a vertical 55-foot Marconi antenna.

Alan turned out to be invaluable. It was hard to remember his age, the Johnsons wrote, "in the face of his great contribution to the cruise as a radio operator." Eurich was not only dedicated at working the rig, he had an amazing ability to fix anything that went wrong. "Almost every day of the entire cruise we were in communication with home," wrote the Johnsons. The fixed schedule was with amateur station W1ZB in Springfield, Massachusetts, or W1FTR in Hartford, Connecticut. "At the worst hours Al would listen patiently for the sounds he wanted to isolate out of all the terrible noises that crowded the earphones," the Johnson's recalled. "The failure to make immediate contact only made him stick more doggedly in the radio corner and return again and again at intervals of an hour to try and finally succeed. This brought us messages from home, but it meant even more at the other end in giving parents a word from one little ship half a world away."

It is the half of W8IGO's article describing the voyage that makes one wish to be young again, makes one fantasize such an experience. "Here," he wrote, "was a year and a half of good fun with no worries.... Just think of a chance to forget two winters and slide quietly along in the tropics, literally on the wings of the wind." At Panama they took on more supplies; it was also their jumping off place. "Here we would leave all civilization such as we had known and drop into a world totally new," he wrote.

The *Yankee* put in to the Galapagos Islands, Easter Island, Pitcairn, Manga, Reva, and

Tahiti.They called at island groups almost unheard of: "The little atoll of Pen Rhyn; ... to Christmas and Fanning, where wrecks of Spanish galleons from the Philippines bound for the mother country laden with treasure lay."; and southwest toward the Phoenix group where, Eurich noted, Amelia Earhart may have disappeared, and finally to the American naval station at Pago Pago on the island of Tutuilla. The *Yankee* sailed to Singapore where the schooner put in for alterations and repairs. Sailing again, the winds took the *Yankee* across the Indian Ocean to Capetown, Rio de Janeiro, and finally back to Gloucester by 2:00 p.m. May 1, 1938.

Eurich noted that the competition for a place on the *Yankee's* next voyage would be intense. "If there is any ham healthy and not afraid of work, with some money, here is the bang-up chance of a lifetime," he wrote in closing. He listed the requirements for a good radioman, adding that "he should be able to live congenially with eighteen people in close quarters for a year and a half."

Let us return to those lonely South Seas. Amateur radio could help remove some of the isolation on those islands, many of which are mere dots on the map (there are said to be 25,000 islands in the Pacific.) A man named D.G. Kennedy, his wife and two children lived on one of the Ellice Islands named Vaitapu. It is a microdot on the map, south-southeast of the Marshall and Gilbert Islands, far north of New Zealand and close to the International Date Line. Kennedy had left Vaitapu for New Zealand for an appendicitis operation. While there he met a ham, Frank Bell, OZ4AA, who interested him in amateur radio. He mastered the code, built a transmitter and receiver, received call letters DGK, and began having QSOs (radio contact) with hams in New Zealand and elsewhere. In view of the fact that he and his family were the only white people on the island and received mail just twice a year, the luxury of communicating in code and receiving news of the outside world can hardly be exaggerated.[9]

Another island station, described as "The World's Loneliest Radio," call letters VK4SK, was located on Willis Island in the Coral Sea, about 400 miles east of Townsendville, Queensland. Described as being "about 500 yards long and 250 yards wide," 22 yards above the seas and surrounded by a coral reef, as of 1932 it had been for ten years the home for a year at a time of two radio operators. Their duty was to observe the weather and report it to the mainland, thus giving time to prepare for cyclones at least 24 hours before they arrived. Although they heard news weekly through Roy E. Abbott, VK2YK, a mainland ham station, VK4SK enabled them to break the monotony and loneliness by contacting other hams. After all, for six months at a time they saw no humans but did see thousands of birds, which nest on the island.[10]

One of the lonely places visited by the *Yankee* was Pitcairn Island. It is the best known microdot in the Pacific, the home of descendants of the mutiny on the *Bounty*. Half way between Panama and New Zealand, but far from the principal trade routes, contact with the rugged little island and its diminishing population had been occasional at best. At first the islanders used signal fires to alert ships of their presence; later they used signal flags and then a shuttered lamp using Morse code to convey information. In 1922 the Marconi company gave the islanders a crystal radio receiver and in 1926 (one source says 1928) an unknown New Zealand ham donated a code transmitter that made possible contact with ships within a radius of 150 miles. Originally it was powered with a gasoline driven generator, but when the gas, which had been stored there by a defunct Italian company planning an air route from South America to Tahiti, ran out, power was derived from a battery which had to be sent to New Zealand for recharging. Needless to say, there were long periods when the transmitter was silent. This was Pitcairn's communications situation until 1931.[11]

In 1930 the curiosity of QST's editors was aroused when Ross Hull, who by January 1931 was an associate editor at QST, related to the staff his experiences on a cargo ship that stopped at Pitcairn. He was in the radio room when suddenly he heard a "raspy spark signal, sending blind and asking that the boat stop at the island to swap shirts, medicines, etc., for fruit." The ship weighed anchor a couple of miles off shore "while the islanders rowed out en masse in two multi-oared boats bringing with them quantities of the most luscious fruit anyone ever heard about." The islander's radio operator was Andrew Young. His story, conveyed by Hull to the editors, so intrigued them that they began a correspondence with Pitcairn and finally found a young ham who had spent a week on the island—none other than Alan Eurich, W8IGO (WCFT when aboard the *Yankee*).[12]

The *Yankee* had anchored off the island for about a week while part of the crew, including Alan, went ashore. He soon made the acquaintance of Andrew Young, Pitcairn's radio operator, and became his guest. The radio was in a one-room shack surrounded by palms and other tropical vegetation. It fronted toward the sea and was backed up by high mountains. Watch was kept at noon and from four p.m. until midnight. When Young was not on duty he had plenty of help, for although only he understood the rig and how to maintain it, many other islanders understood Morse code and were capable of handling transmission. Code was almost a second language for Pitcairners. Children tapped out Morse on whistles. Eurich discovered that many of the residents, including the women, could translate code into English with remarkable accuracy.

While Young and some of his fellow islanders were on the *Yankee*, Eurich pinch-hit for him on the radio. He heard GLYQ, the call letters of the S.S. *Rotorura*, announcing that it was coming close by to drop off Norman Young, a Pitcairner who had been to Panama for medical help. It also requested fruit and honey. How to alert the residents? Andrew Young had instructed Eurich to walk out and call "Sail ho!" at the top of his lungs. He complied. The call was repeated all over the island, a crowd gathered around the station, and plans were immediately made for the gathering of fruits and the launching of boats.

Alan had the greatest respect for Young's achievements with the primitive short wave equipment. The Pitcairner had heard ships over 1000 miles away and on the broadcast band had heard KFI in Los Angeles. He had held two-way conversations with ships 400 miles away. But, Eurich added, PITC would have to change or cease to exist in a few years because his equipment was so antiquated that it would soon be impossible to use with present day ship installations.

League officials were struck by the Pitcairners' plight. They were also aware that to work Pitcairn Island was one of the great goals of DX-ers. Why not give the islanders an up-to-date transmitter and receiver? Officials queried radio manufacturers, suppliers, and shipping companies and in due time had received offers to help from twenty-one of them. "Then, and only then, was the size of the job at hand fully realized," wrote Lew Bellem, W1BES, a radio engineer who helped on the project.[13]

First was the matter of power: Pitcairn Island was without electricity, nor was gasoline available; batteries wore out. The answer was to harness the wind to charge enough storage batteries to avert a power failure when there was prolonged low wind velocity. Dynamotors, machines which transfer current of one voltage to that of another voltage, could create the high voltage necessary. Frequencies of the transmitter were fixed at the 600 meters universally used by ships; and second and third frequencies, considering the remoteness of Pitcairn, in the 40 and 20 meter amateur bands. Either phone or code could be used. "All the equipment," wrote Bellem, "should be as simple and foolproof as possible from the standpoint of hooking up and operating, since Andrew Young's experience with tube transmitters was nil and his familiarity questionable."

In due time the wind charger was obtained, along with two storage batteries capable of providing a reserve of power good for eight to ten hours a day. "Every precaution," wrote Bellem, "has been taken in design, construction and choice of parts to preclude the possibility of breakdown. All resistors and fixed condensers have been chosen so as to operate well below their ratings." Eurich, who was consulted via shortwave, stressed the importance of high grade insulation because of the salt air. His advice was followed. Dozens of spare parts were to be included. When assembled the rig was subjected to practical trial in the United States, and the results were satisfactory. The new Pitcairn radio was deemed capable of working the globe.

The next task was getting it delivered. Two shipping companies volunteered their services. One would take the rig, packed in seven cases, to Cristobal on the Atlantic side of the Panama Canal; it would then be transferred across the Isthmus to Balboa and from there shipped to Pitcairn in another company's vessel. It was predicted that with reasonable good fortune, Pitcairn Island would have a new, state-of-the-art transmitter and receiver by sometime in February 1938. "May the amateur in contacting PITC remember that he is treading the sanctity of 147 years of almost absolute isolation. Largely by his conduct will the rest of the world be judged by the islanders," Bellum concluded.

No *QST* indexes for 1938 and 1939 mention Pitcairn Island, but it is assumed that the new radio arrived safely. It is known that radio activity from Pitcairn increased. As of 1973 the operator was a thirty-six-year-old islander named Tom Christian, Fletcher's sixth generation descendant, and Tom's wife Betty. At that time his rig was a Hallicrafters. He told Ian Ball, author of *Pitcairn: Children of Mutiny*, that he could "be ten men and work the set for ten hours a day," so many hams want to work Pitcairn Island. Today, according to the Pitcairn Web site, the Christians are still the radio operators. However, there is now an official British (the Web site says New Zealand) weather station atop a mountain at 870 feet above Botany Bay. Since the summer of 1985 AT&T has established telephone service with the island. It is "routed through New Zealand to Pitcairn ... at a cost of $11.83 for the first three minutes."[14]

But the shortwave radio is still important. Because there is no competition, Pitcairners can use the shortwave to contact relatives and friends elsewhere, particularly in the States. A telephone patch is turned on and a third party can converse with the islanders. The islanders are aware of how DX-ers want to work their little spot in the world, and they have done what they can to help. They have also participated in creating a new DX station on Ducie Island, 325 miles from Pitcairn and just two and a half miles in circumference. They were there in 2002 and again in the winter of 2003. Meralda Warren, the Pitcairner from whom I purchased T-shirts, alerted me by e-mail that she and others were sailing for Ducie Island to do DX-ing from there. Mrs. Warren operates the 12 meter position. More than twenty percent of Pitcairn's population sailed for Ducie to participate in the radio activity.[15]

The 1939 world's fairs were a fitting close to two decades of unprecedented exploration, experimentation, and technological advance. They displayed the world, or at least the United States, at its very best as it tottered on the brink of World War II. The fairs' sponsors sought publicity of any kind, for enormous sums had been invested and only heavy attendance would balance the books. Two publicity stunts that went awry were the sailing of Chinese junks from Chinese ports to each of the fairs.

The vessels were built specifically to sail across the Pacific. One, the *Sea Dragon*, was the inspiration of Richard Halliburton, the well known travel writer. His exploits had thrilled romantically inclined teenagers as well as otherwise mature men and women who had read his books, among which *The Royal Road to Romance* is probably best known. The *Sea Dragon*, writer Guy Townsend writes, was "poorly designed and constructed." Moreover, Halliburton's choice for captain, one John Wenlock Welch, was, Townsend quotes Halliburton as saying, "a regular Captain Bligh." The crew of the *Sea Dragon* included wealthy young men who paid dearly for the opportunity to sail across the Pacific in a Chinese junk—too dearly, as it turned out.[16]

The *Sea Dragon*, bound for the Golden Gate International Exposition at San Francisco, was radio equipped. It set sail from Hong Kong on February 4, 1939, but six days later limped back into port for repairs and modifications. The Philippine cook and two of the young men left the vessel at this time; twelve Americans remained on board. Again, on March 4th, the Chinese junk set out. On March 23rd it was heard from by radio—and never heard from again. Halliburton, his crew, and the *Sea Dragon* were lost at sea.[17]

The other Chinese junk, this one slated to sail to the New York World's Fair, was better conceived and better built. Three adventurers, Jim Peterson, Homer Merrill, and Rex Purcell, VS6BF, had begun preparations at Hong Kong eight months before they sailed. The vessel, dubbed the *Pang Jin*, unlike Halliburton's, was carefully built. "We personally selected every piece of timber, coil of rope, and bucket of paint used in its construction," Purcell wrote. The three men were equally careful in selecting their radio. The transmitter he designed for them was good for both 'phone and CW and was powered by a Johnson Sea Horse motor generator; it also charged a storage battery for the receiver. The antenna "stretched from mast to mast, but since the boom rose above the mast tops it frequently broke as the sails were shifted," Purcell wrote, adding that "under the most trying difficulties, the rig operated consistently and well."[18]

Purcell was well qualified to man the radio. After four years in the U.S. Army Air Corps he spent four years flying for the Philippine Aerial Taxi Company, whose communications were carried on by radio. He did not have an American amateur license but the British government granted him a special one, VS6BF, limited his rig to 50 watts, and stipulated that his license would be void upon arrival at New York. As it turned out, his work with the ship's radio probably saved the *Pang Jin* from sinking in an Indian Ocean cyclone and may have saved their lives when the boat experienced heavy weather and sank in the Red Sea. Purcell adequately described what it was like, working the transmitter in a terrifying storm:

> "QST de VS6BF, QST de VS6BF, QST de VS6BF, ar K." Over and over I pounded the call. The heavy crashing of the junk sounded dangerous and labored down in the shack. Waves cascaded over the deck, thundering and smashing. The barometer was dropping alarmingly and had already passed a figure lower that I had ever seen before."
>
> Finally he made contact. ZS6DY in Johannesburg was standing by. "Go ahead please."
>
> "Chinese junk *Pang Jin* in severe storm off east central coast of Madagascar. Urgently need weather reports and forecasts on direction of cyclone this vicinity. Can you arrange? a.r.k."
>
> "ZS6DY to VS6BF. Will try to obtain weather info for you immediately. Please QRX [stand by] while I check."

Purcell, VS6BF, waited, all the while trying to hold his receiver's frequency setting steady—quite a task in such a storm. Overhead the shouts of the men battling the wind and waves were dimly audible. Conditions were undoubtedly becoming worse. Then the message came through: "Cyclone off east central coast Madagascar plotted as progressing east to west, 28 miles per hour. Weather bureau advises you proceed northwest in order to escape danger zone."

The *Pang Jin* complied, sailing northwest as advised, and reached calmer waters. Because they used the 'phone, the crew often heard the sounds of automobile horns, telephones ringing, a loudspeaker, and once, bath water as a tub was being filled. To these men with unkempt beards and dirty fingernails, the thoughts of a hot bath and the other accouterments of civilization inspired them to keep on because civilization lay just ahead.

Although the *Pang Jin* was better built and better manned than Halliburton's *Sea Dragon*, it still met with ill fortune. Five days out of Aden the *Pang Jin* sank in the Red Sea. In a news bulletin following Purcell's and Polk's article, it was stated that all members of the crew were saved by the Greek freighter S.S. *Olga E. Embiricos*. Nothing was salvaged but their lives and a few personal belongings.

These are just a few adventures of hams working on ships during the 1920s and '30s. They were chosen as examples because they were somewhat out of the ordinary: adventures on a barquentine sailing vessel, a meeting with Pitcairn Islanders, an attempt to sail Chinese junks to the world's fairs. There were dozens of hams on dozens of other vessels in the same era, and we can assume correctly that all experienced adventures, in many of which the radio played a vital part.

Balloons into the stratosphere also excited the public in the 1930s. Auguste Piccard in

August 1932 ascended to 55,600 feet. He had with him a young Belgian scientist and radio operator, Max Coxy, B9. Several messages were sent, the first ones ever from such an altitude.[19]

The 1920s and '30s were also the great years of experimentation with lighter-than-air craft. The navy had the *Shenandoah*, the *Akron*, the *Macon*, and the German-built *Los Angeles*; only the latter escaped disaster. The three American-made dirigibles crashed, the *Shenandoah* on September 3, 1925, the *Akron* on April 3, 1933, and the *Macon* on February 12, 1935.

The *Shenandoah* story is of particular interest to hams. The huge airship, the first to be inflated with helium, warranted an estimated 225 articles in the *New York Times*. It was plagued with bad luck. On the night of January 16, 1924, with wind gusts of up to seventy-seven miles an hour, the monster zeppelin broke away from her mast at Lakehurst, New Jersey, with a skeleton crew of just fifteen aboard. She drifted nearly out of control while her radio operator, Gunner Robertson, desperately put back together the rig, which he had been working on. When he was finally on the air he contacted AM radio station WOR in Newark. From then until the great airship was back at Lakehurst, Robertson tapped messages in Morse code of the great dirigible's progress to WOR, as well as to officials at Lakehurst and to any hams who might be listening in. Someone at WOR decoded every message, which was then broadcast in English to thousands of BCLs who listened to their radios until the *Shenandoah* was again safe in her hanger at Lakehurst. A ham named Bob Dommins "stayed up all night listening to the Naval radio station at Lakehurst on Morse code and WOR." He saw the *Shenandoah* passing over his parent's farm near Freehold, New Jersey.[20]

The *Shenandoah* made many publicity trips over American cities and made one trip across the country to San Diego. There its radioman contacted the U.S.S. *Canabus*, 4,400 miles away in the Pacific, using a low-power shortwave rig on 90 meters. Much of the news about the dirigible's trips told of serious storms the lighter-than-air craft encountered. On September 23, 1925, an atmospheric disturbance over Ohio tore the silver beauty into three parts. All those in the control cabin were killed when it tore from its fastenings and plunged to earth, but twenty-seven crew members in other parts of the broken ship survived.[21]

The *Akron* was the largest of the zeppelins. For nearly two years the great airship cruised over the land, its perambulations being mentioned in the press. On Monday, April 3, 1933, it departed from Lakehurst "on a regularly scheduled flight to calibrate New England radio stations."[22]

The weather turned bad. The report transcribed by the *Akron*'s radioman seemed to indicate that the storm they were encountering was over Washington, D.C. The *Akron* headed out to sea to avoid the disturbance and, unbeknownst to its officers, headed directly into the heart of one of the severest storms to strike that part of the Atlantic in years. When the great aircraft went down, it carried seventy of its seventy-three men to their deaths. Four were rescued, but one of them, Robert W. Copeland, the *Akron*'s radio operator, died after he was rescued.[23]

The *Akron* had a sister ship, the *Macon*. On February 12, 1935, in fog thirteen miles southeast of Point Sur in California, her radioman tapped out SOS and a red rocket was sent into the sky. Two of the ship's huge gas bags had given way, the stern had crumbled, and a plunge into the Pacific was inevitable. The crew had time to don life jackets and launch rubber boats, while navy ships sped to the site of the disaster. Eighty-one of the eighty-three men aboard were rescued. Of the two who died, one was Ernest Edward Dailey of North Bend, Oregon, the airship's radio operator. On May 31, 1935, he was honored by having his name engraved on a copper plaque commemorating brave radio operators. It is located at the Battery, a park at the southern tip of Manhattan Island. In August of the same year the deceased radioman received a navy commendation.[24]

Interest in all kinds of airborne conveyances was so great that officials of hundreds of cities, large and small, brought about the construction of airports. The opening of a new air field was

The airship *Macon* went down in the Pacific, carrying to his death its radio operator (courtesy National Air and Space Museum, Smithsonian Institution).

cause for a great celebration. Although there were all sorts of events ranging from parachute competition (who could land closest to a target on the ground) to stunt flying and wing walking, a basic attraction of the larger shows was an air race in which contestants zoomed around a triangular or square course. Pylons were shaped like pyramids, thirty feet high, covered with black and white bunting and flying a checkered flag at the top of a mast. Placed three to five miles apart, the planes whizzed around them at 150 to 200 miles an hour. Amateurs were placed at each pylon and at the timer's stand. They kept the judges informed of the planes as they raced around the pylons, of any possible illegal actions on the part of the pilots, and of mishaps. At the air races in Miami in 1930 two forced landings and one crash were reported.[25]

Again in January 1931 the city of Miami held a four day All American Air Meet. At least 148 contestants registered for the activities. The crowd thrilled at the races, in which "the planes bunched at the pylons like horses on a track and roared down straightaways at speeds estimated at close to 200 miles an hour"; they watched the "ships roaring past grandstands and standing on wing-tips at pylon turns." The Miami Radio Club participated with a ham at each pylon and another at the central control stand. Operator Wizenbaker, W4AKW, witnessed a plane crash at his pylon, number two. He ran to the wreck to give what assistance he could while another ham, Bowers, W4NB, sent information of the crash to the control stand. In almost no time, because of the radio message, an ambulance was speeding toward the wreck and within a half hour newspapers were issued with news of the accident, based upon the radio reports.[26]

Besides pylon flying there was long distance racing. For the coast-to-coast race in connection with the National Air Races at Los Angeles, September 8–16, 1928, the ARRL broadcast a special bulletin alerting two thousand of the better amateurs in the country to render assistance. The *New York American* (a now defunct newspaper) offered three cups as prizes: one to the ham who received a message the greatest distance from a plane, one to a ham conducting two-way communications by phone or code over the greatest distance, and one to the amateur judged to render the most valuable service during the flight. Plans were made for local stations to cut in on talk from planes in flight, giving the listener "the thrill of listening to a 'radio air interview.'"

The sport of gliding went through a period of national interest during these same years. Amateur radio operators figured prominently in their activities. Gliding enthusiasm centered at Elmira, New York. The city has been described as shaped like a pie plate, surrounded by ridges. From these ridges gliders could take off in different directions, depending upon the wind. The National Soaring Meet was greatly aided by amateurs maintaining contact between the airport and the ridges from whence the gliders took off. On the very first day of the 1928 meet a glider manned by a Major Purcell, fell "with a resounding and blood-curdling crash into trees near the field." Within minutes the hams had sent an urgent call for an ambulance. Less than twenty minutes later the injured major was on his way to the hospital. The crash was the first of two during the meet in which hams were able to contact authorities and have help on the way within minutes.[27]

Hams were even more prepared for the Elmira meet the next year when there were six specified hillside takeoff points. In the two weeks of the meet 1,694 messages were handled, all on 56 megacycles and all on duplex 'phone. In one case, the ham called for an ambulance just as the crash took place; fortunately the pilot walked away unhurt. By 1933 a few of the gliders had radios and hams gave the pilots, who did not have licenses, a crash course in the mysteries of radio; enough for them to travel to Buffalo and qualify for a C license.[28]

And one read of a California amateur who was also a glider enthusiast—until he crashed and suffered "a fractured skull, a broken left leg, a broken back, an infection of the right knee, four fractures of the jaw, cut lips and lost teeth, a month's unconsciousness, and hospitalization until January 20, 1932." But, added QST, "He's on the air again now, but not on wings this time."[29]

# 10

## More Amateurs and Aircraft:
## The NC-4, Byrd, the Dolebirds,
## the Lindberghs, and Others

A book could be written about aircraft activities in the year 1919. The London *Daily Mail* had offered a prize of 10,000 lbs, equal to about $50,000, for the first pilots to fly across the Atlantic. Two British flyers attempted the feat in a biplane powered with a Rolls-Royce engine. Their heavily fueled aircraft got off the ground—and disappeared. Their radio went dead. Fortunately they were picked up alive from the sea and taken back to England. Later, two more British pilots in a twin engine Vickers biplane flew from Newfoundland to Ireland. Also, during the late spring and summer of 1919, the U.S. Navy's blimp *C-5* was moored in Newfoundland prior to a planned nonstop flight to Ireland. Heavy winds tore the great lighter-than-air craft from her moorings and she went out to sea unmanned; her wreckage was sited by a steamer, but by the time it reached the locality, it had disappeared.[1]

The navy also scheduled a heavier-than-air transatlantic flight, not for the *Daily Mail* prize, but for the advancement of air transportation. The proposed route was from Newfoundland to the Azores to Portugal to England. The four identical aircraft proposed for the flight were huge amphibians with three, later four motors. These "Nancy's" as the press called them—their real designation was NC-1, -2, -3 and -4—look today to modern aircraft like dinosaurs look to modern animals. Impressive, with bright yellow wings just four feet shorter than that of a Boeing 707, these primitive pterodactyls of the air age were put together, as author Richard Smith writes, of "wood, linen, and wire, held together by nuts and bolts, wood screws, glue, and paint."[2]

The flight of the NCs—three started out from Newfoundland, two went down but their crews were rescued, and only the NC-4 completed the flight—was characterized by preparations for any emergency. More than five dozen destroyers plus perhaps forty more ships were alerted to the flight. As for radios, never had a navy project been so well equipped. The aircraft all had radios and were expected to keep in touch with the other aircraft and ships.

They still used the spark transmitter, but the vacuum tube for the receiver was by then in use. The NCs were equipped with transmitting sets operated on 500 watts, powered by a wind-driven generator mounted outside the hull. They could not operate without the plane being in motion, so a smaller, twenty-six-pound emergency transmitter, with a radius of about seventy-five miles, was also on board. The rigs were long wave, from 450 meters up. Both radiophone and Morse code were used. Somehow the radiomen squeezed into the equipment-crowded fifth compartment from the front of the plane.[3]

To Ensign Herbert C. Rodd has been attributed much of the credit for the NC-4's success-ful flight. At age eighteen he was radioman on the Great Lakes steamship *Lakeland*. When it ran aground on November 10, 1914, Rodd stuck to his key, bringing to the site a salvage tug which was able to tow the ship to safety. Subsequently he enlisted in the navy. His resourcefulness came to the attention of navy brass and he was assigned to service on the NC-4.[4]

He notified the supply ship *Aroostook* when repairs or parts were needed, requested the recharging of batteries, and contacted ships along the route for the NC-4's bearings. He was con-stantly asking for weather information. On the flight from Newfoundland to the Azores the young man stayed fifteen hours at his station. He continued his good work as the NC-4 flew from there to Lisbon, from Lisbon with a couple of stops to Plymouth. "If there was a 'hero' aboard the NC-4," wrote Richard Smith, "he most certainly was Herbert C. Rodd." His log of the flight constitutes appendix D in *First Across*.[5]

It was an expensive, well-planned, valiant effort, the flight of the NCs. The entire opera-tion, however, including the fates of the NC-1, -2 and -3, pointed out the necessity for radio com-munication, good weather reporting, and the many needs for improvement in aircraft before transatlantic airways could become practicable.

Lindbergh's successful flight in May 1927, from New York to Paris (without a radio) ush-ered in a period of unprecedented national aviation activity. By the end of 1927 thirty people including two women had lost their lives in transatlantic attempts, and seven others had died in preparations for such a feat. Although America-to-Europe flights had been achieved, as of December 1927, no one had succeeded in flying from Europe to America in heavier-than-air craft and only two dirigibles had successfully made the journey. To the question as to the viability of transoceanic flying, the general tenor of discussion among pilots was positive, that it could be done. The public, however, continued to look upon flying, wrote Admiral Richard E. Byrd, "as one of the most attractive forms of suicide."[6]

On June 29, 1927, Richard E. Byrd and his crew aboard a Fokker triplane dubbed the *Amer-ica* roared down the runway at New York's Roosevelt field and were soon airborne—barely so. Among the safety measures Byrd took with him was the radio. The main radio (for he also had a small, waterproof set for use if they were forced down, and a kite to hold up the antenna and also pull the boat along in the wind) was especially designed for the flight by radio experts Mal-colm P. Hanson and L.A. Hyland of the Naval Research Laboratory. "We wanted to prove that we could locate ourselves at sea with radio," Byrd wrote. While the small set was shortwave, the main radio was a long wave set. It had an automatic sending device that could repeat calls at the rate of ten a minute on a prearranged wavelength. Somewhere on the outside of the plane was a radio power unit, a propeller-driven generator.[7]

If they were in fog or clouds—and in fact they were in fog or clouds for all of two thousand miles—they could use the radio to locate their position by "getting lines of direction from ships or shore stations." The radio did work well, making contacts while flying in the clouds two miles up. Byrd gave Lieutenant George O. Noville, the radio operator, credit for the operation, writ-ing that it was a triumph of science, that the whole flight was worth while "to demonstrate this one thing [the radio] which we had been so anxious to prove."[8]

When the *America* came over Paris the city was completely closed in by fog. There was no alternative but to return to the sea and make a forced landing in the water. They did this suc-cessfully near the town of Ver-sur-Mer. The Wright motors had functioned perfectly for forty-two hours, the plane did not flip over, and the men were all able to get ashore. Byrd had proved his point—that trimotored planes could carry passengers and freight across the oceans. The radio had also operated well.[9]

Three outstanding examples of aviation activities in which shortwave and amateurs partic-ipated are the Dole Contest (Oakland, California, to Oahu, Hawaii); the flight of the Fokker trimotor *Southern Cross* from Oakland, California, to Hawaii, the Fiji Islands, and Australia; and the attempt of the *'Untin' Bowler* to pioneer an air route from Chicago to Berlin. They demonstrate the growing recognition of the importance of amateur radio to aviation.

If Lindbergh could fly non-stop from New York to Paris, why couldn't pilots fly non-stop to Hawaii from California, just 2,400 miles away? Following Lindbergh's achievement, James D. Dole, the "pineapple king of Hawaii," offered a prize of $25,000 for the first successful flight from California to Hawaii and $10,000 for the runner-up. (Actually, the feat had already been accomplished twice but this did not stop the Dole contest.)[10]

The press dubbed the fifteen entries the "Dolebirds." (Some reports specify fourteen official and four unofficial entries.) The race got underway on the morning of August 16th, 1927. Prior to the start, two participants leaving from San Diego for Oakland were killed on takeoff and another pilot, taking off for Oakland from Los Angeles, jumped to his death from his disabled plane when his parachute failed to open.[11]

By the late 1920s the Department of Commerce was helping scheduled flying, such as Air Mail, by establishing radio beacons every 200 miles along established airways. The beacons, besides flashing a light 360 degrees, were installed with automatic radios. The system worked as follows: the radiotelegraph letters "A" (dot dash) and "N" (dash dot) were used for the beacons. Two directive antennas were used: the "A" was sent on one antenna and the "N" on the other. When the aircraft was precisely on course, the two letters would be superimposed and appear as a "T," a blending of the other two letters. If the plane moved slightly off course, one letter would be stronger than the other and be recognized as "A" or "N," depending upon whether the deviation was to the right or left. Every half hour the beacon signals stopped while weather information was transmitted. These were voice transmitters and receivers although the radiotelegraph signals "A" and "N" were in code. The aircraft equipped with radios mentioned below made use of these beacons as long as they were accessible, and at sea they could contact ships for position.[12]

Not all the entrants would profit from these aids, however. Of the eight planes that took off from Oakland, just four made it out to sea: the *Woolaroc*, *Miss Doran*, *Aloha*, and *Golden Eagle*. (The *Miss Doran* returned to Oakland but made repairs and headed out to sea again.) Of those four, the *Woolaroc* had a transmitter and a receiver, the *Golden Eagle* just a receiver; but neither the *Aloha* nor the *Miss Doran* was wireless equipped.[13]

The winner was the monoplane *Woolaroc*. After 26 hours, 17 minutes, and 33 seconds, the monoplane landed at Oahu's Wheeler Field. The pilot, Arthur Goebel, later gave credit to the radio for helping in the navigation. Two hours later the *Aloha*, another monoplane, landed safely and thus won the $10,000 second prize. Martin Jensen, *Aloha*'s pilot, later stated unequivocally that "if we had a radio set aboard ... we would have arrived first without any trouble." It seems that at one point the *Aloha* lost three hours circling while waiting until high noon for Captain Paul Schluter, the navigator, to shoot the sun and fix their position.[14]

Amateurs grabbed the opportunity to monitor the contestants' progress as the fragile planes winged across the Pacific toward the Island of Oahu, and hams scored a triumph for their splendid organization which resulted in their furnishing news involving the flights faster than commercial radio.[15]

Their preparations began with two amateurs, one in Oakland, California, C6ZR, and one at Lihue, Island of Kauai, Hawaii, 6AGL. These two, who had worked out schedules prior to the Dole race, would now track the progress of the Dole contestants and inform the world. Meanwhile an outstanding manufacturer of shortwave radios, Ralph Heintz of the firm of Heintz and Kaufman, had installed a state-of-the-art shortwave radio at 33.1 meters on Major Livingston Irving's *Pabco Pacific Flyer*. Heintz also asked one of the West Coast's best amateurs, Fred Roebuck, 6AAK-ex KFYH, to man the firm's station, 6XBB-6GK. He was to keep continuous watch

The *Woolaroc* had the advantage of radio in winning the Dole Race (courtesy National Air and Space Museum, Smithsonian Institution).

over the *Pabco Pacific Flyer's* signals. Roebuck in turn requested hams in the San Francisco area as well as in Hawaii to assist him in listening in for signals from the plane.[16]

Another group of amateurs in cooperation with the Oakland *Post-Enquirer* had put up a rig at the Oakland Airport. Ronald Martin, 6AYC, lent his transmitter. Using call letters 6NO, contact was made with BUC in Honolulu who was in touch with the naval radio stations by phone.

"With a number of Honolulu-California schedules working, and with a shortwave station on one of the racers, amateur radio was 'sitting pretty' and ready to follow the race from start to finish," commented *QST*. But then came disappointment. Major Irving's *Pabco Pacific Flyer* crashed on takeoff. Not to be deterred, the Oakland station 6NO kept a watch on the *Woolaroc's* 600 meters and listened for naval and marine reports on the flyers' progress, giving that news to the press minutes before it was received through commercial stations.[17]

What had happened to the *Golden Eagle* and the *Miss Doran*? Almost immediately a search was begun; possibly as many as sixty naval and private ships conducting the hunt with the addition of naval planes from the carrier *Langley*. Ham operators kept at their sets, "the operators giving up time, sleep, and food to carry out the work." Hams checked with Honolulu and found false a report that the *Miss Doran* had been spotted eighty-five miles from Hawaii.[18]

Then there was Captain. William P. (Lone Star Bill) Erwin and his navigator-radio operator, Alvin H. Eichwaldt. Erwin was the thirteenth ranking American World War ace; Eichwaldt was a mariner, having gone to sea at age fifteen and remained a seaman. The thirty-two-year-old Erwin had entered the Dole race, planning on taking his wife along, but at age twenty she was judged too young so twenty-seven-year-old Eichwaldt accompanied Erwin as navigator. Both knew how to handle a transmitter and use Morse code.[19]

Erwin had started out from Dallas in his Swallow monoplane, the *Dallas Spirit*. His aircraft was one of the two that started the Dole race and had to return to the Oakland Airport. (The other was the *Miss Doran.*) The plane had passed the Golden Gate when a trap door in the bottom of the navigator's cabin, insecurely fastened, blew open. As the wind rushed in it ripped the fabric and created a big hole in the bottom of the fuselage. There was nothing to do but return to the Oakland Airport.[20]

This did not stop Lone Star Bill. He immediately set about repairing his plane and announced his plans to fly to Hawaii anyway. With the two ships down—in those days planes were often called ships—Erwin cancelled all thought of racing to Hong Kong. "There can be no thought of racing or stunt flying while those two ships are down in the sea," he said. "When I get to Honolulu if either the *Miss Doran* or the *Golden Eagle* has not yet been found I am going to refuel and fly back looking for them again."[21]

Before Erwin left Oakland Ralph Heintz and Fred Roebuck, with a few other hams, had prevailed upon him to place in the *Dallas Spirit* the fifty watt shortwave transmitter that had been on the *Pabco Pacific Flyer.* Although it was installed hurriedly the set worked well in preliminary tests, so well that the *Dallas Spirit* was expected to keep in touch during the entire flight with both the mainland and Hawaii. Heintz and Roebuck had urged all hams to stand by on the plane's wavelength to listen for Erwin and Eichwaldt's call, KGGA. Even when they were not tapping out messages a continuous note would be issued from the short wave transmitter.[22]

The overloaded plane warmed up for the takeoff at 2:20 p.m. August 19th. "If we get off the ground we'll be long gone," Erwin said just before he and Eichwaldt entered the plane. At 2:15 the *Dallas Spirit* began the takeoff and by 2:17 it was 150 feet into the air, just short of the end of the runway. By 2:18 it was just a speck in the distance. Twenty thousand onlookers had cheered it on its way.[23]

By the time the plane was passing the Golden Gate, Roebuck, Heintz, and many other hams had tuned to KGGA's wave. The signals from the *Dallas Spirit* were described as being "tremendously powerful"; they were heard up and down the Pacific Coast, inland, and came in clear in New York City, where the *New York Times* receiver, 2UO, copied the transmissions. "For hours the steady drone of the transmitter brought news of the progress of the plane, interspersed with the very human comments of Eichwaldt," wrote Frates and Budlong for QST. (The "steady drone" was an audible, continuous sound when Eichwaldt was not sending a message.)

Heintz and Roebuck, listening in at the Oakland Airport, registered the messages beginning with the first, "Going strong. We are passing the docks. Will see the Lightship soon." As the day wore on their messages continued to be cheerily light-hearted: "2:50 p.m. We are flying at 900 feet and under the fog, with thirty miles visibility...." "2:55 p.m.... We are turning up 1,650 R.P.M. and making 95 air speed. All instruments working fine...." "3:30 p.m.—Our ceiling is increasing and the sun is breaking through." "3:33 p.m. Just had a drink of water." "3:49 p.m.... all ok except Bill just sneezed. We are keeping a sharp lookout for the Doran plane and the *Golden Eagle*...." "5:11 p.m.—Just passed the S.S. *Mana* at 5:10 Coast time and dipped in salute. They answered on the whistle. Of course, we could not hear it but we saw the steam...." "6:05 p.m. Please tell the gentleman who furnished our lunch that it's fine, but we can't find the toothpicks.—Bill." "7:10 p.m.—The weather is part cloudy with a smooth sea. Visibility about thirty miles. Have seen no wreckage or anything that might be either of the ones we are looking for. The visibility is still very good."[24]

All the messages were sent ZWT (words sent twice, ZWT being the commercial signal designating this procedure.) At about 8:30, the variation of the received signal's pitch became more pronounced than usual. This indicated bumpy weather conditions and uneven speed, a result of the transmitter's power source being a wind-driven generator. As the revolutions per minute changed, the power delivered to the transmitter fluctuated resulting in changes in the transmitted frequency. This caused the pitch of the received signals to vary.

"To those who could read the story of the varying note, this caused considerable concern," wrote Frates and Budlong, "which was only partially relieved by the jocular and unconcerned comments of Eichwaldt." It was noted, however, that words were no longer being sent twice. One of the radiomen listening was J.O. Watkins of the Federal Telegraph station at Daly City, near San Francisco. He interpreted the change in pitch after about twenty minutes as either engine trouble or as the plane climbing. But at 8:59 the sound suddenly steadied to normal pitch. Then it gradually continued to rise. "It is difficult to describe my feelings as I heard the note begin to wail at a frequency that indicated a wild velocity which I knew no sane pilot would voluntarily produce under such circumstances with a loaded plane and which I felt certain would be caused by a fall," Watkins wrote. Then the wavelength began swinging, indicating that the aerial was twisted or the navigator was changing his position (probably balancing himself against violent movement on the plane and accidentally bumping the radio dial). This continued until Eichwaldt flashed the SOS that the plane was in a tailspin, then sent "but we are out of it." There were a few moments when the sounds returned to normalcy. "(Then) came the second SOS and the announcement of the second spin, the rising and falling of the note telling its own story to those ashore. The second SOS was cut short by the crash." Watkins deduced that the *Dallas Spirit* had gone into a tailspin at 2,000 feet and crashed into the ocean at 200 miles an hour.[25]

Frates and Budlong, in their report for *QST*, paid particular homage to Eichwaldt, to his "cold nerve and supreme courage." They pointed out that during the half hour before the crash, while the ship "was bucking squalls one after another," the radioman continued his unconcerned comments and jokes, yet he had to know they were in great danger. "When the second spin came, and the plane started down to its end, Eichwaldt continued sending in the same even, unhurried manner that he had used during the flight. He stuck to his post to the end, sending calmly and evenly right up to the time the plane hit. With the note rising to a shrill shriek and falling almost to zero—denoting violent movements of the ship—the dots and dashes came through like clockwork until they were actually heard sputtering out as the antenna hit the water." Then *QST* added, "To know that he was heading for his death, and then to stick by the key telling the world just what was happening right up to the last second required courage of the highest order. Eichwaldt," observed Frates and Budlong, "preserved the highest traditions of the radio operating fraternity."[26]

While it was tragic for flying, the entire Dole episode gave a tremendous boost to shortwave radio. Up to this time radios used in planes were long wave at a standard 600 meters. The all-powerful *New York Times* was quick to credit shortwave radio for the paper's contribution to the news. Well should the newspaper, for Eichwaldt's messages, including the final two SOSs, reached the *Times* from an estimated 3,500 miles via the 200 foot vertical antenna stretched from the *Times* Annex to the radio room, all of this amid the high-rise structures of Times Square. The *Times'* rig was described as "a standard three-tube regenerative circuit built especially for shortwave reception." There was nothing novel about it; it was similar to receivers used by hundreds of amateurs. Quite possibly this was a record for receiving a message from an airplane. "That the *Dallas Spirit* employed just a single fifty-watt tube and that the plane was all the time close to the water, the accomplishment was phenomenal," wrote the *QST* writers.[27]

In 1928 a risky attempt at achieving something never before accomplished succeeded. It gave both aviation and radio a great boost. This was the flight of a Fokker trimotor, the *Southern Cross*, from California to Hawaii, on to the Fiji Islands, and from there to Australia. Between Hawaii and Fiji the plane would fly over a greater expanse of ocean, 3,138 miles, than any plane had ever before attempted. The Australian flyer responsible for the flight was Charles Kingsford-Smith, a determined glory hunter. "Smithy," as he was known to his friends, was a rare combination of fly-by-the-seat-of-your-pants daredevil and careful, meticulous planner. According to his mother, "Smithy" had loved flying ever since, at age five, he had jumped from the roof of a shed with an open umbrella and had nearly been killed.[28]

The crew consisted of two Australians, "Smithy" and Charles T.P. Ulm, his relief pilot, and two Americans, Lieutenant Commander Harry W. Lyon, a merchant marine skipper as navigator, and James Warner, the radio operator. Warner had recently been discharged after sixteen years in the U.S. Navy and was working as a clothing salesmen when Kingsford-Smith offered him the position as radioman; he had never been aloft prior to the takeoff at Oakland.[29]

Over a period of ten months Kingsford-Smith and Ulm studied every overseas attempt, both successful and unsuccessful, and planned their journey so meticulously that they had removed just about every possibility of failure save for the obvious risks involved in flying with the aeronautical and wireless technology of the time. Their shortwave radio was built by Heintz of San Francisco; it was similar to the one on the *Dallas Spirit* and on Sir George Wilkins' flight to Spitzbergen. It operated on 33.5 meters and was powered by wind generators on the wings. The plane was also equipped with a 600 meter transmitter for contacting ships at sea, with an auxiliary distress arrangement, and with a receiver. As "Smithy" told the press in Hawaii, "I can't say too much for our radio. It was a vital item of our equipment and it means more to transoceanic flying than any other one item I can think of with the exception of extremely careful preparations."[30]

Hams all over the world listened for the *Southern Cross*'s call letters, KHAB. At the beginning, during the flight to Hawaii, there was a brief brouhaha when Warner, the *Southern Cross*'s radio operator, warned amateurs that the *San Francisco Examiner* would prosecute hams giving out messages sent to 6ARD, the *Examiner*'s radio, with which Kingsford-Smith had a contract. But Warner soon realized that if every amateur shut off his receiver the world might not hear of their progress, so he began sending messages QST (non-addressed), which gave all amateurs the right to receive the messages. This was a good move on Warner's part. "Reports from the wire services in San Francisco indicated that amateur reports were beating those of 'non-amateur' stations, Navy radio and the commercial companies," reported J. Walter Frates.[31]

Amateurs cooperated with each other in keeping track of the plane. Hawaiian hams kept in touch with San Francisco operators, giving them messages that may not have come in clearly to the mainland stations. KHAB's signals were also heard in Alaska. A South Dakota ham was in touch with an Australian ham and so received information about the ending phase of the trip. "When commercial and 'non-amateur' stations had either long since given up the ghost or were having difficulties on the Pacific Coast, amateurs in San Francisco and Oakland were still listening to the steady drone of the transmitter and comfortably copying positions until KHAB reported itself beyond the Loyalty Islands and within a few hundred miles of Brisbane, where daylight intervened," reported Frates. Then the burden of communication was taken up by Australian and New Zealand hams. (The Loyalty Islands are in the South Pacific near New Caledonia, which, in turn, is northeast of Australia.) Frates noted that on their flight from Hawaii to Fiji the crew were in touch with amateurs on lonely islands such as the cable station at Fanning Island, in the central Pacific south of Hawaii, where 1AJ was in communication with them.[32]

Kingsford-Smith gave both his American crew members great credit. He considered Lyon the best navigator ever, and said he would use him for any flight anywhere. As for radioman Warner, he was "a man of no mean skill and courage. Imagine listening for a message in a little corner in the bottom of this airplane with her great engines roaring out in front, and sending and receiving perfect messages from 4,000 miles away."[33]

The romance of the adventure was brought home to millions by way of shortwave and the nation's press. When they flew into black storm clouds, gained altitude in an attempt to get above a storm, and worried about gas consumption, they let the world know. When the *Southern Cross* was bucking a storm, radio operators "heard the plane's generator shriek wildly upward. The signals were rendered barely audible and then died out altogether." There was fear that the plane had crashed into the sea, but soon it was heard from again. "Always at the end of their radio messages came a cheery word from the men. With courage that never faltered they defied with jest the anger of the wind that threatened to hurl them into the waves below," reported the *New York Times*.[34]

The *Southern Cross* completed its flight to Brisbane successfully. Lyon and Warner sailed back to America while Kingsford-Smith and Ulm made the rounds of Australian cities. Smith continued in the flight business, trying to establish air mail and transportation routes between Australia and New Zealand and even from Australia to England. The saying that "he who lives by the sword dies by the sword" is applicable to this competent, courageous man. He eventually disappeared flying off the coast of southern Burma.

As was stated at the end of the report on the flight of the *Southern Cross*, "In the not-far-distant future when aerial fleets will be making transcontinental and transoceanic passengers and freight flights and will carry on their radio communications on a shortwave band similar to the marine one of 600 meters, a great deal of the credit for the work will be due to the pioneering efforts of the amateur, who devotes his time and equipment unselfishly for the advancement of the art and without a pecuniary interest."[35]

Forgotten now, hardly mentioned in the annals of aviation progress, was the failed flight of a big two motor Sikorsky amphibian, the *'Untin' Bowler*. Colonel McCormick, publisher of the *Chicago Tribune* wanted the *'Untin' Bowler* to forge a commercial air route from Chicago to Berlin. The flight from Buffalo to Chicago, from where the flight was to begin, was headlined with many pictures. During a storm over Ohio the crew contacted an amateur in South Bend who obtained a weather report for them. The colonel announced prizes amounting to $400, impressive in 1929, to the amateurs who conveyed the most messages from the *Bowler* to the *Tribune*. Amateurs in Nova Scotia and Labrador were also arranging to "watch by ear" the plane's progress as it took off from Port Burwell at the northern tip of Labrador for Mount Evans, Greenland. Preparations for the trip were described in detail. These included descriptions of the long wave radio to be operated by the flyers, no radioman being along. However, a *Tribune* reporter was aboard to file stories by way of the radio.

The ARRL appealed to all members of the league to cooperate, to work the *Bowler* on each of its hops, and hundreds did. The aircraft's route took it from Chicago to Milwaukee to Buffalo to Remi Lake, Ontario, and from there to a point just beyond Rupert House at Ungava Bay. This is southwest of the Hudson Straits and east of Hudson's Bay; from there the aircraft was projected to fly southeast to Europe and Berlin. At Ungava Bay the pilots were marooned for two days amidst flowing ice. The *'Untin' Bowler* finally made it to Port Burwell, some forty miles farther. The tide rose forty feet. Ice floes, some of them small icebergs, threatened the plane. In spite of their efforts to save it, the ice finally tore it from its moorings and carried the amphibian three miles out to sea, where it sank on July 12. The crew finally reached the rail head at Churchill, on the western shore of Hudson's Bay, and from there came down through Winnipeg to Chicago.[36]

Even though the *'Untin' Bowler* did not make it to Europe, let alone Berlin, Colonel McCormick chose to present the promised prize money to the amateurs who had received, and then transmitted to the *Tribune*, the most messages from the *Bowler*. He gave several other hams certificates signed by Maxim of the ARRL and himself. Communications by radio were highlighted in many of the news items and amateur radio scored favorable publicity. In spite of the failure of the *'Untin' Bowler* to complete its projected flight, the general consensus as revealed in the newspapers was one of optimism. The time would come when regular air service would exist over the great circle route from Chicago to Berlin.[37]

As would be expected, the military was deeply involved in experimentation during these years of frenzied air activity. In January 1930, the army decided to test the air corp's ability to be operational under polar conditions. One of the purposes of the flight was to obtain first-hand

The hams with the most calls to the curiously named *'Untin' Bowler* received prizes from Colonel McCormick. Here is the Sikorsky Amphibian with *'Untin' Bowler* imprinted at the front of the aircraft's cabin (courtesy I.I. Sikorsky Historical Archives).

knowledge of the value of shortwave radio communication in operations over long distances and in remote regions. New equipment, such as heated face masks and gloves, were also to be tested. Although the flights were all in the lower 48, the winter, although at first mild, turned ferociously cold and the operation was anything but a junket. Commanded by Major Ralph Royce, pilots in open cockpits of the eighteen Curtis-Hawk pursuit planes, with two (some articles say three) transports, all from Selfridge Field, near Mount Clemons, Michigan (just north of Detroit), were to fly to Spokane, Washington, and return by a slightly more southern route; the total distance was about 3,500 miles. All the planes were equipped with skis instead of wheels. All told, forty-three individuals were involved.[38]

The experiment was something of a landmark for hams. The air corps approached the ARRL requesting ham cooperation and received it with enthusiasm. One of the transport aircraft, a Ford Trimotor, was equipped with shortwave equipment with call letters AB6, to work at 32 and 54 meters. How well could AB6 maintain contact with other stations? The ARRL enlisted its most reliable relay members to keep in touch with the Arctic patrol; also taking part were the army's signal corps and General Electric's experimental shortwave station at Schenectady. Amateurs taking messages were instructed to transmit them, via their relays, to the league's message center in Hartford and the army's message center in Washington, D.C. This demanded that specific schedules be arranged among the relayers, and that they abide by them rigorously. After being received in Hartford and Washington, the information would then be relayed out to amateurs again, who were to deliver it to the nearest newspapers. This was publicity for the army air corps—and inadvertently for amateurs. The ARRL notified members along the routes of the First Pursuit Group and requested that they keep in touch. All fifteen hundred amateurs involved in the relay system were informed of the maneuvers and invited to participate.[39]

The one flaw in all their plans was the slower speed of the transport planes, including the Ford Trimotor equipped with the shortwave equipment. It was also plagued with bad luck and fell days behind the pursuit planes winging their way toward Spokane. When the radio plane reached Minneapolis it was directed to await the return of the pursuit group and rejoin it on its way back to Selfridge Field.[40]

With the radio plane out of service Major Royce had to fall back upon amateurs. At every city where the group landed he was in contact with hams who cheerfully conveyed messages to Selfridge, Hartford, and Washington, and in return received weather reports and other messages for the pursuit group. Several hams stayed at their rigs for many hours, relaying messages successfully when other means of communication were impossible. Although no fliers lost their lives, there were forced landings, moderately serious mishaps and motor breakdowns that grounded a plane or two. Once they were flying so low that a barn loomed just ahead; they zoomed upward and just cleared it.[41]

The trials and tribulations of the First Pursuit Group's battle with freezing temperatures was heavily covered in the nation's press with amateurs sharing much of the news. The ARRL received accolades from the assistant secretary of war, who wrote that he was "perfectly amazed at the results produced by your enthusiastic and able members." Sgt. K.D. Wilson of the group's radio AB6 also thanked amateurs for their "splendid cooperation." From Selfridge Field came the comment that "with the little time to prepare, the amateurs did well."[42]

The best descriptions of working shortwave radio aboard an airplane were written by an intelligent, courageous woman. When her husband announced plans to forge an air route to the Orient via Canada, Alaska, Siberia, and down to Japan and China, he informed her that she was coming along as his crew, operating the radio. She would have to take an examination for a third class license. This involved theory as well as fifteen words per minute with Continental Morse code. "Now Charles," protested Anne Morrow Lindbergh, "you know perfectly well that I can't do that.... I had to be tutored to get through elementary physics in college!" Her husband said they would study together. He brought home "a small practice set of buzzers and keys," hired a radioman to tutor them, and together they studied code and the theories of radio. They passed the exam for third class radio operators, Charles earning a higher grade than Anne.[43]

Anne spent hours at the buzzer, achieving a modicum of confidence. Her first contact was with a prearranged operator on Long Island. "Who—is—at—the—key?" asked the operator. Anne wrote down the words nervously, for she still listened for letters, not whole words. Then she tapped out "Anne—Lindbergh—how—is—this—sending?" "Pretty—good—but—a—little—heavy—on—the—dashes," came the reply, "just—like—my—wife's—sending." Anne concluded that "there was still a good deal for me to learn."[44]

Her professionalism came by way of on-the-job training. On July 27, 1931, the Lindberghs took off in a beautiful Lockheed aircraft, dubbed the *Sirius*, powered by a 600 horsepower Wright Cyclone engine. Black with orange wings and pontoons instead of wheels, it was a strikingly handsome plane, even by modern standards. Seats were tandem, Anne's behind her husband's. First they flew to Washington, D.C., to pick up official papers. Then they began their journey, their first stop at North Haven, Maine, near the Morrow family's summer residence. Anne knew that WOA at North Haven was awaiting her message. This was no longer practice, this was the real thing! She had to contact a station from a flying airplane. Working the transmitter and receiver in the cramped space of a cockpit was a lot different, she quickly learned, from practicing transmitting and receiving in her bedroom.

What she had to master was a complex rig. It was the transmitter-receiver used by Pan American Airways on its South American flights. To change frequencies she had to choose from six transmitting coils. The antenna, with a metal ball at the end to weigh it down, had to be reeled manually out and in, so many turns for each frequency. One switch on the rig always gave her a 400 volt shock. That afternoon on the way to Maine was hardly a success: static, "a kind of stage fright," and the complexities of the rig all worked against her.[45]

The next leg of the flight was to Ottawa. "I had my first successful day with radio," Anne

**Anne Morrow Lindbergh mastered Morse code well enough to earn certification and be the radio operator for her husband (courtesy National Air and Space Museum, Smithsonian Institution).**

wrote. "I was in contact with one or the other of two stations every fifteen minutes of the trip. I was able to send out our 'Posn' (position) and 'Wea' (weather) regularly and to hear in return the comforting 'dit-darr-dit' (r) which means 'received ok.'" By using the abbreviations she inadvertently was telling her readers that she was rapidly picking up QST English. According to her biographer, at some point on the Orient flight "a radio operator, impressed by her skill in taking and receiving messages, informed her over the radio, 'No man could have done better' She considered this the highest accolade anyone could give her."[46]

Once she had mastered the art of fitting the correct transmitters coils, replacing them in storage, flipping the right switches, and tapping out messages, Anne discovered that receiving messages demanded four hands when she just had two. On the shortwave the dots and dashes veered, so one hand was necessary to turn *that* dial, a second hand to adjust another dial, a third hand to hold the pencil and a fourth to hold the writing pad. Her "right hand took messages on a pad balanced on my left knee," she wrote, "while my left hand crossed over and controlled the dials on my right." She must have sighed. "There was no use getting around the fact that the radio equipment was installed for a radio operator," Mrs. Lindbergh commented. "I would simply have to learn to be a radio operator."[47]

She learned fast. Anne did an expert's service when they needed contacts to get to Baker Lake, Atkavik, Point Barrow, Nome, Kamkchatka, and Nemuro. Often the *Sirius* was enveloped in fog. Approaching Point Barrow she was in constant touch with the Barrow radioman. "And even when I couldn't send," Anne reminisced, "—when we were flying too low for the antenna to be out the right length to send but I could drop it down to receive—he [the Barrow operator] went right on sending weather and information. It is wonderful and heartening to get when you are cold and isolated, flying through the fog." When they landed at Barrow she identified the radioman by the khaki mackinaw he was wearing. "I was so gratified to him I had to shout to him immediately and thank him," she wrote.[48]

The trip to the Orient was in 1931; two years later, in July 1933, the Lindberghs were again in the air, this time on a five-and-a-half-month, 30,000 mile survey of possible air routes from the United States to Europe. Anne was again to be at the key of KHCAL, the plane's call letters. She tells of her preparations for the long survey, including practicing on the radio. This indicates that she never made a hobby of radio, so as the date of their 1933 expedition approached she had to brush up on her skills. In her writings she reveals a new confidence. "It feels good to be in the ship again," she wrote in her diary upon their taking possession of the *Sirius* in California, where it had been repaired following an accident in China (and now with a more powerful motor and a new propeller). "That's where the key was, that's where the transmitter was.... I feel comfortable and safe, that lovely red wing below me." (The radio equipment, the same as used on the Orient flight, would be installed later.)[49]

Her skills were tested to their limits. "On this trip I expected to do much more radio work than flying," she wrote. The extent to which she describes her activities with the transmitter and receiver, both in her extensive article in *National Geographic Magazine* and in her *Diaries and Letters*, make clear that radio was her primary task throughout the long Atlantic survey. She describes the positions of the rig in the cockpit. Again and again she details her sending and receiving, along with some amusing asides. She complains of her fingers getting so cold that she could not send well. On other occasions she mentions "bad radio" or "terrible radio communication." So dominant was her task that it entered her dreams. "I wake to find myself repeating dots and dashes, and feel I am fighting something—what? Darr darr darr—darr dit darr—darr dit daar.... Everything turns into dots and dashes in my mind."[50]

The day they flew to the Cape Verde Islands was uneventful, Anne wrote, "except for half an hour which was, for me, one of the most thrilling of the summer." She was sending out CQs (General Call) on short wave. Listening at 24 meters at 11:30 GMT (Greenwich Mean Time), she wrote,

> I heard very plainly WSL at Sayville, Long Island, also sending out a CQ.... I decided, rather recklessly, to call him.
> "WSL—WSL,—WSL—de—KHCAL—ans (answer) 24." I tapped it out easily, confident that my 15-watt transmitter could never reach him. Silence—and then the CQs at the other end stopped. Then clearly:

"KHCAL—KHCAL—KHCAL—de—WSL." He was answering! I was hot and cold from excitement and did not dare touch the set lest it should break the contact.

"QRK" (I receive you well, your signals are good), we went on. "QRU"? (Have you anything for me?

QRU—casually, like that! ... I would let him know ... that we were not just around the corner at Atlantic City.

"Lindbergh plane—en route Cape Verde Islands—min pse (please wait a minute)...."[51]

According to her biographer, this contact established a world record of over three thousand miles for radio communication between an airplane and a ground station. Anne also mentions contact with WCC in Chatham, Massachusetts, nearly four thousand miles away. When the operator indicated that he wanted to interview her—"answer answer few few questions questions first radio interview from airplane," she cut him off.[52]

She had become adept at changing frequencies, using the equipment and contacting stations from the cockpit, but she was still a novice when copying code. Probably she also sent it rather slowly. In her delightful *Listen! The Wind*, which covers just ten days of their survey, she writes that "Receiving was still quite difficult. Except for familiar words and expressions, I could not yet translate the messages in my head and was forced to write them down letter by letter as the sounds came in my ear. My pencil and not my mind apparently did the translating.... When the pencil stopped I would look down and see what I had written."[53]

KHCAL, with Anne in charge, continued to operate when she flew with her husband. Gradually these flights declined as the years went by, Charles flying solo more and more, participating in the air war in the South Pacific, and then, deeply involved in the environmental movement worldwide, opting more and more to be a passenger on commercial airlines.

Amateurs also served during airline disasters. On December 28, 1934, a Curtis Condor, a huge two-engine passenger biplane flown by American Airlines, was lost in a snowstorm in the Adirondack Mountains between Syracuse and Albany. Just four persons were aboard: the pilot and co-pilot (brothers named Ernest and Dale Dryer), another pilot, J.H. Bron, en route to his home in Boston, and one paying passenger, a Department of Education official, R.H. Hambrook, whose ultimate destination was Washington, D.C.[54]

In the case of the American Airlines Condor, one engine went out and the weight of icing taking place on the wings was too much for a single motor to carry. The plane flew lower and lower. When trees were sighted pilot Dyer stalled the plane, nose up, and allowed it to smash into trees, which broke its speed and allowed it to fall to the ground. Although the wings were sheared off and a propeller bent, the plane remained intact and did not burn. After it was located, the pilot said that it could hardly be seen from as low as 1000 feet altitude. All escaped serious injury although the pilot's jaw was broken.[55]

Nevertheless their situation was precarious. Two and a half feet of snow had fallen and the temperature was well below zero. They were without adequate clothing and food. The radio faded steadily and finally gave out. "We fought to stay awake," said Hambrook, the one paying passenger. "Sleeping meant freezing.... Saturday dawned but we heard no sign of life. We were too tired to care." They improvised a shelter from broken plane parts and tree boughs. Sunday morning came. They heard the droning of a plane, but did not see it. Then on Sunday afternoon they heard a big Condor in the air. One of the crashed plane's pilots threw a bucket of gasoline on the fire to make it flare up, and they were seen.

A group of amateurs from Schenectady had organized what they called a "General Electric" expedition to provide communication between search planes and the airports from which they had flown. The hams had set out with three cars and a truck loaded with portable gear. At first their base was at the small airport at Gloversville, New York. With gas-driven generating equipment they maintained communications between the airports at Albany, Buffalo, Newark, and

Boston and with the pilots of the many planes involved in the search. When, late on Sunday afternoon, the plane's location was determined, the hams loaded their portable gear and in a few minutes were making their way over ice-covered roads and then over a woodcutter's trace which brought them to a cabin close to the wreck; the downed aircraft was about four miles away through hip-deep snow. Within fifteen minutes they had the portable radio working. The radio was active until afternoon helping other search parties get back to their bases. "Some of the operators had not slept for 36 hours and others had not eaten for 26," reported *QST*. "Excitement provided the necessary stimulant."[56]

The fifty hours between the crash and their rescue had taken its toll on the occupants. Three were taken to the hospital, the pilot with frozen feet and hands and signs of pneumonia. His brother had tried to walk out with the rescuers but ended up being carried on a toboggan; he had a broken jaw and an eye injury. The pilot-passenger was suffering from exposure and exhaustion. Only the paying passenger was in good shape. He boarded a plane at Utica for Washington that Sunday afternoon but was forced down at Newark and had to spend the night there.

These are just a few of hundreds of examples of amateurs cooperating and participating in the enthusiastic advance of aeronautics taking place during the 1920s and 1930s. Improvements of both advanced in tandem, and helped bring both arts—flying and radio—to where they are today.

The army air corps' experiment was just a smattering of the experiences of amateurs in cold places. The Arctic and the Antarctic beckoned.

# 11

# Amateurs and Polar Exploration: Phase One

People reading the daily newspaper and watching television today may have less in-depth knowledge of the polar regions—the Arctic and the Antarctic—than did their counterparts living between 1900 and 1940. During those years polar exploration was headline news. There were Peary and Cook, both claiming to have reached the North Pole. There was Captain Robert Scott's demise in Antarctica, Roald Amundsen's success, and the incredible story of Shackleton's ill-fated attempt to cross the Antarctic continent. In the Arctic in the 1920s there was Amundsen again, crossing over the North Pole in a dirigible, the *Norge*, and the attempt of the Italian Umberto Nobile to duplicate Amundsen's achievement in the dirigible *Italia*. Americans had their own Rear Admiral Richard E. Byrd, flying over the North Pole (like Peary's and Cook's claims, an achievement disputed with strong arguments pro and con) and his undisputed flight over the South Pole. Australians had every reason to be proud of the scientific work accomplished by Sir Douglas Mawson's expedition in the years 1911–1914.

Today incidents of great adventure, hardship and suffering no longer characterize man's conquest of the polar regions. In the post–World War II period permanent research settlements have been established in Antarctica. Today Russian ice breakers periodically cross the North Pole while doing the everyday job of breaking a watery path for flotillas of freighters carrying cargo from northern Russia to the rest of the world. The Northwest Passage has become a shipping route. Transatlantic flights cross the Arctic many times a day, often but not intentionally crossing over the North Pole. In the 1999-2000 season, an estimated 10,013 tourists traveled south on 116 trips to Antarctica, and two to three million tourists visit above the Arctic Circle annually.[1]

Take a world globe and study its top and bottom—the Arctic and Antarctica. Just in case one's knowledge has slipped away since geography was studied in grade school, the Arctic is an ocean, not a continent. The top of the globe shows the northern reaches of Asia, Europe, North America and Greenland extending into the ice-covered Arctic with peninsulas and especially archipelagos—numerous rocky or snow-covered islands—dotting the otherwise barren expanse. The vision of the ice-covered ocean should be one of pressure ridges, chasms, leads (the name for open water) and icebergs. Trudging across such an ice-covered body of water, men either pulling sledges themselves or using dogs, was not only an incredibly hard task, it was also dangerous. Ice floes could break away and the explorer, his sledge and dogs find themselves on a deteriorating piece of ice gradually floating away from camp. There was also the constant danger of plunging into the frigid waters or into a chasm. Pressure ridges could occur as often as four per mile.

And let us not forget that this all had to be accomplished in freezing temperatures and often in darkness.

For six months of the year the polar regions are primarily in night. The months from late April until September, or October until March, witness little more than a glow on the southern or northern horizons (Antarctic or Arctic). The stars and the moon are bright if there are no clouds, and displays of the aurora borealis or, in Antarctica, the aurora australis, have men waxing ecstatic with their beauty. But much of the time the weather is bad with high winds, terribly cold temperatures, and snow churned up so thick by the winds that one can hardly see three feet ahead.

Unlike the Arctic, a sea surrounded by land, Antarctica is a continent averaging six thousand feet in altitude. It is encircled by the most agitated waters in the world. In this region where the Indian, Atlantic, and Pacific oceans meet, storms are almost the norm. Great icebergs drift north. Ice floes that get thicker and larger as they approach the continent often prevent ships from approaching the mainland. Today aircraft do most of the work of hauling in and out supplies and personnel. The animal, bird, and sea life is abundant although it once was much more so before whaling and sealing reduced the numbers astronomically. Most people affiliate Antarctica with penguins, and indeed there are thousands of them in a number of varieties.

Recently there has been renewed interest in the experiences in Antarctica of Scott, Shackleton, and Mawson, and in the continuing question of whether Peary or Cook actually reached the North Pole. Yet, largely forgotten were scientific expeditions with experts conducting all manner of experiments—meteorological, astronomical, biological, zoological, geological, oceanographic, magnetic, ichthyological and, in the Arctic, anthropological. For the most part, their efforts were not marred by tragedy. Their contributions to the scientific knowledge not only of the polar regions but of the earth itself, and its atmosphere and its climate, were considerable. Many of these scientists grasped at radio as a wonderful way of maintaining communication with the rest of the world. (Of course, scientific work continues unabated today.)

The role played by radio in polar exploration is best presented by describing an Antarctic exploration that took place prior to radio's appearance. This was the *Belgica* expedition of 1898–1899. At the time, sixty years had elapsed since an expedition had attempted to advance beyond the southern ice barrier. Due to a lack of knowledge of what was needed, the *Belgica's* preparations were lacking in many ways. Its leadership and principle funding was Belgian, although the crew and scientific personnel included several nationalities, including the American surgeon Frederick A. Cook and a Norwegian named Roald Amundsen who would return and be the first person to reach the South Pole.[2]

As for Frederick Cook, he became an enigmatic figure in the history of polar exploration. Cook claimed to have reached the summit of Mount McKinley in Alaska but this achievement was successfully disputed. He emerged from the icy wastes and announced, on September 2, 1909, that on April 21, 1908, he had reached the North Pole; on September 6, 1909, Peary announced that *he* had reached the Pole on April 6, 1909. Again Cook's claims were disputed; Peary clearly won the argument although later researchers would question his claim also. Possibly Cook was a Dr. Jekyll–Mr. Hyde kind of man, because his narrative of the *Belgica's* struggle tells a convincing story of a competent physician who was very concerned with the well-being of the men on the iced-locked ship.[3]

Today a man of Cook's interests would probably be a psychiatrist. Much of his book describes the moods and attitudes of the men. When they were frozen in and faced with months of darkness on the beleaguered ship, Cook analyzed their physical and mental states. "This part of the life of polar explorers is usually suppressed in the narratives," he wrote, "[but] an almost monotonous discontent occurs in every expedition through the polar night." He expands on the subject: "The curtain of blackness which has fallen over the outer world of icy desolation has also descended on the inner world of our souls. Around the tables, in the laboratory, and in the forecastle, men are sitting about sad and dejected, lost in dreams of melancholy from which, now and then, one arouses with an empty attempt at enthusiasms." Fortunately just one man, a seaman, went insane during the *Belgica's* ordeal, and even he recovered when the sun arrived again.[4]

To Cook, physical troubles were even more dangerous. Men lost their appetites and their digestive systems did not function correctly. Their pulses became irregular and their blood pressure elevated; they could hardly walk a hundred feet without breathing heavily. Although none were over age thirty-five, all turned gray-haired. "The gait is now careless," he wrote, "the step non-elastic, the foothold uncertain." The death of a young man named Danco accelerated the malaise.[5]

Not mentioned by Cook was the psychological element present in the reality that the *Belgica* might never break free of the ice. During severe storms, or times when the wind blew just right, the old ship creaked and cracked and the fear was that it would be broken up. If this happened, supplies did not exist to enable the eighteen men (for one had died) to make their way over the ice to safety. In fact, the crew in their last mid-summer gazed with agony at a lead (an area of ice-free water) just yards from the *Belgica*, yet the ship could not move to it. They tried sawing up the ice and setting off explosives, but in the end it was a favorable wind that saved them. On their journey north they were marooned again for nearly a month, but finally the *Belgica* and its men reached safety at Punta Arenas in southern Chile.[6]

Surely later explorers into the polar regions were aware of the mental, physical, and emotional troubles that men endured during the long winters' nights as described by Cook and others of the *Belgica*. One would think that one of the first moves would have been to install wireless on their vessels and if possible, at their land bases also. Explorers, after all, are expected to be interested in using any and all devices that might help ensure success. Yet I find no reference to wireless in the campaigns of the ill-fated Sir Robert Scott, whose last expedition, which did reach the South Pole, was in the years 1910–1912. Neither did the victor in that race, Roald Amundson, have wireless along. Nor did Peary and Cook have wireless when they claimed to have reached the North Pole.

Shackleton did not have a transmitter on his ill-starred ship *Endurance*, which was crushed by ice in the Weddell Sea. He does mention, however, the failure of the receiver he was known to have on board. Radio was on the *Aurora* (which had been Mawson's ship), which was to meet Shackleton's overland party in the Ross Sea (if his party had succeeded in landing and crossing the continent). The ship tore loose from its moorings on May 6, 1915, and was adrift for ten months and trapped by ice. On board the *Aurora* was a radio operator named L.A. Hooke with a transmitter and receiver given the expedition by the people of Sidney, Australia. This was a long wave transmitter, the shortwave "revolution" being still five to seven years away. Long waves could not carry nearly as far as short waves.[7]

Although his wireless was considered capable of working stations just two hundred miles away, Hooke persisted night after night in trying to work civilization. The closest wireless, he thought, was the one at Macquarie Island, but the government had removed it in the interests of economy. For war reasons a transmitter at Awaru in New Zealand had also been removed. Finally, with a quadruple aerial eighty feet above the deck, on March 25, 1916, Hooke worked stations in Tasmania and New Zealand, at least nine hundred miles distant. In terms of wireless, this was a fluke, but such extraordinary contacts do occasionally occur. News of the plight of the *Aurora* surpassed war news on the day it was issued, and the contacts greatly aided the *Aurora*, crippled with a broken rudder, in returning to civilization.[8]

Why didn't *all* polar explorers bring wireless along? Finances could have been one reason. Expeditions were almost always hard up for funds. Radio was still in its early stages of development. It was heavy. It was not yet trustworthy. Possibly it struck them as superfluous and non-essential.[9]

The first Antarctic explorer to make use of wireless was the Australian Sir Douglas Mawson, one of the most respected of all polar explorers, and rightly so. His Australian Antarctic Expedition in the years 1911–1914 was radio equipped. It is clear that Mawson, whose aim was not to reach the Pole but to gather all kinds of scientific information, understood the possibilities in this new means of communication, primitive as it was at the time. He was aware of the "madness and dissidence" on the *Belgica* and "knew that melancholia and madness could be contagious."[10]

It is not surprising, then, that among the 5,200 separate boxes, bundles, and items stored on board Mawson's ship, the *Aurora* (the same vessel later damaged with Shackleton's expedition), were "oregon poles"—wireless masts. Some of them were to be installed on Macquarie Island, 850 miles south-southeast of Hobart, Tasmania, where a small contingent was to conduct scientific work along with a radioman to handle a state-of-the-art Telefunken transmitter and receiver. (Mawson does not say, but in all probability this was a long wave rig, shortwave radio not yet having made its mark on the art.) The rig was to serve as a relay station between Mawson's base in Antarctica and civilization in Tasmania. The remaining oregon poles remained on the ship bound for the Antarctic base.

At Macquarie Island all did not go well with the radio technicians and their equipment. Arthur J. Sawyer, the radioman assigned to the island, recorded in his diary that Hannam—Walter Henry Hannam, who was to be the wireless man at Mawson's Antarctic base—had "opened all the barrels and boxes, therefore exposing almost everything to the rain, the generator lying out with an overcoat over it." But progress was made. A flat hilltop, dubbed "Wireless Hill," just three-quarters of a mile from the main dwelling, was chosen for the radio installations, including the antennas and a hut. Sawyer, whose occupation was as wireless telegrapher with the Australian Wireless Company, appears to have done a competent job.[11]

After two weeks the *Aurora* headed south to establish the main base at a point the expedition dubbed Cape Denison on Commonwealth Bay in Adélie Land. They soon discovered that it was the windiest, stormiest part of a frigid continent. (Study the bottom of a world globe: it will be seen that they were due south of Australia in land claimed by that country.) Then the *Aurora* sailed fifteen hundred miles west and dropped Frank Wild and six other carefully chosen men on Shackleton's Ice Shelf. At this western base they constructed a small hut. Wild had a radio with him, but it was just a receiver; he makes no mention of it in his report beyond that they erected a "small antenna."[12]

Back at the main base, with considerable difficulty and with the aid of a derrick, the oregon poles and radio equipment were landed and the task of erecting the masts proceeded slowly. "'Dead calm, up with the wireless masts!' was the call when there were a few hours of stillness.... Everyone hastily dashed for his burberrys [outer clothing], and soon a crowd of muffled figures would emerge ... dragging ropes, blocks, picks and shovels.... The first thing to do was to establish good anchorage. Then the oregon masts, section by section, were erected and securely stayed by stout steel-wire cables.... Fumbling with bulky mitts, handling hammers and spanners, and manipulating knots and bolts with bare hands while suspended in a boatswain's chair in the wind, the man up the mast had a difficult and miserable task." A twenty-two year old named Bickerton, taken along as a motor engineer, did most of the work. One day he was frostbitten while climbing to a block on the topgallant mast. At full height the masts were ninety feet above the ground. It took several months, beginning on April 4, to get them up. Meanwhile a smaller hut, to serve as a work room, was attached to the main dwelling. There the wireless was placed. A good foundation was laid for the small motor and generator that would power the wireless equipment.[13]

It was October before the rig was ready for service. Hannam was able to send messages to Sawyer on Macquarie Island, and many were received there, but contact could not be made from Sawyer back to Hannam at the hut. It could have been due to the intense static caused by gales and drift snow. Whatever the reason for the failure to receive code from Sawyer on Macquarie Island, Mawson blamed Hannam, saying that "wireless was the biggest failure of the expedition." Hannam in his diary complained that Mawson expected too much. After all, he protested, it was the first attempt at wireless in the Antarctic. The discussion was muted when in October the winds blew down one of the masts. Several months later the mast was successfully erected, and the radio functioned well.[14]

All members of Mawson's expedition could now look forward to spring, summer, and autumn in Antarctica. In due time exploratory journeys with sledges and some with dogs set out on planned journeys taking them as many as 300 miles away and return. All would have to cross treacherous ice and snow, crevasses and pressure ridges, glaciers and occasional unstable ocean ice. It is a positive commentary on Mawson's administrative ability, his judgment of men, and his careful planning, that all but his own assignment were carried out without the loss of a man. Only Mawson's resulted in the death of his two companions, Dr. Xavier Mertz and E.S. Ninnis, one disappearing in a chasm; the other died in his sleeping bag. That Mawson was able to make his way back to the hut alone is also testimony of the abilities of this remarkable man.

Mawson's return after the loss of his two companions resulted in his arriving at the main base just hours after the *Aurora*, which had waited several days in hopes that he would show up,

These are the "Oregon poles" erected at Mawson's camp in Antarctica (courtesy of South Australian Museum Mawson Collection).

steamed away. Captain Davis had brought the ship to Cape Denison on the 13th of January, 1913, when just nine members of the expedition were at the hut, the others still being on their assigned expeditions. By January 18th two of the three parties had returned; only Mawson and his companions were still out. The captain landed supplies sufficient for another winter, then cruised east and back, hoping to see Mawson's party, and, failing at that, sailed away on February 6th. He was actually following Mawson's penciled instructions of what to do in case he (Mawson) did not return. Just a few hours after he sailed, Mawson appeared at the base camp.

At this point radio came in handy, although it did not make a difference in operations. Hannam had embarked on the *Aurora* to be taken home. He had changed places with the *Aurora's* radio operator, twenty-seven-year-old Sidney Jeffryes, who had previously applied for work with the expedition and had been rejected. Unbeknownst to Mawson, he had been taken on as wireless operator on the *Aurora* for its second voyage to Antarctica. When it reached Cape Denison Jeffryes volunteered to stay on, thus relieving Hannam. Mawson thought it should be possible to contact Captain Davis, who was barely eighty miles away. Jeffryes contacted Hannam, who was listening for him on the *Aurora* by previous arrangement at 8, 9, and 10 p.m. The contact was made. Mawson ordered the captain to return to Commonwealth Bay. Then the elements entered in. The wind blew hard; Captain Davis could not approach Cape Denison satisfactorily and even if he could, the wind was so strong that the motor boat from the ship to shore and return could never make it. Mawson sent the captain a message giving him the choice of staying on in hopes of the weather alleviating, or steaming for the west base and picking up Wild and his men.

The captain made his decision. It was necessary that he reach Wild and his party at the western base while the *Aurora* could approach it. The captain knew that he had landed sufficient supplies to last the main base for another year. So seven men were left there, among them Jeffryes, Hannam's replacement, and Mawson. Meanwhile the *Aurora* was able to pick up Wild and his men. Now, like it or not, Mawson had to spend another winter at the hut with six other men.[15]

Although sources do not state so specifically, Jeffryes, "with his enthusiasm for wireless and his messianic eyes," appears to have been a better radioman than his predecessor. From February until July, 1913, he did excellent service. "He worked the rig every night, listening for signals and calling at intervals," Mawson wrote. "The continuous winds soon caused many of the stays of the wireless mast to become slack, and these he pulled taught on his daily rounds." Twenty-two-year-old F.H. Bickerton, the engineer, kept the gas engine and generator working consistently. It was on February 15th that they first heard Macquarie Island, which was sending a weather report to Hobart, Tasmania. On the 20th calls were exchanged with Macquarie Island and after the 21st the exchange of information became more or less regular.[16]

Having contact with the outside world was indeed a treat. By the 23rd Mawson had sent a message to Lord Denman, governor-general of the Commonwealth, detailing news of the expedition, including the loss of Ninnis and Mertz. Through Lord Denman a message was forwarded to King George V requesting permission to name a tract of newly discovered country King George V Land; a return message gave permission. Sympathy messages were sent to Ninnis's and Mertz's relatives. On the receiving end, the men learned of Scott's death in Antarctica and of Amundsen's success in reaching the South Pole. In March they heard that the *Aurora* had reached Hobart safely and that Wild and his western base party had been taken off the base in good condition. One can imagine Mawson's gratification in contacting his fiancé in code, and hearing from her in return.[17]

For several months Jeffryes was the most valued man at the base. He and Bickerman worked every night from 8:00 p.m. until 1:00 a.m. sending and receiving messages. It was a tedious, harrowing job, what with the continual wind, the crackling of St. Elmo's Fire, the dogs barking, sounds made by others in the hut, the fading of the signals and even the tapping out of messages with hands suffering from below freezing temperatures. "Jeffryes," Mawson wrote, "would sometimes spend the whole evening trying to transmit a single message, or conversely, trying to receive one." Not just Macquarie Island, but occasionally New Zealand and Australian cities were worked, as were ships at sea, including naval vessels. Every night Jeffryes sent out a weather report.[18]

Early in July—mid-winter in Antarctica—the instability that Mawson may have detected, causing him initially to reject Jeffryes' application, surfaced. His increasing eccentricity had been noticed, but was attributed to the stress he was under, constantly working the wireless, trying to send and receive messages with what just a decade later was considered primitive equipment. The first indication that he had passed beyond acceptable behavior was on July 10th when he asked Mawson for some poison to destroy a recurrence of venereal disease, yet there was no indication that Jeffryes then or ever had had v.d. Then he became paranoid, wanting a list of the accusations made against him. He threatened the others with jail when they returned to civilization for planning to murder him. Mawson ordered one of the men to watch Jeffryes constantly, fearing that the wireless operator might become violent. On July 27 Jeffryes tendered his resignation.[19]

The situation, under the circumstances of just seven men living through the Antarctic night in a small dwelling, was a serious one. Melancholia and insanity could spread, as it had in the case of the marooned *Belgica*. The fate of the second year inhabitants hung in the balance. Trying to cope with the problem, Mawson wrote a statement which he read to the men, formally gathered around the table. He complimented Jeffryes on his excellent work, explained that he might possibly be allowed to live in a nearby ice cave, save for the problem of conveying food to him, and then pointed out that he really could not resign from his position—that to do so would mean that he could not return to civilization on the *Aurora*, because he would be no longer a member of the expedition. He flatly stated that Jeffryes was ill.[20]

Jeffryes veered between eccentricity and madness in the following months. One day he was found transmitting in Mawson's name, stating that only Mawson and he (Jeffryes) were well, and might have to leave the hut because the other five were ill. So Bickerton, who was terribly slow at code, took over, sending instructions that no more messages were to be accepted from Jeffryes because he was insane. Upon the return to Australia Jeffryes was placed in insane asylums, although his relatives insisted there had never been a sign of insanity in him prior to his trip to Antarctica.[21]

Few cases of insanity among radiomen have been recorded. Even with Jeffryes, it must be said that while he was sane he made a distinct contribution to Mawson's Australasian Antarctic Expedition. He was an excellent radioman, glued to his rig for five or six hours every night, listening carefully, trying to draw out messages under terrible conditions. To spend an entire evening at the single task of getting just one message delivered to civilization speaks well of the man's dedication. Staying on duty just to get the news of the outside world for the others to read when they arose in the morning was in itself a major contribution. While during his insanity Jeffryes had threatened his six comrades, he has to be credited with making the long winter night more tolerable for his six hut-mates until his mind gave way.

Two men who were with Peary's expedition in 1909 in which he claimed to have reached the North Pole made polar exploration their life's work. One was Donald Baxter MacMillan, the other was Captain Robert "Bob" Bartlett. Although Bartlett is the better remembered, and both made use of wireless, MacMillan's use of the new method of communication was more energetic. As we shall see, he had a wireless operator along on his Crocker Land Expedition of 1914–1917. Although the rig failed, MacMillan had such faith in the medium that he provided for an ambitious radio presence in his 1923–1924 expedition and on all his subsequent forays into the Arctic.

Donald MacMillan was born in 1874 in Provincetown, Massachusetts. His father, Captain Neil MacMillan, had perished in the Arctic when Donald was just a boy. In due time Donald graduated from Bowdoin College and did some graduate work at Harvard. He came to the attention of Arctic explorer Robert E. Peary, who took him along on his last trip, the one on which, in 1909, he claimed to have reached the North Pole.[22]

From that time on MacMillan devoted himself to scientific Arctic exploration. He made twenty-eight trips into the vast region between 1910, when he was in Labrador, and 1948. He was captain of his own vessel, the 115 ton *Bowdoin*, which had been built to his specifications. He was professor of anthropology at Bowdoin and he lectured widely. His two books, *Four Years in the White North* and *Etah and Beyond*, are well written and full of the kind of information that the layman wants to know, but which is so often left out of such writings.[23]

At about the time Mawson's Australasia Antarctic Expedition was coming to a close, MacMillan's Crocker Land Expedition was heading for the Arctic. The name was derived from Peary's statement that from a far point in his explorations he had seen land to the northwest. He had named it Crocker Land. MacMillan's aim was to verify the existence of such an entity and explore it.[24]

While exploration was predominant in MacMillan's aims, this expedition, and subsequent ones he led, were entirely scientific in purpose. This time he had raised funds from friends, from the American Museum of Natural History, the American Geographical Society, the University of Illinois and a dozen other colleges and universities. His expedition, which was planned for a two year stay in the Arctic—and four members, including MacMillan, actually spent four years there—was well outfitted. When the *Diana*—MacMillan had not yet built the *Bowdoin*—left the Brooklyn Navy Yard on July 2, 1913, it was loaded to the gunwales with supplies. Their home base in the Arctic, named Borup Lodge, was situated at Etah, just north of the present American Thule Air Force Base. [25]

MacMillan was aware of the dangers confronting men, both physical and mental, during the long Arctic night. Interspersed throughout his two books are comments on how he kept his men in good health. It should be pointed out that the Arctic was peopled by the Inuit (Eskimos) who lived there year round; thus, unlike Antarctica, there were contacts with people other than t members of the expedition. MacMillan had learned to dress his men as the Inuit dressed, to build igloos, and in many ways to live as they live. Thus it is a fact that life in the Arctic was somewhat more amenable to explorers than in Antarctica. In addition, he had found that excessive sleep, likely to occur during the Arctic night, was dangerous for men's health. To keep his men physically robust he had them setting traps for foxes. Every day, regardless of weather, they bundled up and headed out for their traps. He wanted them to have many interests such as photography, meteorology, zoology, ethnology, practical astronomy, and growth of the sea ice. He commented on how men were puzzled when he interviewed them for asking if they could do readings, liked to participate in acting skits, could sing, etc. These were attributes necessary to keeping men occupied when inside their lodging during the long Arctic night.[26]

So it is hardly a surprise that early on he realized the value of wireless for men's morale. Among the personnel of seven scientists on the Crocker Land Expedition was a navy radioman, young Jerome Lee Allen. He had received training at the Navy Wireless School and had done work at the Bureau of Standards and the Naval Radio Laboratory. When the *Diana* floundered on rocks Allen contacted the American Museum to charter the *Erik*, out of St. Johns, to carry the expedition and its cargo on to its destination. Once there it was Allen, who was an electrician as well as a wireless expert, who set up the engine and dynamo to supply power not only to the wireless but also for the housing, which was equipped with electricity. Indeed, their base, Borup Lodge, as it was called, was extremely well designed. The presence of electricity made a real difference.[27]

Unfortunately Allen did not have a robust constitution and became ill during the first winter. Nevertheless he worked diligently. "Jerome Allen deserves the very highest praise for his indefatigable efforts to establish communications with home through his wireless apparatus," MacMillan wrote. With Inuit help the young radioman strung an aerial from the hill behind Borup Lodge across the river valley to the heights to the east. "Yet after all this effort not a rewarding buzz was heard!" When this did not work, Allen constructed huge box kites to which he attached an aerial, but the winds were so blustery and unpredictable that he met with failure again. Then he had the rig transferred two miles across water to Starr Island, his theory being that the hills surrounding the base house prevented his contacting other operators. But again he failed. His ultimate humiliation came on May 23rd of the first year: five sealed bottles containing messages were thrown into the sea in hopes that someone would find and read them—a not so vague indication of the lack of faith in wireless.[28]

According to MacMillan's biographer, "if something did not work, trying one remedy never was enough; he [Allen] would try six, and not only had the keenness to think of them, but the patience to apply them. He sat up frequently until one or two in the morning, studying anything from algebra to English." He did well also as a motion picture cameraman. At one time MacMillan sent him 120 miles down the coast to a bird-rookery at Saunders Island. "Much of the 12,000 feet of film brought back by the expedition is the result of his patience, energy, and skill," MacMillan commented. Allen was one of those going home at the first opportunity. This was when the relief ship *George B. Cluett* arrived on the 15th of September, 1915, more than two years after the *Diana* had left the Brooklyn Navy Yard. But the *Cluett* got frozen in at Parker Snow Bay, just ninety miles south of Etah, and the men had to go south by sledge to finally reach open water and a boat to take them home.[29]

MacMillan and three other members of the expedition did not return until August 1917. Two attempts had been made to fetch them from Etah and both had failed. Now it was Captain Bartlett's ship *Neptune* that picked up those remaining, including MacMillan, and returned them to civilization four years after they had left the Brooklyn Navy Yard.[30]

As for the Crocker Land Expedition, it failed to discover Crocker Land—it does not exist—but except for that, MacMillan's expedition was quite successful. He listed twenty-one accomplishments, and they were substantial. The failure of wireless was a disappointment, but in retrospect not surprising. The attempt was with long wave transmission, which was unpredictable in the years prior to the widespread use of the vacuum tube and the regenerative receiver. The far more efficient shortwaves were just beginning to be appreciated. There was little chance, except for an atmospheric fluke such as experienced by the operator on the *Aurora* in the Antarctic, of carrying a message more than a few hundred miles. Even so, it is a little strange that Allen was unable to work anyone. Just chalk it up to bad luck.[31]

In 1920 MacMillan was again in the Arctic, this time to Hudson's Bay. In 1921–22 he headed an expedition to Baffin Land, and in 1923–24 he was off to Northern Greenland. By this time he had his own boat, the *Bowdoin*, built to his own specifications. "She is more vessel to the foot than you ever laid eyes on before," it was said, "built in Maine where they know how, and exceptionally sturdy to withstand the crushing force of 40-foot-thick ice." It was reported that long copper strips were being attached to the hull to serve as a ground for the radio equipment.[32]

The changes wrought in radio between the time of MacMillan's four year stay in the White North, 1913–1917, and 1923 were, for want of a better phrase, breath-taking. The vacuum tube was being standardized and produced by the hundreds of thousands; the regenerative receiver was in common usage; and continuous wave transmitting was quickly making the old spark transmitter obsolete. Moreover, hams, restricted since 1912 to 200 meters and below, had discovered the virtues of shortwaves. In the early 1920s they were DX-ing all over the globe.

Commodore MacMillan was aware of all this; after all, he had brought along a radio operator on his expedition of 1913–1917. On his 1921 expedition, with a receiver, he had heard time signals from Arlington. These were on long waves; now it remained to see how shortwaves could do in long-range transmitting. So it is not surprising that early in 1923, prior to his departure for northern Greenland, he dropped in at league offices at Hartford. He informed Maxim that he would be carrying on two-way communications with amateurs from his base within 700 miles of the North Pole. As reported in QST, the use of radio in the polar regions was a morale thing, although the phrase was not used in those days. "In the past," QST reported,

> members of every expedition into the Arctic were confronted with a condition more terrifying than six months of darkness, more awe-inspiring than miles of floating ice, more difficult to contend with than a temperature of 60 degrees below zero; namely, loneliness and the thought of being cut off from the rest of the world. Ever since the first dash for the pole men have overcome hardship of every description only to break down finally in the face of something they could not see or feel. The loneliness of the Arctic has been their greatest handicap.[33]

MacMillan explained that in the *Bowdoin*, which would be home to the seven men on the expedition, one room would be devoted to radio. It would house a "real ham transmitter, hence it will be impossible to hear the wind whistle, the Eskimo talk, or the seal bark, but fellows, we can hear amateur messages direct from the Arctic Circle," QST informed its readers. The rig would be state-of-the-art Zenith, made in Chicago. The station call would be WNP (Wireless North Pole). Installation on the *Bowdoin* posed no problems, but the antenna was another matter. Either a portable mast or wires from a cliff to the iced-in ship would be necessary. The schooner would also have a state-of-the-art AM receiver which, it was hoped, would bring in programs from American commercial broadcasting stations.[34]

MacMillan asked the help of the ARRL, which was delighted to come to his aid. He knew of the relay system fostered by the league and envisioned using it to transmit messages, regardless of what amateur might hear them, to expedition member's friends and relatives. Another task falling to the amateurs was delivering, in person or by telephone or telegram collect, a news story of about 500 words every week. They were to convey this free of charge to any newspaper that was a member of the North American Newspaper Alliance. (QST carried a list of members for use of the hams.) To receive credit for doing this, the amateur fortunate enough to hear the story and convey it to a newspaper was instructed to send it also to ARRL headquarters with necessary information as to how the message was received. Moreover, the league agreed to furnish a highly qualified ham to accompany MacMillan on his fifteen month foray into the Arctic. For his part, MacMillan agreed to send via the radio a weekly list of calls heard, so that "you fellows will know from week to week what stations have been heard in the vicinity of the pole." His expedition, it was said, would "be the first party bound for the north to keep in touch with civilization by radio."[35]

In a month's time the radio rig was built and shipped to the *Bowdoin*'s home port of Wiscasset, Maine. Rather surprisingly, it was rather a medium wave than a shortwave set, with 220 meters most likely to be used, with 185 and 300 also available. Along with a shortwave and a long wave receiver, it was installed with the help of M.B. West, the Zenith engineer who had designed the transmitter. Arriving at about the same time was a ham from Bristol, Connecticut, Donald H. Mix, call letters ITS, the operator chosen by the league to accompany the *Bowdoin*. Described as "tall, raw-boned, sandy-haired, freckle-faced, [with] a big mouth, a wide grin, a good fellow," the only criticism MacMillan had of him was that he was too quiet! Mix, the youngest member of the expedition, was an amateur whiz who had established himself as a leader at working far-off places—DX-ing—and was also known as an expert at maintaining the equipment. All evidence is that he was a good choice.[36]

Enthusiasm for the expedition was enhanced by MacMillan's message that appeared in a box in QST. He expressed his "deep appreciation to all members of the American Radio Relay

Donald Mix, radio operator on Schooner *Bowdoin*, 1923–1924 (courtesy The Peary-MacMillan Arctic Museum, Bowdoin College).

League for their cooperation in making possible the first real radio work in the Arctic regions." He stressed what radio would mean "to the morale of a crew in the frozen North.... Undoubtedly the greatest hardship of the Arctic explorer is to be cut off entirely from his own world. The isolation," he stressed, "has spelled disaster for many an expedition." He reminded *QST* readers that "an attempt at communication to and from the Northland has never before been attempted in a systematic, well organized way, with the cooperation of such an organization as yours." Captain MacMillan was optimistic of the results. Frozen in during the long Arctic winter, he wrote, "we shall look forward eagerly to our radio talks with you ... who [will] be listening eagerly to 'copy' WNP—Wireless North Pole."[37]

WNP was on the air before the *Bowdoin* left the harbor at Wiscasset. Mix had erected a "little four-wire flat-top [antenna] just 23 feet long, but about 70 feet above the water" and it served quite well. Stations as far off as Arkansas had been worked. The first story for the North American Newspaper Alliance had been sent out and received; it appeared in members' newspapers on June 27th. Then the *Bowdoin* was not heard from for several days because the ship was under sail instead of power and the transmitter could not be used without a power source.

When again operational, Mix did not have much luck in contacting amateurs, but he did

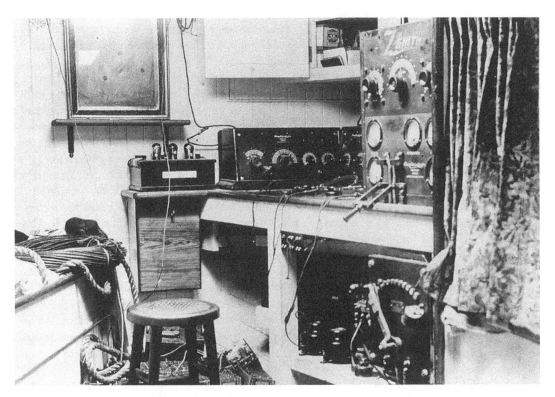

Radio room with Zenith rig aboard the *Bowdoin*, 1925 (courtesy The Peary-MacMillan Arctic Museum).

receive their calls quite frequently; he also received time signals from Arlington every night. Transmitting was still difficult. MacMillan allowed the antenna to be stretched to a short bowsprit that was added to the *Bowdoin*. It was a failure; when the vessel plunged into a hollow in the sea, the bowsprit was gone, leaving the antenna a complete wreck.

Even today summer is not as good for radio as winter. July 1923 was particularly bad. "If one gets a good hour in ten days, he should feel elated," remarked *QST*. However, Mix was already hearing West Coast stations. The log the ARRL compiled of amateurs who had worked WNP, and had informed the league, listed thirty-three contacts in July. All reported bad QRN (atmospherics), several had been able to take down 200 or 300 words of the 500 word newspaper article but not the full 500 words; worse, just seven hams appear to have contacted WNP with their transmitters. Late in July, from off the Labrador coast, Mix was able to send a message to President Harding, who was on his ill-fated vacation to Alaska. "Greetings from the crew of the *Bowdoin*," read the message; "all hope that you are ... enjoying your sub–Arctic trip." The greeting, received in the east, was relayed via amateurs across the continent to the president.[38]

Not until August 17th did the expedition arrive at Refuge Harbor, ten miles north of Etah, its winter quarters. By their reckoning they were at 78° 31', 689 miles from the North Pole. Mix transmitted the news of their arrival, and their good health, to whoever picked up his message.[39]

No transmission was heard again until August 27th at 11:35 p.m. By coincidence, the ham who worked Mix was R.B. Bourne of station 1ANA of Chatham, Massachusetts—Don Mix had contacted a fellow New Englander. *QST* reported that "Mix's transmitter cut a swath clean through the aurora borealis and the Arctic sunlight ... success for the installation in the *Bowdoin* is assured, and with the coming of snappy winter air, WNP will boom down with a kick."[40]

Of little interest to amateurs but of great interest to the men of the *Bowdoin* was the success of their long wave AM receiver. Much pleasure was derived from the Wednesday night radio programs broadcast from the Edgewater Beach Hotel in Chicago. They were specially arranged for the *Bowdoin* men by Zenith's E.F. McDonald.[41]

On October 25 the sun bade goodbye to the little schooner frozen in the ice at Refuge Harbor. Not until February 28, 1924, a hundred eighteen days later, would it return. Reception in September was disappointing because the "snappy weather" was slow in coming. Contact had been slightly better along the Pacific fringe than in the east, where just four stations had heard WNP, and their reception was unsatisfactory. League officials desperately wanted the experiment with shortwaves to be successful. If western stations were more successful than eastern ones, so be it. Above all, it was necessary that reliable, steady contacts be made. Now that the Arctic night had set in, league members were hoping that just such reliable contacts would be established.[42]

But they were to be disappointed. Mix tried to make contacts at 180, 220, and 300 meters, but failed far more times than he succeeded. Yet one permanent, reliable contact was made. This was 9BP, an amateur station worked by Jack Barnsley of Prince Rupert, British Columbia. That settlement was the western terminus of the Canadian National Railway, 500 miles north of Vancouver and 100 miles south of Ketchikan, Alaska; it was about 2200 miles from the *Bowdoin*. Beginning in September he had received scores of messages from MacMillan's crew. These he had relayed to their friends and relatives through the league's traffic system. He also conveyed news stories from the *Bowdoin* to the North American Newspaper Alliance. Reciprocating, Barnsley conveyed such news items to the *Bowdoin*'s men as the Japanese earthquake and the outcome of the Dempsey-Firpo fight. Barring poor atmospheric conditions, he was working with Mix on a daily schedule.[43]

Reception was poor even in October, with Barnsley being the only reliable amateur maintain-

Schooner *Bowdoin* and supply vessel *Radio* in front of the station site (courtesy The Peary-MacMillan Arctic Museum).

ing regular traffic from the *Bowdoin*. Whether on the West Coast, the East Coast, or in between, while a good many heard Mix, few were able to make contact with him. Possibly, it was conjectured, this was due to the magnetic pole; possibly it was due to the aurora borealis. Conditions improved noticeably in November, with Mix being heard by some hams in the Northwestern ARRL division, five in the Dakota division, and by five Canadians. But the reception remained undependable. Barnsley was the only ham to take full news messages, which he trafficked to papers subscribing to the newspaper syndicate. Some kind of distance record was established, however, when Mix talked for fifteen minutes with a Hawaiian station—an estimated 4600 miles away.[44]

In December the Radio Corporation of America (RCA) approached the ARRL for help in transmitting to the *Bowdoin* men a Christmas message from President Coolidge. That distinguished gentleman had filed the forty-three word greeting with RCA believing its radiomen could deliver it. They could not, so they asked the league for help. "*Would we!*" was the reply, but the transmission proved difficult in the extreme. Hams had "the darndest streak of hard luck that has been seen for many a day ... like a radio set on the night you invite your friends over to stage a little demonstration and—you know." Hartford was struck with one of those freak nights when little gets through. Twenty-two stations were trying to contact Mix with the message but none succeeded. Not until the night of January 1–2 did Barnsley, the old stand-by back from his vacation, connect with Mix and convey the message. "Some beautiful little streak of freak conditions had almost ruined us," lamented *QST*. In a reciprocal vein, Mix had transmitted a 1500 word Santa Claus message to America's children "from the North Pole." It came through with no trouble at all.[45]

In spite of the difficulties in delivering Coolidge's Christmas message, December was considered a good month for two-way contact with the *Bowdoin*. Mix heard the hams contacting him very well, but his communications with them were not entirely satisfactory, fading being the most common problem. This was in spite of a new antenna stretched from the ship's masts to some hills on shore. One theory was that the high winds that rocked the antenna could be the reason for the poor reception. All told, Mix reported that he QSO'd (made contact with) stations 51 times in November, 25 times in December, 35 times in January, just 8 times in February, and just once in March.[46]

On April 14th (possibly the 13th) Mix was heard; then the transmitter was turned off because the supply of oil was getting low. That last contact was heard by Everett Sutton, 7DJ. Mr. Sutton was fifteen years old, living in Port Angeles, Washington. He took a string of messages from Mix including the information that the *Bowdoin*'s crew were all well, but steady daylight plus the oil situation would probably result in silence until late summer. About 7DJ, *QST*'s K.B. Warner noted, "Ether burners please note that his set has but one '5-watter,' ... and works on a 2-wire aerial 50 ft. high, using the clothesline in the backyard for a counterpoise; his receiver is a single-circuit regenerator with a one-step audio." Warner added, "this lad and this station have done what a hundred of our best have failed to do, and we think it the most outstanding incident in the WNP communication this year."[47]

Mix's transmitter was not turned on again until August 1, when the *Bowdoin* was homeward bound. Of 16,000 words cleared, Mix reported that 8,000 were cleared by Jack Barnesby; of 13,000 words received, Barnsley was responsible for 9,000 of them.[48]

At one p.m. on September 20, 1924, the *Bowdoin* entered Wiscasset harbor before a crowd of several thousand well-wishers and the whistles of several locomotives side-tracked for the special purpose of the welcome. One of the speakers of the welcoming group was Hiram Maxim. Mix had, after all, been the first amateur to operate a shortwave rig in Arctic waters.[49]

The amateur experience with the *Bowdoin* should be considered a success, but not an unqualified one. The long days of sunlight, the aurora borealis, winds affecting the antenna, the still primitive transmitters and receivers, all prevented the steady, reliable communication that was hoped for. Moreover, the "shortwaves" at 180, 220, and 300 meters were far higher than would soon be common; meters down to 20 were forthcoming with proven abilities to carry long distances. The failure to deliver Coolidge's Christmas message on time demonstrated the weaknesses still existing in the new art of shortwave sending and receiving. It is to be noted, however, that there was national interest in the triumphs of shortwave. The articles from the *Bowdoin*, sent via shortwave and distributed to the seventy members of the North American Newspaper Alliance (not all of whom ran the stories), received considerable public attention.[50]

Donald MacMillan was just one of the explorers into the Arctic in the 1920s who was interested in radio communication. Roald Amundsen was there in 1922 with his vessel, the *Maud*, which was frozen in for several months. The *Maud*'s transmitter sent out two messages a day, but they were heard almost exclusively in Europe. This tendency—for messages to be heard in Europe but not in the Western Hemisphere—was the subject of considerable research. A Canadian vessel had as one of its assignments an investigation as to why there were so many radio transmission problems confronting Arctic expeditions. In a very real sense, however, all amateurs operating in the Arctic contributed bits and pieces to the general knowledge of the air waves, the aurora borealis, the problem of twenty-four hours of daylight, and finally, how to overcome these barriers to reliable communications.[51]

By the mid–1920s polar explorers were aware that radio represented just one of the many technological advances that might expedite their exploring endeavors. Motorized ground vehicles—today known as snowmobiles—were being designed for use in the frigid regions. The potentials of airplanes and lighter-than-air craft—dirigibles—whetted their imaginations. What if they could explore the vast wastes of the Arctic and Antarctica with these new means of transportation? Why couldn't they fly over the North and South Poles? Radio could improve the safety factor by informing rescuers of the location of a damaged or crashed vehicle. The challenge remained while the danger, they felt, was receding.[52]

Aircraft, radios, and internal combustion engines of the time were indeed crude by today's standards, but to people of the 1920s and even 1930s they were wonderful developments. Mankind just could not wait to put them to use. And yet, the aircraft's structural design was still experimental. There were no computers to do intricate compilations about chance and performance, no wind tunnels to test a plane's behavior under various atmospheric conditions. Compared to today, metallurgical knowledge was rudimentary: would the many components of the engine hold up under varying conditions for many hours at a time? Who today would want to trust his or her life to such an airplane, to a transmitter or receiver using fragile vacuum tubes, a rig powered from batteries that had a limited life?

But to the explorers of the 1920s and 1930s, state-of-the-art was sufficient. The risks were great, but so was the potential for fame. For scientists the promise of solving questions about the earth and climate and wildlife were sufficient to warrant the risks. Around the North Pole there were a million miles of uncharted—what? MacMillan had spent four years, 1913–1917, in a search for Peary's Crocker Land, which proved to be a mirage. Yet the theory remained that amidst the vast, unexplored Arctic region, somewhere the ocean gave way to a substantial piece of land. For some courageous men the risks were warranted by the mystery that was combined with a desire to establish firsts—first plane over the North Pole, first dirigible over the North Pole, first aircraft over the South Pole, first expedition to explore an unknown portion of the icy regions at the ends of the earth. In the 1920s especially, but well into the '30s, polar exploration was big news. It commanded newspaper headlines while the public interest inspired the funding of private expeditions.

Radio, aircraft and lighter-than-air ships into the polar regions are the subject of the next chapter.

# 12

## Amateurs and Polar Exploration: Phase Two

In the 1920s and '30s there were Commander Donald MacMillan's forays with the *Bowdoin*, Captain Bob Bartlett's trips into the Arctic with the *Morissey*, and the activities of Lincoln Ellsworth, Roald Amundsen, Umberto Nobile, George Hubert Wilkins, and Richard E. Byrd in the polar regions. These are just the better known of many expeditions into the ends of the earth.

Radio was by now grasped as the great security blanket. By contacting help it could save the lives of explorers who had met with disaster. All that was needed was an amateur who could fix a damaged rig under the worst of circumstances, and was adept at receiving and sending Morse code or making a few technical changes to use voice. An added incentive was the medium wave AM receiver, so that broadcast programs from civilization could be heard.

The MacMillan expedition of 1925 involved ambitious plans in which amateur radio would play an important role. Commander MacMillan proposed to explore the million square miles surrounding the pole, determining once and for all whether there was land in the region. To carry out his project he had raised money from private sources—a task at which he was becoming extremely adept—including funding from the National Geographic Society, publishers of the popular *National Geographic*. He also received assistance from the navy. It agreed to furnish him planes, pilots, and maintenance personnel. In doing this, he was ahead of Lieutenant Commander Richard E. Byrd, who with Captain Bob Bartlett had been planning a trip to Etah using navy planes to carry out their explorations. Byrd would have his expedition, but it would be with MacMillan. Bartlett was left out.[1]

MacMillan's expedition would consist of the *Bowdoin* and a second ship, the S.S. *Peary*. This Scottish whaler, which would carry the planes and Byrd and his staff, was to be under the command of wealthy Eugene F. McDonald, Jr., president of the Zenith Radio Corporation. Byrd appears to have been given *almost* complete control of the aeronautical activities, meaning the uses made of the three Loening amphibians taken along. The description of his activities implies that he had permission, though it was not spelled out, to fly to the North Pole. Yet ultimately MacMillan had the final say over the flying. Toward the end of the explorations he forbade Byrd to make one more flight.[2]

Clearly, MacMillan, McDonald, and Byrd did not get along. The conflict appears to have been mostly MacMillan and McDonald vs. Byrd, who felt that MacMillan was too cautious, too slow, and was clearly not sold on the practicality of heavier-than-air craft in the Arctic. The conflict with McDonald was over the radios to be used in the aircraft. McDonald had brought along state-of-the-art shortwave transmitters and receivers. He planned on experimenting with them in many ways, including installation in the planes. But the navy's policy was to use long wave rigs in their planes because they were reliable for close-by two-way communication; a navy long wave spark transmitter was installed on the *Peary* to maintain contact with the aircraft. The *Bowdoin* was equipped with state-of-the-art shortwave for working the rest of the world. When the planes were in the air the navy forbade the use of the shortwave transmitters.[3]

The shortwave sets brought along by McDonald were designed to work on 20, 40, 80, and 160 meters. Amateur radioman John Reinartz, a naval reserve officer, 1XAM—1QP, helped build them; he was along as the *Bowdoin*'s communications officer. A 32 volt storage battery charged by a Delco gas engine generating outfit powered the set. The antenna was forty-five feet long, of stranded gold-plated wire, running from the ship's deck to the cross trees of the main mast. McDonald's short-wave sets for the planes, which he was not allowed to install, would use a dry cell battery and operate at 40 meters, and could be set to work either code or phone. Of course it was a serious blow to McDonald's plans to be denied the privilege of experimenting with these small, compact units.[4]

Amateur radio gained much favorable publicity from this expedition. This was because the National Geographic Society and the navy made use of shortwave daily reports received by amateurs which were then conveyed to the press. The *New York Times*, for example, ran more than twenty articles between July 3rd and September 19th. Many tell how the message was received—from what amateur, giving his name and call letters—and how it was relayed to the National Geographic Society or the navy. Readers following expedition activities were made aware of the amateurs' contributions by repeated mention of them in the newspapers.

Although hundreds of hams heard messages from Etah, only a fraction established two-way contact with WNP. Even if the amateur did transcribe a message but did not make two-way communication, he or she was to forward the message by mail or radio to the league as well as to the National Geographic Society and the navy. The ham was then credited with assisting the expedition. One of the reliable contacts was Arthur Collins (call letters not given) of Cedar Rapids, Iowa. Fifteen years old, Collins was as reliable as a dedicated adult. He made radio history by corresponding at twenty meters. The young man adhered rigidly to league instructions by telegraphing or telephoning collect the messages to the National Geographic Society or to the navy; or direct to a private person.[5]

In a lengthy article headlined "Schoolboys Aided MacMillan's Work," the *New York Times* related that "most of the amateurs who have been playing a part in communications with the *Bowdoin* ... are high school boys." It told how Collins had received a message that Byrd's airplanes were in the air, conveyed it to the proper authorities, and it was in the newspapers before the planes had landed. The story also emphasized the relay aspect of the communications: The amateurs were praised unequivocally.[6]

The first ever press service interview in the Arctic took place on August 6, 1925. It was conducted by a reporter with the Associated Press in Arlington Heights, Illinois (near Chicago). The entire interview, conducted in international Morse code, took an hour and a half. Shortwave broadcasting of the human voice was also tried, with moderate success. By mid–August Collins was hearing John Reinartz's voice and the crew singing 'America.'"[7]

When the *Bowdoin* and the *Peary* returned to the States, the general impression was that the expedition had been a complete success. *The National Geographic* ran an article listing their scientific accomplishments. The *Times* emphasized radio's contribution to polar exploration. "Down in the radio room," Reinertz reported, "someone is continually on duty, sending messages that all is well at home. There has never been a day when we have been out of touch with the world." So interested were people in this means of communication that the newspaper ran a schematic of Reinartz' transmitter with explanation of how it worked "so that amateurs who want to build the set can do so."[8]

Yet John Reinartz apparently failed to live up to his obligations. He was suspected of failing to send some messages. More serious was his failure to act on the grounds that he was seasick at a critical time involving the *Bowdoin* and the *Peary* in stormy seas on the return voyage. He was even suspected of sabotaging some of the equipment. He was removed from his position "for cause" and replaced by one of the operators on the *Peary*. His failure is a real puzzle, for Reinartz was an innovative radioman with a good reputation. His actions harmed his career. The reasons are inexplicable and at this late date must remain so forever.[9]

The final word of the amateur contribution appeared in the October issue of *National Geographic*, with a million readers:

> A unique feature of the MacMillan Arctic Expedition was the daily contact maintained with civilization by means of the shortwave length or high frequency radio transmission, and an important aspect of the work was the cooperation throughout the United States of some 1,200 amateur radio operators of the American Radio Relay League.... [T]hese radio amateurs, who receive no pay for their services, have worked long hours in the night receiving messages of many thousands of words, addressed to the National Geographic Society and the Navy Department, and given by them to the news-papers and press associations.[10]

The year 1926 was Arctic exploration's vintage year. Two attempts would be made to fly over the North Pole: Richard E. Byrd in a Fokker triplane and Roald Amundsen in the Italian dirigible *Norge*. They would take off from Spitzbergen, Norway. Hubert Wilkins was exploring the Arctic from Point Barrow, Alaska. Amateur radio was involved in all three expeditions.

Byrd's attempt was a private one, the navy brass having frustrated his desire for a naval Arctic expedition. However, he and Floyd Bennett had received leaves of absence to make the attempt. With money raised privately and a ship, the *Chantier*, lent him by the Federal War Shipping Board at a dollar a year, Byrd, with a thoroughness that always marked his plans, set about preparing for the expedition. On April 5, 1926, the ship left New York Harbor. Although most of the crew were green landlubbers, they were intelligent, dedicated men and they learned quickly.[11]

Once at sea a stowaway appeared. He was Malcolm P. Hanson, an employee of the naval radio research laboratory. Just prior to departure he had worked on the expedition's radios forty-eight hours without sleep. When he did lie down he slept a full twenty-four hours and awoke to discover that the *Chantier* was already at sea, or so he said. Byrd's explanation of his appearance differs slightly. "Now the expedition owes a very great deal to Hanson," Byrd noted in his diary on April 7th. "He has worked day and night for weeks on high frequency radio sets for ship and plane. It seems that he had three or four more days work to do and deliberately stowed away and so sacrificing himself for the good of the expedition. He did not want me to know he was aboard when we left and so have the responsibility for his act. My only concern is to get him out of his scrape." On April 15 he noted that "Hanson the stowaway is working night and day on the radio. He gets only three hours sleep on an average, a night. He has a terrific proposition [challenge] with the high frequency [shortwave] radio. The field is so new he is meeting with many unsuspected difficulties. Whatever results he gets I am extremely grateful to him." Later Byrd recorded how the amateur rigged up a sub-radio station on the poop deck to get away from the interference he encountered in the radio room located amidships. "It was frightfully rough and windy last night but not a word of complaint from Hanson and too he got good results," Byrd noted. "Good for Hanson." To the radioman's disappointment, he was ordered home, leaving the ship at Trondheim with three cheers from the crew ringing in his ears. Already the radio was being used for nearby contacts; through it Byrd was kept informed of the progress, or lack of it, of Amundsen and Wilkins. Meanwhile the *Chantier* with the Fokker trimotor (it used Wright Aeronautical engines), dubbed the *Josephine Ford*, continued on its way to King's Bay at Spitzbergen.[12]

According to K.B. Warner, Byrd's radios consisted of a 500-watt, 500-cycle transmitter on board the ship, operating on 42 meters, call letters KEGK, and a 50-watt crystal-controlled transmitter on the *Josephine Ford* also operating at 42 meters, call numbers KNN. Confidential press messages addressed to the *New York Times* were to be an important part of the traffic. Amateurs were urged to send via telegraph, press rates, collect, any messages they picked up for the *Times*. Byrd also said they had with them on the aircraft a shortwave radio operated by a hand dynamo, should they be forced down. That KNN worked at least during the first half of the North Pole trip is clear from Byrd's diaries and the messages to the *Chantier* at Spitzbergen and those received in the States detailing the progress of the flight. They were telephonic rigs. On his flight to the Pole Byrd kept in touch via radio for the first part of the flight:

> Send a radio [message] back that we are 85 miles due north of Amsterdam Island. Got over ice pack just north of land....
> Send a radio that we are 240 miles due north of Spitzbergen. Then pull in your wire [antenna].
> Radio that we are 230 miles from the Pole.... Radio that we have reached the pole and are now returning with one motor with bad oil leak. But expect to be able to make Spitzbergen.[13]

Then for several hours prior to the plane's arrival over Spitzbergen nothing was heard. The operator aboard the *Norge*—Amundson's dirigible—thought he had heard a faint signal, but apprehension was growing rapidly; then the *Josephine Ford* appeared.[14]

The date was May 9th, 1926.

The United States went wild over Byrd's claimed achievement. Spoken congratulatory messages were sent to KEGK, Byrd's radio on the *Chantier*, via General Electric's powerful shortwave station at Schenectady, 2XAF, operating at 32.79 meters. Among those who spoke to him was Malcom P. Hanson, the expert who had installed the rigs. The hyperbole equaled and possibly exceeded the excitement of the first trip to the moon in 1969.[15]

Roald Amundsen hid his chagrin at Byrd beating him to the Pole, but this did not stop him from continuing post-haste preparations for the huge, Italian-made dirigible *Norge* to take off. It would leave from Spitzbergen, as had Byrd, and attempt to cross over the Pole and continue on to Alaska. Besides Amundsen among the ten aboard was Umberto Nobile, the Italian dirigible maker; Lincoln Ellsworth, who put up the money; and Frederik Ramm, a *New York Times* correspondent.

On May 11, 1926, the *Norge* was "let go" and quietly rose into the air. Two of the ten men who were crowded into the little gondola, Captain Berger Gottwaldt and Storm-Johnsen, were "in the little radio cabinet" (Gottwaldt was considered the radioman). Amundsen was so impressed with radio that he discussed it briefly in his narration of the trip. "It is very wonderful after all, this wireless," he wrote. "We elders, who have known nothing other than cables will certainly never overcome this feeling." While the *Norge* was still in the air Captain Gottwaldt was decorated by Norway's king with a gold medal for his work with the airship's radio and Ellsworth received two congratulatory messages.[16]

Their radio, built by the British Marconi company, was long wave, working between 500 and 1500 meters; it had a power of 200 watts and was considered to have a range of 1000 to 2000 miles. "This is too bad," commented QST, "as such waves probably will prove useless during the time when they may be most needed." The power came from a generator in back of the tiny radio room; it was operated by a small propeller sticking out from the side of the ship. The antenna, 300 feet long, was dropped through an insulator on the floor of the cabin; it had a sinker attached on the end to weigh it down. When the *Norge* ran into a storm, ice coated the antenna, which dragged on the ice and broke several times. The windmill generator also froze, so that the receiver ceased to function. All contact was lost. Then, unexpectedly, the radio, at least the receiver, worked. They did not get the call letters of a station they heard but thought it was located in Nome; from it they were able to determine their position. Eventually they landed in Alaska at the small community of Teller, about ninety-one miles northwest of Nome. As with Byrd's silence in the last few hours of his flight, the *Norge* was not heard from for many hours; only Wilkins's operator at Point Barrow thought he heard the dirigible's faint signals. The *Norge* had journeyed 3,393 miles and had been in the air seventy-two hours.[17]

QST editorialized about the *Norge*'s radio conditions:

> [It] illustrated by the conspicuous failure of its communications the great value of shortwaves to such exploring parties.... Wavelengths were 600, 900, and 1400 meters, although 900 seems to have been used exclusively during the passage. Radio contact with Spitzbergen seems to have been satisfactory up to the time the party neared the Alaskan coast. [Then the generator and the aerial iced up.] For two days an anxious world had no news of her. Twenty-four hours after her landing at Teller her radio officer had succeeded in overhauling an ancient spark station belonging to the Lohman reindeer ranch at that place and finally got word to Nome that the party was safe. What a pity that Amundsen did not carry high frequency! This should be a lesson for all time.[18]
>
> They had crossed the Pole on May 12, 1926.

As with Byrd's accomplishment, Americans, indeed the whole world, were excited over the long flight of the *Norge*. Frederick Fram, the *Times* correspondent aboard the dirigible, sent by radio the first messages ever from the North Pole. They were not picked up in New York, however. They were heard much closer to home, along the west coast of Norway, relayed by wire to Oslo and sent by cable to New York. The question remains of why the *Norge* did not have a short-wave transmitter rather than a long wave one. The merits of shortwave working long distances and using far less power had been positively confirmed, yet not all radio engineers of the time were convinced of its reliability.

In 1926 Captain George Hubert Wilkins, an Australian adventurer, came to the United States to raise money for an exploration of the million miles of waste between Alaska and the North Pole. He proposed to conduct all his explorations from the air; his expedition would be the first to operate without ships. Wilkins raised $25,000 from the North American Newspaper Alliance, had the support of the American Geographical Society, and raised other funds from the people of the city of Detroit. As for the alliance, it had sent its own journalist to accompany the expedition, a likable man named Palmer Hutchinson.[19]

For his work Wilkins obtained a trimotor Fokker with Wright Whirlwind engines and a single engine Fokker powered with an improved Liberty engine. He also had attached to his party, at the request of the Detroit manufacturers, a Snow Motor section. It consisted of primitive snow-mobiles which were to cross the 560-plus miles from Fairbanks to Point Barrow carrying with them the wireless apparatus and other supplies, including extra gasoline. Radiomen, mechanics, and scientists made up the expedition personnel.[20]

While Fairbanks in 1926 was, by Alaskan standards, a small city, Point Barrow was just a village. It consisted, according to radio operator Howard Mason, "of seven white people, two hundred Eskimos and about eight hundred dogs.... The little village would be an oasis in a desert," he wrote, "if there were a desert."[21]

F.H. Schnell, traffic manager of the ARRL, was enthusiastic. As with MacMillan in the previous year, amateurs were expected to hear reports from Point Barrow and convey them, as before, to the North American Newspaper Alliance. Two expert hams from Seattle, Howard F. Mason, 7BU, and Robert "Bob" Waskey 7UU, were to accompany Wilkins and work the radios. Both the aircraft carried radios designed and built by Malcolm Hanson, the same man who was with Byrd. A portable transmitter and receiver had also been developed by Burgess laboratories which, it was planned, could be used by the explorers in case of a forced landing. Amateurs at Madison, Wisconsin, were to man the Burgess Battery Company's station on a twenty-four hour a day basis; it was planned that they would receive most of the dispatches sent for the Alliance.[22]

Mishaps accompanied the expedition. In Fairbanks, where the planes were assembled,

**Hubert Wilkins's Fokker aircraft in Alaska (courtesy Ohio State University Photo Archives).**

Hutchinson walked into an airplane propeller and was killed instantly. The single engine Fokker, dubbed the *Alaskan*, crashed on landing. No one was hurt. Then the trimotored Fokker, named the *Detroiter*, flown by another pilot, likewise crashed on landing. Again, no one was hurt, but the plane was badly damaged. The estimated time for repairs to the *Alaskan* was three weeks; for the *Detroiter*, six.[23]

Meanwhile the snow-sled caravan had started for Point Barrow. After just sixty miles, at a place called Tolovana, they had to abandon the snow motors because they were so fuel-guzzling that no gasoline would be left to be stored for the planes at Point Barrow. Bob Waskey replaced Mason, who returned to Fairbanks. (He would later fly to Point Barrow.) From Tolovana they advanced via dog sledges for the next several hundred miles. Waskey made use of a little battery operated set of his own design to contact Fairbanks every night. Mason heard him on the portable set to be used in the planes. Plans were that after Mason left for Point Barrow, the ham in charge of the Signal Corps station at Fairbanks, would man the rig; his name was Hemrich. Saturday night reception was troublesome: people turned on their electric pumps and heaters for their Saturday night baths, and QRM was terrible.[24]

It took the overland party, which included Waskey and the radio, seven weeks to reach Point Barrow. They ran short of grub and had to kill some of their dogs before they killed enough game to keep themselves and the remaining dogs with food. They had to leave the heavy gas engine generator, which was to furnish power for the permanent radio, 160 miles from Point Barrow. A rush party was sent to retrieve it. Meanwhile Wilkins had been moving supplies to Barrow in the repaired single-engine Fokker. On the first two trips Hanson's set did excellent service. Operating on 46 and 61 meters, it derived its power when in the air from a wind-driven generator. On the ground Wilkins used a hand turned generator which required the work of four men taking turns, while he worked Fairbanks with the dots and dashes of a rank amateur. "Those signals were," he wrote, "...the first messages ever sent and received by wireless from Point Barrow."[25]

In 1926 the flying activities, from Fairbanks to Barrow and return and for exploration, were all carried out in the single engine Fokker. On the first exploratory trip in the *Alaskan* Wilkins and his pilot, Carl Ben Eielson, flew extensively over the ice of the Beaufort Sea, searching for land but finding none. The story run by the North American Newspaper Alliance of their return to Point Barrow emphasized the achievements of their shortwave radio, which "made radio history." With the antenna on the wing and a sending set weighing just fifty-eight pounds, Wilkins was heard by the Fairbanks radio operator.[26]

Wilkins made three round trips between Fairbanks and Barrow in the *Alaskan*. After the second trip from Fairbanks he broke his right arm. They reached Point Barrow on the third trip but an accident ruined the propeller. Wireless made it possible to inform the men at Fairbanks of their troubles, and for the Fairbanks men to wire unhappily that the Fokker trimotor, although repaired, did not run well and in their opinion could never make it to Point Barrow. Wilkins and Eielson finally flew back to Fairbanks with the damaged propeller. Heavily loaded with gasoline, the *Alaskan*, with a new propeller, crashed on take-off at Fairbanks for its fourth trip. This time the machine was wrecked beyond repair. By now, however, the trimotor was pronounced airworthy and they flew it to Point Barrow. A few days later they learned by wireless of Byrd's and Amundsen's trips over the North Pole; they even saw the *Norge* in the distance. Waskey, the ham at Point Barrow, flashed the news to the North American Newspaper Alliance and gave them a great scoop. He followed up by reporting the *Norge*'s landing at Teller. This was much to the chagrin of the *New York Times*, whose party, sledging overland, were still thirty-five miles out of Point Barrow with their portable station KDZ. The year 1926 had brought great luck to Byrd and Amundsen and Nobile, but for Wilkins, time had run out. He stored his planes and equipment at Fairbanks, received permission from Detroit to continue another year, and called a halt to the 1926 expedition.[27]

In 1927 Wilkins had two Stinson-Detroiter biplanes, the DN-1 and DN-2. In time Howard Mason was installed with the radio in an office at the Point Barrow whaling station (Waskey was not with the 1927 group). Wilkins and Eielson quickly prepared for their first exploratory flight of 1927. They chose one of the Stinsons, the DN-2. Mason had checked over the wireless installation and pronounced it in good condition. It was one of those built by Malcolm Hanson and had a hand-driven generator with an attachment for automatically sending several different combinations of code letters, the code arranged with A.M. Smith, the *Detroit News* reporter who was a member of the expedition.[28]

It was forty-two degrees below zero on the "fine morning" they took off and headed north, heavily loaded with gasoline and emergency supplies. "Have tried all lengths of wire," Wilkins wrote as he was flying, "but antenna current on wireless is scarcely visible on meter. Doubt that messages are leaving, but have sent every half hour since leaving Barrow." The plane acted up. It kicked and coughed and sputtered occasionally. Wilkins and Eielson found what looked like smooth ice and snow and landed without damage. They examined the plane, worked on the motor for more than two hours, Eielson getting badly frostbitten fingers for his reward, and took off again, the motor running nicely. They knew they were low on gas but it must have been a heart-stopping moment when the motor just quit. Out of gas, literally on wings and prayers and Eielson's good piloting, the plane silently came down and made a fairly soft landing, although it was damaged. At least they could get in touch with Point Barrow by wireless, or so they hoped. "I had little faith in the wireless machine as the antenna current meter failed to register," wrote Wilkins. "But we repeated the short-clipped message: 'Went out 550 miles. Engine trouble. Forced landing three hours. Sounded 5,000 meters. Landed out of gas 65 miles N.W. Barrow.'" Later he made use of the automatic code machine, inserting the letters 'KO' meaning 'Engine Trouble.'" Were they heard?[29]

According to operator Howard Mason at Point Barrow, the "OK" signal came through well, every half hour, announcing that they were doing well. After about two hours the signals became weaker. "It gave me a rather sinking feeling," he later wrote, "because I knew the old set was not kicking out like it should and that with them putting 100 miles between us every hour we would soon lose their signals, which we did." A few weeks later Wilkins explained why the signals became weaker and finally disappeared. Whenever he cranked the hand generator, the heat liberated from his body caused hoarfrost to form on all metal work in the cabin, including the radio set—it was that cold. This prevented the transmitter from functioning properly. As long as Wilkins sat still, this did not occur. "Rather an unusual excuse for a radio set not working," Mason commented, "but true."[30]

That afternoon it clouded up and a blizzard raged, with no communication from the DN-2. The storm, the worst of the season, continued unabated. Two days elapsed with no signals from the plane. Then some faint dashes were heard on the DN-1's wavelength. At least it was known that they were alive. An hour later the sounds were heard again, "about R4 this time." The automatic contactor on the hand generator was sending two letter combinations that stood for "forced landing at sea," "out of gas," "plane damaged." Some "jerky hand sending" followed, indicating that they were about one hundred miles from Point Barrow. The messages were repeated in entirety several times, then silence.[31]

Search flights winged over the Arctic, but the possibilities of finding two men and an airplane on an ocean of ice that continually moves due to wind and currents made the likelihood of discovery remote. Then one night about one a.m. there was an unusual howling of the dogs (Eskimo dogs do not bark) at Point Barrow as an Eskimo came running in with a letter from Wilkins. He and Eielson had walked across the ice for sixteen days, finally reaching Beechey Point. They had left the transmitter with the plane. The next day Alger Graham, the other pilot, in the DN-1, brought Wilkins and Eielson back to Point Barrow.[32]

Wilkins wanted still more exploration. The transmitter located at Point Barrow was installed in the DN-1. Mason received a few radio parts from Hemrich, the operator at Fairbanks, and set about creating a new transmitter for the Barrow base. "The powers that be" he wrote, "...were quite alarmed to watch me build a transmitter out of the few parts at hand, improvised with condensers made of pieces of five gallon cans, etc., and were sure that it was not going to work." But it did. As for the DN-1's flight, it was without mishap and the radio worked well. Now it was getting late in the season. The crew was relieved when Wilkins ordered them to pack up and prepare to return to Fairbanks and the "lower forty-eight." On June 18, 1927, all members arrived safely in Seattle.[33]

While sitting in his San Francisco hotel room one morning early in 1928 Wilkins glanced out the window and saw "the most efficient looking monoplane I [had] ever seen.... The machine was broadside to me. I marked its beauty of streamline, angle of incidence, and attack of the wing in level flight.... It had no flying wires; no controls exposed—nothing but a flying wing." Wilkins was gazing at the first Lockheed Vega built. Indeed, if a Vega were to land on an airfield today, people would undoubtedly say that it was a post–World War II model. Instead of the almost apple-box frailty of other planes, this one was sleek, streamlined, and missing exposed wires and struts.[34]

Wilkins raised money, purchased the second Vega built, hired his favorite bush pilot, Ben Eielson, and made preparations for a flight from Point Barrow to Spitsbergen. He had San Francisco's two radio engineers, Heintz and Kaufman, install a thousand dollar rig in the plane. "To tell the truth," wrote Wilkins, "I had small confidence in the reliability of shortwave wireless and was reluctant to allow its installation to interfere in any manner with the flying qualities of the machine.... Each year ... we had provided wireless equipment and each year at critical moments it had miserably failed us." This year he bought the best, supervised its installation, and proposed to work the radio himself. Though they were rank amateurs, he and Eielson could do a little sending of the International Morse code. If the plane load had to be lightened, the radios would go, Wilkins said. "There was no reason to suppose that our wireless equipment should be any more reliable than other equipment on the plane and there was as much chance that our wireless power would fail as there was that our engine would quit." So much for his faith in either the plane or the radio. This man was indeed an adventurer—but a most careful one.[35]

The rig proved quite dependable. While at Fairbanks they were in "more or less regular communication" with Fred Roebuck, 6ARD, at San Francisco. While at Point Barrow they used the Presbyterian Mission's Delco engine to drive the generator necessary to work their short wave. While in the air Wilkins set the wireless aerial to get the best resonance. He could send messages, but the receiver had been left at Barrow with Leon C. Vincent, the school teacher. (Wilkins probably did this because the noise of the plane totally drowned out anything from the receiver, which also added weight to an already overloaded aircraft.) Vincent was to listen for the Vega's signals and thus follow its progress and be aware of trouble should something happen. On the flight the generator began acting up, which caused Wilkins to be skeptical that any messages got through. Even so, at twenty minutes to the hour, on the hour, and sometimes a quarter after the hour he sent information. Near Spitzbergen he tapped out, "Now within a hundred miles of Spitzbergen. We are in bad situation. Heavy clouds about us.... All open water below." Then two sharp mountain peaks came into view, and he was able to send "Spitzbergen in sight. Spitzbergen in sight. We will make it." They landed first at Dead Man's Island near Spitzbergen, were isolated for five days due to a storm, took off and shortly thereafter landed twenty-five miles away at Green Harbor.[36]

A wireless was there and they were able to notify the world of their success. Then they examined the generator. A bearing on one of the armatures had burned out. They did not know when this happened. Later—so much later that it had to be added as an appendix to Wilkins's book— Leon C. Vincent sent a letter informing him that he had indeed heard him faintly from 4:00 p.m. Sunday, April 15th, to 3:50 a.m. Monday, April 16th. The buzz that he heard the last time was right on schedule, so he assumed, correctly, that all was going well. It would seem that the Vega was heard for nearly 12 of the 20 hours of the flight. The plane had covered 2200 miles, 1300 of which had never been explored before.[37]

As a result of this trip Wilkins apparently became convinced of the reliability of shortwave transmissions. Shortly after the flight he wrote, "It has been definitely demonstrated that a shortwave communication can be maintained throughout the year from the Arctic latitudes as far south as the equator, and, once sufficient wireless stations are established, it will be possible to use radio directional apparatus and to receive weather reports that would make Arctic travel as safe as elsewhere."[38]

One of Wilkins's first messages was to Dr. Bowman of the American Geographical Society: It was a puzzle to those who did not know of their prearranged messages. If Wikins saw mountainous land on the long trip he was to send " Black Fox Seen"; if he saw flat land, the message was "Blue Fox Seen"; and if none at all, "No Foxes Seen." He sent the latter message.[39]

The part amateurs and radio played in the "firsts" accomplished by Byrd, Amundsen, and Wilkins have been described here in considerable detail because of their fame. However, there were other scientific expeditions into the Arctic during the 1920s and '30s. QST's section entitled "Expeditions" gave the whereabouts and the accomplishments of the various enterprises. It also listed their shortwave call letters, often the name of the amateur manning the rig, the meters being used, occasional descriptions, even photographs of the transmitter, and listed the call letters of hams in the States who had worked them. All of the groups had shortwave radios and amateur operators along. Without doubt many thrilling events took place involving amateurs and their shortwave rigs that have gone unrecorded.

For at-home amateurs in the States the public interest in the Arctic and Antarctic added greatly to the fun of being a ham. To hear code from a ship in the ice-covered ocean just a few hundred miles from the North Pole was cause for elation, even worth reporting to the local newspaper; to indulge in two-way conversation gave a ham bragging rights at the next meeting of the local amateur club.

Following are the radio operations of a few more, but emphatically not all, of the expeditions to the polar regions during these years:

In 1926 one important scientific group which was equipped with shortwave radio and depended upon American amateurs for communication with the outside world was the American Museum (of Natural History) Greenland Expedition. It sailed at about mid–June for Etah in the *Morrissey* under the command of Captain Bob Bartlett and was to be gone about four months. Its charge was the collection of material for the museum. It was funded by George Palmer Putnam of G.P. Putnam's Sons, publishers. The radio operator was Edward Manley, SFJ, of Marietta, Ohio, and an ARRL member. Atwater Kent and the National Carbon Company supplied money for the radios, and offered "an Arctic trophy" to the amateur who handled the most traffic from the expedition. As usual, messages were to be forwarded to G.P. Putnam's offices in New York City. The *Morrisey's* call letters were VOQ. "This was Captain Bartlett's first experience using shortwave radio in the Arctic," wrote operator Ed Manley, "and it made things different from his trips with Peary when they were often out of touch with civilization for as long as two years."[40]

On this expedition the *Morrissey* had one serious mishap. On the night of July 25 off of Northumberland Island it went on the rocks at high tide. There was a real possibility of abandonment. Because of shortwave radio the accident was reported not only in the States but also in New Zealand, "which interested me greatly," wrote Manley, "showing how small the world is to short wave radio." One ham, identified only as 9CP, kept in touch constantly on the night of August 2nd "taking the longest message sent out, the story of the wreck." Manley suggested that 9CP should receive the prize, a narwhal tusk, offered by Putnam for the most consistent contacts with the *Morrissey*. (The Narwhal is a small Arctic whale with a circular tusk up to nine feet long.)[41]

In 1928 the *Bowdoin* was iced in at Anatalok Bay, Labrador. This year a supply ship, the *Radio*, accompanied it, with the call letters WOBD. It was heard until the ship turned south, leaving the *Bowdoin* and WNP alone in the Arctic. Cliff Hemoe, the *Bowdoin's* operator, wrote of being capsized and almost drowning as he rowed a boat from the expedition's housing to the schooner where the shortwave radio was located. When the ice was strong enough he drove to the *Bowdoin* using dogs and a sled. "Not many fellows go out to their radio shack in a dog team!" commented Clark C. Rodiman of the ARRL, who wrote up Cliff's experiences. In the *Bowdoin* he had to "sit down to a cold pair of headphones and a fifteen-below-zero schedule." Sending with mittens damaged the smoothness of his "fist," but he was still able to send and also receive and write down messages. Most messages were being sent on 20 meters.[42]

As 1928 was Wilkins's year of success it was Umberto Nobile's year of catastrophe. Late in May, from the same hanger at Spitzbergen from which the *Norge* had been launched two years before, Nobile took off in the *Italia* to reach the North Pole also. He did reach the Pole, but on the return to Spitzbergen the dirigible fell onto terribly rough ice. One man died and six were carried off in the broken dirigible, never to be seen again. Nobile, with a broken leg, arm, and crushed chest, along with eight other survivors and his little dog Titina, were left on the ice.

Fortunately one of the pieces of equipment that had fallen from the airship was a shortwave transmitter and receiver. Radio operator Giuseppe Biagi, seeing the inevitable crash, had wrapped his arms around the small shortwave set and closed his eyes. When the crash came he fell to the ice with the radio crushed into his stomach. Once an appraisal had been made of their situation, and the radio was found intact, Biagi immediately used the set to notify anyone listening of the disaster. They did not know if they were heard. They could hear from their ship, the *Città di Milano*, sending messages indicating that it was making rescue preparations. Trouble was, their whereabouts in the vast Arctic ocean was unknown.[43]

Nobile and five colleagues—the other survivors having headed for land, two of the three finally being rescued by a Russian plane—remained on the ice in a small tent. Biagi, one of the most stable and optimistic of the survivors, kept working his radio. Finally what Nobile called "the miracle of the radio" happened. The message was picked up by a young Russian amateur in Archangel who notified the Russian authorities. Soon they were also in touch with the *Città di Milano* and with Russian and Norwegian planes. It was the Russians in the ice-breaker *Krassin* who rescued them. Afterwards Nobile was ill treated by his government. Rivalries involving General Italo Balbo played a central part in this. Nor did fascism condone failure.[44]

Before we leave the Arctic, it is worth noting how radio, and especially shortwave radio involving amateurs, became a necessity for Arctic business. In 1929–1930 the *Nanuk*, a schooner owned by a fur trader, left Seattle for a trading voyage to the Siberian side of the Arctic. Among its crew members was radioman Robert J. Gleason, a young fellow in his early twenties. The *Nanuk*'s radio was long wave, fixed to work at 600 meters, the disaster calling channel. Shortwave could be used only while ships were in port. Gleason brought along his own homemade five-watt transmitter to be used when legal. Both the long wave transmitter as well as the shortwave worked well, and Gleason used both, the shortwave often illegally.[45]

The venerable Hudson's Bay Company belatedly became aware of the virtues of shortwave transmissions. In the 1930s it began using twelve-watt Morse code sets at its remote posts. Power was by windmill-operated generators with twelve-watt storage batteries that, writes HBC historian Peter C. Newman, "were nursed like colicky babies, kept warm by claiming the place of honor behind each staff house stove." Eventually the network grew to fifty-four stations."[46]

Activities continued in Antarctica. One very important ham possessing a poor "fist" was Rear Admiral Richard E. Byrd. When communication from his lone observation hut became more garbled than usual, suspicions arose that all was not well with him. As a result a rescue mission set out for his hut in the darkness of the Antarctic winter. It arrived just in time. Byrd was found in very poor physical condition due to carbon monoxide poisoning. Without radio communication he probably would not have survived his lonely ordeal.[47]

This took place during Byrd's second expedition to Antarctica (1933–1935). In the first expedition (1929–1930) he flew over the South Pole. For the first expedition and even more for his second one, Byrd relied heavily on radio. In 1929–1930 he had with him several radiomen including naval lieutenant Malcolm P. Hanson, who had stowed away on the *Chantier*. Radios would seem to have been everywhere. Byrd's two ships, the *Eleanor Bolling* and the *City of New York*, had rigs and operators who kept in touch with the new headquarters, dubbed Little America. Eleven dog teams pulling sledges were used to carry supplies from ships to the base. To know if they ran into trouble, radio announced their departure and the base radioed their arrival. Meanwhile Hanson was getting the three 65-foot radio towers set up at Little America and installing the permanent rig there, WFA. Radios were also being installed in the planes, along with one-cylinder gas engines that could run a rig in case of forced landings. At least some sledge teams on scientific missions were also equipped with radios.[48]

Operators were not allowed to contact hams at the time of official schedules and were forbidden to work hams when gas engine generators were using up too much fuel. Even so, Little America's radio operators contacted many amateurs who transmitted messages from expedition members to friends and relatives. When those hams tapped out "Please QSL Card" the expedition's operators tapped out that they were "at latitude 789:34 and longitude 163:30 west, the southernmost radio station in the world." Then they added, "Put a tack there on your map if it goes down that far. We are 2100 miles from the nearest post office or other human habitation. The next boat will reach us in 1930, and we will be pleased to send cards then.[49]

Polar exploration did not end with the return of Byrd's Second Antarctic Expedition in 1935. Lincoln Ellsworth was in the Antarctic in 1936 and 1939, exploring new lands from the air. Other scientific expeditions were periodically in the Arctic. The Second World War highlighted the importance of polar regions, and after the war, government-sponsored expeditions with costs hardly a factor made scientific investigation of the regions—both the Arctic and the Antarctic—relatively safe. Rare were the radiomen who were not also amateurs. They played a part—a most important part—in the twentieth century exploration of these regions.

As their written letters and articles indicate, most amateurs who went out with polar expeditions relished the experience. As Howard Mason, with Wilkins, said, "Shortwave radio in arctic exploration has been proven. And the amateur is the man for the job. If you get a chance to join an expedition you no doubt will find it a lot different from what you are used to, and different from what you will be expecting. There will be a lot of hard work, many obstacles that must be overcome, and a lot of hard tack and corned willie as well as ice cream and cake. But if your health is good, and you are sure you can stand the cold weather, go to it. You will have the time of your life."[50]

A more accurate picture of the life of a radio operator on a polar expedition had to await the article by A.G. "Gerry" Sayre, W2QY: "Ham at 30 Below" in the January 1939 QST. Gerry, who was on the MacGregor Arctic Expedition that sailed from Newark, New Jersey, on July 1, 1937, gave the negative as well as the positive sides of the experience. He suffered severe seasickness as the vessel struggled northward, and described the storms, the cold, the struggles with the ice, the fire that broke out close to 3,000 gallons of gasoline in the hold, and the loss of 4,000 feet of antenna wire overboard. The little gas engine for generating electricity was hard to start in sub-zero temperatures. The radio room had to be preheated or condensation resulted in blow-ups. Having explained all this, he *really* plunged into a negative description:

> When your bunk gets damp and mildews, when your clothes get sticky and clammy as you hang them up, when even your duffle bag of clean clothes mildews through, when the tubes collect "blizzard blankets" and the cartons become moist and damp, when your tools and other metallic objects rust in the damp salt air—then you begin to appreciate the precautions required....
>
> Yes, there are two sides to the picture, all right. I certainly had plenty of illusions shattered on this trip. Don't you believe all the bunk they paint about the glory and glamour of expedition life. Just salt it down that expeditions are about nine-tenths good, hard work and about one-tenth bunk. My private opinion, after having spent a little time "Up North," is that it is not worth while for an amateur to want to waste a year of his life out in the great unknown, away from the warmth of his family fireside, from clean rooms and hot baths and three square meals a day, from friends and companions, from all that one has come to enjoy and appreciate of civilization....
>
> Think of having to put up an antenna that has blown down, when it is pitch dark twenty-four hours a day and the temperature is from 10 to 30 below, with a howling Arctic wind so strong it almost bowls you over ... or of living closely confined with a group of men with whom you continually rub elbows all day and night for months on end, or of monotonous meals providing the same scanty fare every day, day after day ... and then, if you still want to go, this is all I have to say:
>
> *More power to you!*[51]

# 13

## To the Rescue: Amateurs in Emergencies and Disasters

Working fellow hams, experimenting, doing "DX," and being active in the local amateur club constitute for many thousands of amateurs the fulfillment of ham radio's offerings. They are aware, though, that the government in granting them the authority to operate amateur stations has implied certain obligations on their part including obeying the regulations and cooperating with government agencies. Responsibilities also may include activity in the radio relay system and comity in carrying on their hobby. But above all else, their obligations demand total cooperation during emergencies and disasters. When telephone and telegraph lines are down, roads are blocked and railroads are unable to move equipment, the amateur is the only source of communication from the place of disaster to the outside world.

It is pleasant to report that from the beginning of ham radio, when such calamities have occurred, a respectable number of amateurs have always pitched in. Often going without food and drink and sleep for hours on end, sometimes risking their lives, they have remained at their rigs, carrying messages to the Red Cross, sheriff's offices, the army, the navy, the coast guard and other organizations involved in bringing aid, as well as notifying families and friends that their loved ones in the danger zone are safe—or, unhappily, dead, hurt, or missing. These hams have earned the right to their licenses.

The federal government and the ARRL have both recognized service in disasters as a primary justification for the existence of amateur radio. First among the five principles justifying amateurs, according to the Federal Communications Commission (FCC) is this:

> Recognition and enhancement of the value of the amateur service to the public as a voluntary noncommercial communication service, *particularly with respect to providing emergency communications* [italics mine].[1]

Ham contributions in times of emergencies and disasters have been substantial. QST usually takes brief note of their accomplishments in minor disasters. With major disasters the journal devotes extensive articles to the roles played by amateurs, granting them the recognition they deserve. But when one goes to newspapers, magazines and books, very little appears. In times of disaster newspapers applaud faithful telephone operators and to a lesser degree telegraphers, praising them for their devotion to duty when their own lives were endangered. But where are the reports giving credit to the amateurs? Newspapers use the amateurs' expertise and transmitters to send stories from ground zero, yet they rarely give credit to the radiomen responsible for this service.

When they do get publicity, it is most likely to be as a notice in QST, where more often than not the ham is referred to only by his or her call letters. Therein lies the problem. How does one place flesh and blood on call letters? Was the ham male or female, teenager, mature young adult, or elderly hobbyist? Middle class, rich or poor? Married or single? Working a sophisticated manufactured rig or a home constructed one? Portable or stationery? Turning call letters into flesh and blood individuals poses a problem indeed.

Amateurs' attitudes toward their contributions in times of distress also play a part. QST editorialized that "such accomplishments are all part of the day's work in amateur radio." Just a blip in the day's work: the modesty may be commendable, but the result is that the history of many a disaster omits the vital contributions of amateur radio operators.[2]

The importance of ham communications can only be appreciated by an awareness of the susceptibility of the United States to natural disasters. We have an average of ten hurricanes a year, six of which reach dangerous intensity. We have floods in every state. We have tornadoes. We even have earthquakes and volcanic eruptions and landslides. Forest fires are endemic from spring through autumn, and ice storms and disastrous blizzards from fall into spring. People suffer and die in these disasters and property worth millions or even billions of dollars is destroyed. Yet these fateful emergencies rarely command notice in history books. Save for the people involved, who never forget, the nation soon loses interest as other crises come along. The amateurs' contributions disappear with the decline of interest in the disaster.

As for the amateurs, whether they have received their just dues or not, their activities in aiding in times of disasters have grown more and more complex. From the earliest decades this necessity for cooperation and coordination of hams with other relief agencies was clear. Although there was almost total lack of organization during the terrible Dayton, Ohio, flood of March 1913, amateurs working independently were credited with maintaining communications when all other means were out of commission. The public already knew of the wonders of wireless from the Republic and Titanic disasters and took note of what the amateurs achieved at Dayton.

March 1913 was the month of Woodrow Wilson's inauguration as president. If the elements are a portent of future troubles, then Presbyterian Woodrow Wilson should have prayed mightily, for the month of his inauguration was one of unusual weather violence. Until about the 13th the nation's weather was normal. Then, as Professor Ferdinand J. Walz of the Weather Bureau's District Three wrote of the Ohio Valley, "Both in the matter of storms, winds, and excessive rain it [March 1913] takes its place in the climate history of the Ohio Valley as generally being the wettest and stormiest and easily the most destructive from high winds and floods of any month for which there are authentic records."[3]

Although as many as thirteen counties were flooded, Dayton suffered the most—by far. That city's geographical situation was perfect for disaster. Three tributary streams—Wolf Creek, Stillwater River, and the Mad River—all flowed into the already swollen Big Miami at Dayton. The Miami's channel was three hundred feet narrower where it left the city than when it entered it. At 7 a.m., March 25, 1913, torrents of water all met inside the city and brought on its worst flood in history.[4]

That amateurs played a part in rescue attempts and in informing the rest of the country of the seriousness of the disaster is known. That the public was acutely aware of their participation was made clear by articles in the Chicago Tribune. However, the Tribune failed to single out individual hams and relate their experiences; its tribute was generalized.[5]

Not until the 1920s and the revolution in low meter–high frequency transmissions and receptions (shortwaves) did land-based amateur radio come into its own as a critical necessity in disasters. The individuals participating, besides being expert radiomen, possessed a sense of civic responsibility. Some were prepared for trouble. They may have had on hand extra batteries or even a small one-cylinder gasoline engine to create voltage for their rigs when the power went out. A few hams even had portable outfits that, although in those days still pretty heavy, could be transported.

It was not always a ham working alone. Early on, amateurs offered their services to aid communities devastated by natural disasters. In 1921, when the town of Hatch, New Mexico, was just about wiped out by a flash flood, amateurs cooperated in bringing help. R.W. Goddard, 5ZJ, was a member of a committee appointed by the Las Cruces Chamber of Commerce to send aid to the stricken town. He enlisted college hams who operated the portable wireless station at the New Mexico College of Agriculture and Mechanical Arts. Within a couple of hours their rigs were loaded on a trailer and rolling behind Goddard's "flivver" on their way to the town of Rincon, forty-one miles north-northwest of Las Cruces and on the edge of the destruction. There they were able to notify Las Cruces of the need for food, clothing, blankets, tents and cots. The Salvation Army and the Red Cross of El Paso pitched in, and relief was sent to the disaster area.[6]

Hams aided transportation. Help was given to the Colorado and Southern Railroad in November 1922. The C&S (a subsidiary of the Chicago, Burlington and Quincy Railroad—the CB&Q) ran trains north from Denver to Billings, Montana. When a "blue norther" hits the region, even today, train and automobile service can be precarious and schedules interrupted. The snowstorm that swept south through Wyoming early that November was a real doozie. By midnight the snow was fifteen feet deep in cuts along the C&S right-of-way. Trains number 29 of November 4th and number 30 of the 5th were stranded somewhere below Casper. But no one knew where they were, for the telephone and telegraph lines were all down; the dispatcher in Denver could not be contacted.[7]

At Casper officials asked amateur 7ZO to try to contact Denver. He failed, but finally contacted a Kansas City ham who said that the lines were open from that city to Denver. That amateur took the message to the Kansas City CB&Q wire operator, requesting that he forward it to the Denver dispatcher. "Advise if you wish help from the northern end."[8]

The CB&Q telegrapher in Kansas City had evidently been born sometime in the sixth century. He flatly refused to believe that a radiogram regarding the matter could have been received. Everything possible was done but this dumbbell refused to take the message. Wells got back on his rig and contacted an amateur at Fort Crook, Nebraska. That ham contacted the CB&Q operator at Omaha, who contacted the operator at Scott's Bluff, Nebraska, who in turn relayed the message to Denver. Within hours Casper and Denver were once more in contact. A rotary snow plow was sent to free train number 29 from the drifts. Later the Denver dispatcher had this to say about the role played by the amateur service:

> Not only did it locate the two trains with fifteen carloads of people but it also prevented the loading of over forty carloads of stock at Wendover, Wyoming, which would otherwise certainly have gone out into the storm and been caught by it. The fate of cattle crowded into open cars during such weather is not a pleasant one to contemplate.

QST ended the story by complimenting amateurs:

> It is the same story, fellows: when the stress comes it is the amateur, the ARRL radio man that is ready for the emergency; the ARRL man is the one who loses sleep, pay, and food, that he may help those who are in difficulties. It is he who through heart-breaking delays, interference, static and fading, sticks to the job and does his unrewarded best. He asks for no publicity and the story has to be taken away by force before he can tell our own members about it.[9]

Until the 1930s, QST observed, contributions by amateurs during disasters were largely "performances of individual heroism—deeds of valor performed single-handed or by comparatively small groups." By the mid–1930s, however, the populace had become aware of the services of amateurs, and expected them to step in. The concept of amateur radio taking over communications services during emergencies became accepted.[10]

The growth came slowly and involved many agencies. The army, national guard, navy, coast guard, Red Cross, forest service, Weather Bureau, sheriffs and police departments created nets whereby communications were maintained in the face of disasters. Amateurs played a major part in these measures because they were an integral part of the operations. But, it might be asked, are radio operators for the army, navy, coast guard, sheriff and state and local police really amateurs? QST gave a reply to this query as follows:

> Amateurs have many affiliations; much "amateur" emergency activity is nominally directed by such authorities as the Signal Corps, Naval Reserve, National Guard, even the Coast Guard and Army Engineers, as well as municipal and semi-private agencies. But in a realistic sense these groups are all affiliates of amateur radio; they are dependent upon amateurs for operating personnel; and to a major extent they utilize amateur frequencies and operate under amateur status. Their sole connecting link is amateur radio. The amateur is the yeast that leavens the whole batch of bread; without him the individual groups could not exist.[11]

By 1938 the league was working hard with other organizations to bring a viable system out of a somewhat chaotic one. An emergency coordinator (EC) was designated for each district and, above him, an EC for each section. As years went by these two systems became more complex and highly coordinated with other agencies or institutions involved in emergency or disaster rescue work.[12]

The effort to create workable nets and smooth cooperation with agencies involved has continued until today. In many communities an active system is in place. To keep the system prepared, trial crises are staged. The organization employs the net, smooths out rough spots, and makes sure that it will work in case of a real emergency. The local amateur club is always at the forefront of the operation, its amateurs dedicated to making the system work well. They know that emergencies will still occur.

The forest service deserves special mention in the development of useful, smooth-working nets. Forest fires are inevitable in the nearly 300,000 square miles of national forests. Time is of the essence in fighting these conflagrations. This is not just because the faster a fire is located, the greater the possibility of snuffing it out before it gets out of control. Changes in wind direction and fire speed can jeopardize those on the fire line. Only by having rapid communication with them by men overseeing the fire can the crews be saved from disaster. If they can be quickly notified of the danger they can beat a hasty retreat or take refuge in a body of water or a cave or some natural barrier. Even today tragedies take place in forest fire fighting, as the news tells us nearly every year.

The forest service tried the semaphore system, heliographs, passenger pigeons, and, shortly after the First World War, Curtis Jennies and British de Havillands flying fixed schedules over specified areas to notify authorities of fires. The pilots, initially without radio, could only alert ground personnel of a fire by dropping a small parachute with a red streamer attached, or else land and make a phone call. None of these methods worked very well.[13]

At first the most reliable means of communication was the telephone. The service had an agreement with AT&T to use its wires at a 50 percent discount provided the service did not compete with the company's regular business. Trouble was, all wires, whether those of AT&T, private local systems (there were many at that time) or the forest service, had to be kept operational. A forest fire could destroy them, as could ice storms, winds, and floods. Moreover, using a telephone during an electrical storm, one ranger wrote, "was about as hazardous as reaching for a rattlesnake in a gunny sack." So the forest service was always seeking new, reliable means of communication. Radio was seen as a possibility even in the days of bulky, primitive spark transmitters and squealing receivers.[14]

The first forest service experiment with radio took place in the summer of 1916 when Ranger William R. Warner mounted his horse and began his weekly trip of thirty-eight miles to the little town of Clifton, Arizona. About midway he noticed an antenna at a local ranch. Upon inquiring he found an amateur in the person of Ray Potter (call letters not given), a high school student. Immediately sensing the potential in radio, Warner was aided by Potter in ordering equipment from a mail order house. At a cost of $115.45 including young Potter's labor, the rig was made operational. On November 26, 1917, Warner was able to contact Clifton. In doing so, he demonstrated three things: the antenna did not need to be in sight of the receiving station; code could be mastered quickly (it took Warner just a few weeks to do twelve words a minute); and the cost was not excessive.[15]

By 1921 airplanes were equipped with transmitters. The forest service had young amateurs—average age 16 to 25 years—operating the ground receivers. For lack of funds, however, the system was not used the next year. Meanwhile, in 1919, a forest service telephone technician known as "Ring Bell" Adams in the Pacific Northwest experimented with army radios. From Mount Hood to the Zig Zag Ranger Station, a distance of twelve miles, he made good contact once he had figured how to make his antenna withstand winds of up to eighty miles an hour. Another experiment between the ranger station near Lolo Hot Springs and the one at Beaver Ridge, Idaho, just twelve miles distant but over "mineral zones" and the Bitterroot Mountains, likewise proved effective. The rigs, which were phone stations, used a lot of juice: 270 number 2 Burgess Edison Storage Cell batteries were needed. Delivering them to the station was a real problem. Once the pack train carrying them had a laggard pack horse at the tail end. It started pulling back, disrupting the whole train, which rolled down the mountain about 100 feet. All the electrolyte was drained out of the batteries.[16]

In late August a fire surrounded Beaver Ridge and threatened the radio camp. The operator, a man named Cutting, placed the rig on a raft, floated it to the middle of a lake, and then retreated from the fire. When the danger was over he returned to the lake, fetched his rig and was back on the air. Adams pointed out that at about the same time in the Clearwater Forest, a telephone line was destroyed by fire and it took several days to restore operation, whereas the radio was back in use in a few hours.[17]

Forest service headquarters in Washington, D.C., manifested mild interest in radio transmission, enough so that beginning in 1928 the service allotted minor funds for a radio laboratory. Until 1951 it was located in the Pacific Northwest. Prior to the Second World War personnel never exceeded eight people including a stenographer. They were poorly paid but they were extremely dedicated, and their contributions were impressive.[18]

One reason for this was that forest service needs differed from consumer needs. Because there was little or no market for compact transceivers, rigs tough enough to withstand a trip down a hill with a fallen pack horse, simple enough "for a mule to operate," and sufficiently inexpensive for a tight-fisted government bureau to accept, the forest service had to design its own equipment. With dedicated amateurs such as Harold K. Lawson, William Apgar, Foy Squibb, and Dwight Beatty (and there were others), the laboratory set out to supply the service with equipment designed for its special needs. What was required had to be simple, rugged, and reliable. By 1930 three types of rigs were in forest service operation: the portable radio, the semiportable radio, and the temporary or field base station.

The rigs they built were always tested by forest service employees at the agency's lookouts, regional headquarters, and even on the fire line. For example, by 1931 Foy Squibb had developed a portable transmitter-receiver that weighed just twelve pounds and used just one and one-fourth watts. To test it, Squibb selected an "average man" from a work crew and gave him "an operating demonstration, including a set-up and take-down, [and] one hour of practice." At a signal, this "average man" "set up the equipment in 18 minutes and sent a coded message requesting 8 men, with location and type of fire. He then waited while the receiving station copied the message, phoned it to headquarters, received a reply, and retransmitted the message to the road man, at which time he disassembled the equipment. The total elapsed time was 44 minutes." The transmitter was capable of being heard 15 miles away by voice and 20 miles away by code.[19]

Sets were tested climatically at locations from sea level to 14,000 feet, from deserts to the most humid forests. Lab workers tried to determine how many "rivets and gussets would be required to make a chassis withstand the shock and weight of a packhorse tumbling over the side of a mountain." They even dropped a chassis from the laboratory roof onto the cement driveway. And they loved their work. In later years Squibb and Lawson acknowledged their experience with amateur radio and "voiced their pleasure at the opportunity to turn a hobby into a vocation."[20]

Always the forest service had two problems: the first was making sure that AT&T did not accuse the service of preempting its business with radios, and the second was similar to ship captains, lords of their vessels, who resented that radio made possible orders from the mainland. Grizzled rangers resented radios that conveyed to them instructions when they were out in the timber. Ultimately the use of telephone lines declined drastically, and in time, and as another generation came along, "grizzled Ranger" resentment also declined.[21]

Then came experiments with vhf (very high frequency) and uhf (ultra high frequency). Early vhf was left to amateurs, who enjoyed experimenting at 5 and 10 meters; the commercial world failing to see the usefulness of these frequencies. But amateurs, including those employed by the Forest Service Radio Laboratory, knew that the higher the vhf, the less power needed—the fewer batteries to carry. They knew that with vhf the antenna length could be just 10 percent of that needed for low frequency. Also, it was believed that vhf was only useful with line-of-sight, but that could be fine for forest service workers on a fireline. Parts did not exist for vhf, but hams—and the forest service radiomen were all hams (or just about all of them were)—all had junk boxes of parts, and in no time at all parts were being used, or re-made, to be useful in vhf work. The initial vhf units were "unique products too complicated in parts and labor for profitable duplication by large-scale manufacturers."[22]

Dwight Beatty and Foy Squibb of the forest service built vhf radios "as if they were one-of-a-kind units intended for their own personal use." In the best amateur tradition, their experimentation was based on a few articles read here, conversations with other hams, a few of their own ideas thrown in for good measure, and a lot of work. Their aim was a practical set for use along the fire line. Their work precluded involvement by manufacturers, for the only demand for such sets was in the forest service—a few thousand sets at most. "The techniques of amateur radio enthusiasts [had] no place in [the] board rooms, production lines, or sales territories" of such companies as RCA, Westinghouse, or Zenith.[23]

And they succeeded, eventually designing a workable smoke-jumper's back-pack vhf radio that weighed just six pounds. The practicality of their work is shown by forest service use of radios. By the end of the 1940 fire season nearly 4,000 radios had been ordered by the regions, 90 percent of them in the portable class (under 21 pounds) and 2000 of vhf, using less than 2 watts of power. The service had 25 frequencies allotted to it in the 2,000–3,000 kc range (100–150 meters) and 75 frequencies in the 30,000–40,000 kc range (7.5 to 10 meters). The use of telephone lines by the forest service declined conversely with the use of radios.

Was the radio laboratory worth the cost to the forest service? Emphatically, yes. As Gary Gray states in his excellent history of radio in the Forest Service:

> For 20 years, the staff at Portland [headquarters of the regional office which supervised the laboratory] was handicapped by lack of resources and leadership. Yet, even after a virtual shutdown during Word War II, it still took private enterprise 20 years to catch up with the Laboratory. The Laboratory staff never went above 8 employees, yet they designed, modeled, tested, produced, inspected and shipped an array of uhf and vhf portables, semiportables, and fixed-based units second to none.[24]

The army and navy have been equally aware of the contributions of amateur radio. The army established amateur radio stations at most of its national guard locations and saw to it that nets existed with other army units as well as with local amateurs. The navy has been cooperative with hams and has encouraged their activities. It can also be said that nearly every community in the United States of any size has an amateur radio club, unaffiliated except with the ARRL, which considers the maintenance of a net system for emergencies and disasters a major reason for its existence. All are aware of the contributions of amateurs in emergencies.[25]

Even in the early 1920s, when the organization of emergency and disaster facilities was just beginning, amateurs did volunteer service of great value to the population affected. Sometimes hundreds of hams were involved. An example took place on February 3rd to 5th, 1924, when the northern half of the United States was swept by a blizzard that temporarily ended communication by wire and telephone. Chicago for a brief time had just one line open to the rest of the world, a single press wire to St. Paul, Minnesota.[26]

During this crisis the net control station (or NCS) was a Chicago amateur selected by the Chicago Radio Traffic Association (which coordinated activities of amateur clubs) to handle the emergency. This ham, call letters 9AAW but name not given, worked a twenty-four hour schedule (the usual quiet hours imposed upon amateurs were suspended for the time of the emergency). In Minneapolis one of the half dozen or so most prominent amateurs of the 1920s and '30s, Don Wallace 9ZT-9XAX, was among the hams handling traffic in and out of that city. "He acted as a radio Paul Revere," reported K.B. Warner, "calling amateur minute-men into action."

Other operators were singled out for their services in Pittsburgh, Des Moines, Iowa City, Milwaukee, La Crosse, Defiance, Ohio, and other cities. Information of an air mail plane, downed at Fort Des Moines, was relayed by a ham to the plane's base in Iowa City. That ham then took down 1500 words of press from another ham and sent it to a California amateur, giving that state the first information about President Wilson's funeral. A typical service in most disaster cases is the need for repair parts and tools to handle the emergency. In the disaster of February 1924, Twin Cities officials needed such aid. Hams conveyed this need by way of an Ohio operator to a Connecticut amateur who then telegraphed the request to the proper manufacturer or service company; the reply, that equipment and parts were being shipped, came back to the Twin Cities again by way of an amateur using 200 meters.[27]

Amateurs aided locally with private messages. A Dana, Illinois, ham summoned medical aid for his town from nearby Streater, and connected a nurse from that town to a doctor in Longpoint, Illinois. The amateur "live[ed] on a farm three miles from town and delivered those messages by flivver during a blizzard!" Another helped a man learn of his wife's condition: she had undergone a serious operation in a Chicago hospital. A ham in the husband's home town contacted a Chicago amateur who checked with the hospital and conveyed good news the same way back to the worried spouse.[28]

It was trouble enough for amateurs to sit at their rigs for hours on end while flood waters swirled around them, or winds and sub-zero temperatures buffeted them in their shacks, but besides these discomforts they had to use their ingenuity to keep their rigs operational. The transmitter was often damaged by water or wind or other elements, and the radioman had to use his expertise to maintain a workable set from parts available. The most common problem was an antenna down, either from ice or wind or both. While the storm continued the amateur had to mount a makeshift substitute, often risking his life in the process. If electricity was out, he had to make his way to the nearest radio or hardware store to purchase batteries to power his rig.

Major disasters include the Florida hurricanes of 1926 and 1928. The entire area from Fort Lauderdale on the north to south Miami was filling with people—300,000 of them by 1926. The big blow of September 11–22, 1926, a category 4 ("extreme": winds of 131 to 155 mph), was considered to have cost $111,775,000 in 1926 dollars and taken 243 lives. Until hurricane Andrew in 1992, a category 5 storm, it was the costliest hurricane in United States history, and in loss of life greatly exceeded Andrew, which caused just 14 fatalities.[29]

Meteorologists called it a Cape Verde type storm because of its place of origin off the coast of Africa. On September 12 it was 1,000 miles east of the Leeward Islands; by the 16th it had passed the Turks Islands north of Hispaniola (the Dominican Republic and Haiti) with winds of 150 miles per hour and was advancing at a speed of 19 miles per hour. By 11:30 p.m. on Friday, September 17, squalls were raking the Florida coast. According to one authority, "over the next twelve hours, Miami was pummeled by the most severe hurricane conditions in Weather Bureau History. The eye of the storm swept directly over the city," with winds of 128 mph and at 7:30 a.m. raging at 138 mph for a full two minutes. A minimum of 8 inches of rain fell and a storm surge of 14–15 feet in Biscayne Bay poured onto the land. People died because they were ignorant of the power of a hurricane, nor did they realize that the brief sunlight and calm of the eye of the storm was just temporary, that in a short time—35 minutes—the lull would end and the other side of the storm would come roaring in. Many were sightseeing close to the sea or bay and lost their lives when the storm surge came ashore.[30]

Electricity was out. The Tropical Radio Company station was Miami's principal commercial radio facility. At about 3:00 a.m. people began arriving at the site seeking shelter; two hours later, at 5:00 a.m., all five of the station's transmitting towers—four loomed skyward 439 feet and the fifth 360 feet—crashed to the ground with a defining roar. No one was hurt, but that ended all broadcasting. All telephone and telegraph lines were down. The small town of Moore Haven, on the southwest side of Lake Okeechobee, suffered most. Dikes gave way, a surge of water 15 feet high swirled across the land, and an estimated 150 to 300 people died. The tempest, once across the peninsula, crossed the Gulf. On September 20 it struck Pensacola, where communications were also interrupted.[31]

Amateurs rose to the occasion. They were the only contacts with the outside world for several days and nights while crews worked diligently to get telephone and telegraph lines operating again. Operators all over the country listened for contacts from hams in the hurricane area. "Messages asking for help, words of assurance to friends and relatives, requests for news and reports of damages were all handled by amateur radio as soon as contact was established," reported QST. Some hams at the site appear to have been awaiting the opportunity to come to the aid of their beleaguered community, having thought out in advance the steps they would take.[32]

At Miami was amateur station 4KJ, operated by John V. Heish. He obtained additional B batteries and made contact at South Jacksonville with a member of the U.S. Naval Reserve, Gifford Grange, 4HZ. Using 40 meter wavelengths, they established an hourly schedule. Heish's first message out of Miami was a request from the sheriff asking the governor of Florida for military aid. Soon Heish and Grange were sending messages for the government, the Red Cross, requests for doctors, food, and milk, orders for building materials, and many messages from Miami people to their relatives requesting aid or notifying them of their safety. Heavy atmospherics (QRN) interfered with the transmissions, but the two amateurs stuck to their rigs. Messages received were delivered by the sheriff's department while messages sent to locations to the north were filed by the receiving amateurs with Western Union as per instructions from the Miami sheriff. Death messages and some replies of importance were handled expeditiously.

After the first day Grange and Heish cleared regular traffic by morning, noon, and night schedules, which left the rest of the day for handling important messages to other points. Grange received the first reports of probable loss of life and property in Miami and learned from a St. Petersburg ham of the damage done there. Other amateurs were equally active. One ham went to Miami with the Florida national guard contingent and kept the troops in touch with the adjutant general's office at St. Augustine. A Tampa station was in touch with Grange and gave important news for the press. At least a score of amateurs in the general area of the hurricane were on the air, helping in any way they could. They handled messages for Southern Bell and did relaying of messages from Miami to Mobile to Memphis; International News Service at Chicago received press messages through amateur radio. Many relays by amateurs in different parts of the nation took place, but were not documented. Indeed amateurs from all over the country were at their stations waiting to help if called upon, and many did participate.[33]

Gifford Grange was credited by the navy for his good work. "[I]t is largely through your efforts that communication was finally established between Miami, Florida, and the outside world during the recent hurricane in the Florida area," read the citation from the commandant, 7th Naval District, Key West. The commendation concluded on a personal note: "I wish to extend to you my appreciation and thanks for your services which reflect not only upon you, but upon the Navy." It was signed by Admiral E.W. Eberle.[34]

The 1926 Florida hurricane, devastating as it was, was but a rehearsal for what was to come two years later. Hurricane San Felipe, so named for the saint's day when it was first noted, September 13, was also a Cape Verde storm. It began on September 4th or early September 5th, 1928, as a wave off the coast of West Africa and grew to becoming a category 4 storm. In the two years that had elapsed since the 1926 blow, amateurs had become more aware of their usefulness and they had learned more about preparedness.

Communications difficulties first affected Puerto Rico and the Virgin Islands on September 13th. Naval radio was the common method of communication, but both stations were knocked out by the storm. The San Juan station got back on the air first, using the amateur band. Amateurs heard its message and promptly contacted the Navy Department. The radio at St. Thomas in the Virgin Islands was also out, but one of the operators was a ham. He contacted a young amateur in New Jersey who also relayed essential information to NAA, the naval radio station at Arlington. The governor of the Virgin Islands was able to keep in contact with naval authorities and direct aid efforts via amateur radio. It is to be noted that until just a few months prior to the hurricane, amateur radio had been illegal in the Virgin Islands. The amateurs involved in both of these islands received commendations from the Navy Department.[35]

Information about the storm was limited. Ships gave good information but they were doing their best to get out of the storm. When the tempest crossed the island of Guadalupe it left 660 dead; when it crossed Puerto Rico in just 24 hours it left 275 dead.[36]

Then, on September 16th, Hurricane San Felipe struck Palm Beach, about 75 miles north of Miami. Telephone and telegraph service were out after 2:00 p.m. and that community was out of contact with the rest of the world. About 40 miles west of there lay Lake Okeechobee, 31 miles across, covering 730 square miles, next to Lake Michigan the largest body of fresh water entirely within the United States. Earthen dikes on its northwest, west, and especially southwest sides prevented its waters from spilling over into miles of flat, low, rich agricultural lands. Thousands of farm workers, mostly blacks, worked the farms. When the hurricane reached the lake, it created a water surge fifteen feet high which simply swept away the dikes, inundated this vast vegetable garden, and officially killed over 1,800 people; unofficially it is believed to have caused the deaths of up to 3,400.[37]

Many Florida amateurs were prepared for the big storm, and two of them stood out for their contributions. One was Ralph Hollis, a driver for the Palm Beach Fire Department, with call letters W4AFC, and Forrest Dana, W4AGR, a civil engineer. Aware of the impending storm, they set out at 1:30 a.m. to buy B batteries for emergency power, borrowed some storage batteries, and set up their single operation at the fire station, which also served as police headquarters. With call letters W4AFC they established contact with W4IX, an amateur at Tavares, Florida (about 200 miles north-northwest of West Palm Beach), and through him conveyed information to the rest of the world. During the day they kept schedules with other stations. At 4:00 p.m. the antenna blew away. During a lull they moved their rig to the police station at the safer end of the building. They reinstalled their equipment under dangerous conditions, succeeding in re-erecting their antenna while bricks and boards whisked by them. They were soon operational again. From 7:30 a.m. Monday, September 17th, until 3:30 a.m. Thursday the 20th, they kept at their station exchanging vital information with other amateurs, reporting the disaster to the outside world, and notifying the army's net control station, WVA in Washington. Through these contacts the first reports were delivered to the Red Cross, and relief, both from the army and the Red Cross, started on its way.[38]

By Thursday afternoon telephone and telegraph lines were up again. By FCC regulations denying amateurs the rights to compete with these commercial enterprises, Hollis and Dana were finally relieved. From the 17th through the 22nd they had been off the air just fifteen hours. For their work they received kudos from Palm Beach and from the army, which mentioned their "clinging to their self-imposed task with a purposeful tenacity worthy of the highest praise.... Mr. Dana lost his home, his automobile, and all personal effects in the storm. Both men ate when food could be received and neither slept in a bed during their self-appointed vigil."[39]

A day or two into the disaster the Tampa post of the American Legion established amateur station W4CV at Belle Glade, which is on the southern tip of Lake Okeechobee. It served the area until regular transmission lines were restored on September 30th. The Palm Beach station maintained schedules with the Belle Glade station until that date.[40]

Some measure of Hollis's and Dana's contribution can be ascertained by noting that they handled approximately 10,000 words of press matter and sent 16 messages to the Red Cross and local authorities, besides 162 personal messages; the station received 20 messages for the Red Cross and local authorities, and about 200 messages from W4CV at Belle Glade. W4AFC trans-mitted the first direct message to the Red Cross via the army station. Both were able to get mes-sages through to the disaster area via W4AFC when all other means of communication were down. We must be aware that other amateurs were helping: QST gives the call letters of 17 ham stations that helped Hollis and Dana. A U.S. Naval Reserve rig operating out of a truck lent by the Miami Herald also did good service.[41]

Florida in 1928 had less than 1,400,000 residents compared with more than 15,000,000 today. The Lake Okeechobee area was just being developed into an enormous vegetable garden. The three newspapers serving it were not even dailies. While the world was learning of the dam-age of Palm Beach and West Palm Beach, it heard little about the tragedy at Lake Okeechobee. In fact, Governor John Martin of Florida did not hear of it, or at least of the viciousness of the situation, until three days after it happened. He learned of it by way of amateur radio relays from West Palm Beach to Moultrie, Georgia, to Asheville, North Carolina, to Lima, Ohio, and finally to a ham in Tallahassee who conveyed the information to him.[42]

QST suggested that hams study the history of disasters and learn from them how to pro-ceed systematically and efficiently. It listed priorities for service: first, messages from the stricken area appealing for aid and requesting food, antitoxin, doctors, and the necessities of life; next it listed the transmission of press messages conveying to the world the extent of the disaster; and finally, messages to and from relatives and friends as to the condition of those in the disaster area. In the latter case, hams from all over the country would be trying to contact hams at the disaster area for information, and it was the duty of the disaster area amateurs to make them stand by while more important messages were taken care of.[43]

The prepared amateur would have considered the sources for a temporary power supply. He or she would already have contacted a store selling B batteries, have storage batteries on hand, or have a small one cylinder gasoline motor to generate electric power. The ham would have con-sulted with other local amateurs to decide in advance the location of the station during the emer-gency. Crudely, this was the beginning of the more formal creation of a net. "Be ready for the emergency call, QRR, when it comes," advised QST. "Jump into the breach with your station if feasible and stand by and avoid interference to those handling emergency traffic if this seems to be the right thing to do."[44]

In the period 1936–1940 the United States weathered a number of serious floods and hurricanes. In all of these amateurs played a significant part by maintaining contact with the rest of the world while other communications were out of commission. Although hams received a modicum of praise for their efforts, the fact that they were not affiliated with great corporations with public relations awareness meant that they rarely received the full adulation they deserved.

Whether or not this was a factor in William S. Paley's decision in 1936 to establish the William S. Paley Amateur Radio Award is unknown. Unlike David Sarnoff, his counterpart of NBC and RCA, who had been a ham and knew and used Morse code, Paley came into radio as a BCL—what hams called a broadcast listener. When he was just twenty-four years old he was introduced to commercial radio by way of a friend's crystal set and headphones. He immediately bought a $100 receiver, a considerable sum in 1925, envisioned the commercial possibilities of commercial radio, and ultimately became the head of the Columbia Broadcasting System.[45]

In 1936 he announced the creation of an annual award to the amateur "who has contributed most usefully to the American people, either in research, technical development or operating achievement." "This action was taken," according to *QST*, "because Mr. Paley felt that the useful service which amateur operators had rendered to stricken communities during the flood disasters in the early part of 1936 was only a single example of the very great contribution they have made to radio communication as it exists today." It is perhaps significant that the first three trophies (1937, 1938, and 1939) went to amateurs who had distinguished themselves by aiding in flood and hurricane disasters.[46]

The 1936 award went to Walter Stiles, Jr., W8DPY, of Coudersport, Pennsylvania, a small town along the Allegheny River. It was early March, and a heavier than usual snow cover lay over the land. Flooding occurred from Connecticut and New Hampshire southward to Pennsylvania; even Washington, D.C., suffered from a swollen Potomac River. Pennsylvania was hardest hit of all, with 84 of the 107 deaths recorded, more than 82,000 buildings destroyed and 242,698 people receiving Red Cross aid. Pittsburgh's Golden Triangle was for a time under 16 feet of water. Damages were estimated at $212.5 million.

Stiles was sure that the flooding Allegheny as well as lesser streams were going to cause havoc in the small towns along their banks. He hastily assembled portable equipment including a gasoline-driven five-kilowatt generator. On the night of March 17 and 18 he kept his transmitter open, conveying information. Then about 9:30 a.m. on the 18th he received an urgent message from the CCC (Civilian Conservation Corps) camp near Renovo, Pennsylvania. The town and camp were both isolated and in need of food, clothing, and medical supplies. Stiles set out over roads harmed by washouts and mudslides to get to Renovo, 68 miles away. He reached the town by 5:00 and by 5:30, according to *QST*, "after pouring water out of the cabinets and setting up the gear on the Y.M.C.A. steps ... was on the air."[47]

Walter Stiles, Jr., had already lost two night's sleep; now he was busier than ever. He remained at his rig for more than twenty-four hours. When a couple of operators from State College arrived to relieve him, he was in a state of nervous exhaustion. In fact, he had worked continuously for 130 hours and had been sustained with some kind of "shots" administered to him by a doctor. Due to his efforts the 4,000 inhabitants of Renovo received food, clothing, and medical attention. "It can be said that he was prepared for the emergency; he recognized the need, he demonstrated incredible degrees of perseverance and technical ingenuity, as well as fortitude and courage, in fulfilling his duty as he saw it," editorialized *QST*.[48]

Stiles was chosen for the Paley Award from among forty nominees. The presentation was at a luncheon at the Waldorf-Astoria Hotel in New York City. It was broadcast over the entire CBS network. More than that, President Franklin D. Roosevelt sent a telegram to Stiles, complimenting him on his accomplishment. "What you were able to do in aid of the flood sufferers emphasizes how important the continued development of amateur radio activity is to the best interest of the nation," the president wrote.[49]

Anning S. Prall, chairman of the FCC, gave the principal speech. He extolled amateurs, noting that there were now 47,000 spread from the forty-eight states to Wake Island. He stressed the FCC's continual concern over their welfare, noting that they stood ready to answer their nation's call in time of war, and were always ready and competent to deal with peacetime emergencies.[50]

The 1936 Allegheny River flood was but a mild prelude to what happened in January 1937. Unusual meteorological conditions resulted in days of unprecedented rain, sleet, and snow in the Ohio River Valley. Sixty billion tons of water amassed in the Ohio River system, with rain falling heavily in some parts of the region. The resulting flood caused havoc throughout the entire Ohio River drainage system. Cold weather and blizzard conditions added to the human misery. A million gallons of oil, gasoline and kerosene released from leaking storage tanks resulted in three miles of the river being on fire. Buildings swept from their foundations and floating away added to the devastation. Flood waters inundated hundreds of villages and cities located along its tributaries. Twelve states and 196 counties from West Virginia to Louisiana were said to be affected by the disaster. At the height of the calamity the Red Cross had 14 concentration camps (in 1937 not an unpleasant name) over an 800 mile stretch, housing more than 790,000 refugees; it had established 105 field hospitals supported by 900 nurses and inoculated a million people against typhoid. The Mississippi River was swollen to flood stage also, but in most places the dikes held. The real tragedy was along the Ohio River system. Rising to the challenge of maintaining communications when all telegraph, telephone, and travel communications were down were the amateurs. The part they played in managing relief efforts during the great Ohio Flood of January 1937 was, in fact, amateur radio's finest hour.[51]

In a larger sense, it was all of radio's finest hour. Commercial stations did service as long as they had power while the Army Amateur Radio System (AARS) and the Naval Communications Reserve (NCR) as well as the coast guard, an occasional American Legion post with amateur members, and police using radio contributed to the relief efforts. What must be remembered is that almost every ham, whether or not a part of a system such as the AARS and the NCR, pitched in and made a contribution. If not a member of the army and navy nets, then the ham was probably linked to a unit formed by the local ham club; someone established a central clearing station and the net was in operation. Even nationally, hams tapped in and offered aid—too much aid! The FCC issued a rare order, that "no transmissions except those relating to relief work or other emergencies be made within any of the authorized amateur bands below 4,000 kilocycles," meaning that only emergency operations would be permitted on the 160 and 80 meter bands "until the Commission determines that the present emergency no longer exists." Sixty stations picked by the ARRL did vigilante duty, making sure the ban was respected. According to DeSoto, hams overwhelmingly heeded the order; QRM (interference) was nearly eliminated.[52]

Yet, as usual, reports tended to ignore the part amateurs played. The Red Cross *Report*, for example, gives the coast guard most of the credit for communications, saying that it coordinated radio communications and manned 76 radio stations including 11 mobile truck units. The *Report* also mentions navy, army engineer, and national guard radio help, with the final phrase that "amateur short wave units also participated." DeSoto estimated that at least 500 ham stations manned by more than 1000 operators took part in relief efforts. Although his estimate may include the hams affiliated with the above services, the Red Cross statement still gives short shrift to a lot of dedicated, independent hams who pitched in and helped.[53]

On January 21, 1937, the military affiliates were ordered into continuous duty. The ARRL through its communications manager, F. Edward Handy, directed the sections communications managers (SCMs) of 11 states to organize emergency nets in the flood-ravaged areas. This was done with remarkable promptness. Other hams throughout the United States and Canada stood by to help if needed. Perhaps 100,000 messages of an official nature were handled, with almost every town and city in the flood-devastated area a recipient of some kind of help via radio. As Clinton DeSoto stated:

> What these prosaic figures add up to in terms of lives saved or suffering averted no one can say. Certain it is that death tolls numbered in the hundreds would have been in thousands had it not been for radio amateurs, that hundreds of thousands of the million and a quarter homeless would have experienced additional agonies of starvation or illness or mental turmoil had it not been for their work.[54]

The devastation was so widespread and the contributions of amateurs so great that DeSoto divided his report by states, beginning with Pennsylvania and ending in Arkansas. It should be stated that throughout, the major contributions were by amateurs with 'phone systems; continuous wave transmissions by Morse code were not as widespread or as useful. Pittsburgh, which had been devastated the year before, was just mildly touched by the 1937 flood. Nevertheless amateurs did valuable service there where a net was centralized and gave hourly reports on flood levels. It was able to inform those down river of the flood's progress; this facilitated evacuating people before they were isolated.

The deluge began taking on aspects of a major disaster in Ohio. When Marietta was cut off from Pittsburgh, amateurs, coordinated by the Marietta Amateur Radio Society's station, filled the communications vacuum. At the small town of Wheelersburg a ham supplied the town's only contact with the outside world for a full week. He operated his 75 meter 'phone at times with power from automobile batteries willingly loaned for his use. He also warned cities down river of thousands of gallons of gasoline floating free on the waters, raising the threat of fire. Occasionally river boats, especially those owned by the army engineers, offered housing to amateurs and their rigs. Of prime importance was the work of the amateur station at the University of Cincinnati, where two 15-kva tractor-driven alternators were set up in case the power went out. It did not, while for more than two weeks fifteen operators worked their central clearing station for the 75 meter 'phone band embracing the entire flood area. Some 2,139 messages were handled, most of them for official business. Thirty-six stations were on their web, with regular schedules kept; a dozen receivers monitored all bands. The army's network gave instructions to its river boats as well as coordinating work with amateurs.[55]

In West Virginia, which suffered from more than nine inches of rain in January, Wheeling was never entirely isolated by downed telephone lines, but with the lines overloaded, hams and operators of the army engineers' network greatly aided the Red Cross and gave needed information to stations down river. That was where the devastation steadily got worse. Parkersburg, a city of 30,000 at the time, was isolated. Amateurs established stations on fire boats to keep the city in contact with the Red Cross and other relief groups. At isolated Point Pleasant, where the Kanawha River joins the Ohio, amateurs there were in touch with the army, the Red Cross, West Virginia National Guard, state police, and state board of health.[56]

At Louisville, Kentucky, the flood crest exceeded all records. During one interval the river was rising at the rate of a foot an hour. Radio warnings had gone out as early as January 21 instructing residents of some low-lying areas to evacuate to higher ground. Thirty square miles of the city were inundated. With 230,000 homeless, Louisville was fortunate in having a good coordinator, Lieutenant W.R.R. LaVielle of the Naval Communications Reserve (NCR). He enlisted and organized the aid of local amateurs, obtained supplies including power (one Delco 32-volt plant arrived by truck from over 160 miles away) and kept contact with officials. In the desire to help, one amateur, DeSoto wrote, "rigged an emergency transmitter that really got down to elementals. It consisted of a Ford spark coil and 6-volt storage battery, circuit *circa* 1907." And, he reported, "the signal was surprisingly sharp."[57]

According to the *St. Louis Post-Dispatch*, after the power had failed 14 radio stations on a shortwave hookup carried on the work of closed broadcast stations, notifying flood-bound families of efforts to relieve them.[58]

The stories go on and on. From Parkersburg down river for 100 miles to Huntington devastation was the worst ever known, with the river wreaking a ten mile wide swath of devastation. At Huntington the hams operating the army's network were on the air for 198 hours, handling 784 messages to relief organizations such as the Red Cross, the American Legion, and the police; in addition nearly 1500 other contacts were made. If the amateur was not in the army or navy, his rig was probably at his home. When flood waters threatened, the ham often moved to a higher place such as the city hall, a church, a school, or upper floors of a safe business building.[59]

Kentucky was terribly hard hit: more than 300,000 were made homeless; there were many deaths and extensive property loss. Red Cross supplies for the city of Ashland were ordered via amateur radio and dropped to the isolated town from the air. At Frankfort, the state capital, 1500 families had to move to higher ground and 2,000 prisoners in the Frankfort reformatory revolted. It was amateurs working with a broadcast crew that kept the governor in touch with the situation. Carrollton, Kentucky, a town of 2500 midway between Cincinnati and Louisville, was totally isolated. A ham named W.O. Bryant, W9NKD, answered the call for communications, was taken there by boat with a couple of aides, and for ten days offered the only contact between the town and the Red Cross, the army engineers, and the coast guard.

It was said that many of Paducah's population of 40,000 owed their lives to R.O. Moss, W9CHL, who during a 120 hour stint was instrumental in obtaining the men and boats to evacuate thousands of residents from the rising waters. He handled 1552 messages with the help of a crew of 25 who delivered them by boat, car, or however possible. When power failed at his home, he transferred to the Heath High School and spent more than 120 hours more on duty. "Maintaining the closest cooperation with all local agencies, keeping the 160-meter band humming with traffic to and from Paducah, W9HCL performed one of the outstanding individual communications jobs of the emergency," wrote DeSoto.[60]

Likewise suffering were the residents of Indiana river towns. Many amateurs were involved. Two General Electric radio cars were sent from Schenectady with operators and did valuable service. At Evansville, Indiana, a city shaped like a crescent, Lieutenant Ben J. Biederwolf of the NCR operated his own station, N9VAI, and coordinated in excellent fashion the work of the many amateurs in the region. At Fort Wayne, Indiana, special mention was made of the NCR group which also coordinated activities efficiently.[61]

Then the Ohio became the southern boundary of Illinois. There the area known as Little Egypt had become a virtual sea fifty miles across. Isolated islands above the flood waters peppered the scene, but many a village, many a farmer were isolated and dependent upon rescue by boats—and radio. The best known town in Little Egypt is Cairo, described during the catastrophe as "an island in flood waters." With incredible energy its inhabitants, with help, had constructed an earth and timber bulwark above their levee. "Cairo," reported the *St. Louis Post-Dispatch*, "was virtually an island ... as backwaters closed in around the city, cutting off all rail transportation.... Streets are deserted. A heavy silence grips the city. A few men patrol the top of the seawall, keeping watch on the oncoming water. The city is a city of grim, able-bodied men—more than 4,000 of them converted by the emergency into levee defenders."[62]

A communications system covering the region was established, primarily making use of 160-meter 'phone stations located strategically in the region. The national guard established a number of these stations in smaller communities; some of the operators with their rigs had come to help from as far away as Chicago. Cairo was at first without a shortwave station, but two amateurs came in and worked ceaselessly until the coast guard arrived with men and rigs.[63]

It has been said that amateurs from all over the nation were willing to pitch in and help.

Close by but not in the flood zone were hams at St. Louis, up the Mississippi from the Ohio-Mississippi convergence. The St. Louis Red Cross office made use of amateur radio stations as a means of communication between Missouri and Illinois flood areas. "Chains [networks] of experienced amateur radio operators have been established between principal towns, relaying information to field quarters of the Red Cross," it was reported. Amateurs in St. Louis and East St. Louis received messages for the Red Cross, conveying them by telephone, automobile, or bicycle to the Red Cross office; pleas for food, clothing, and shelter also were delivered to the proper agencies. Hams delivered personal messages of families separated by the flood. One ham handled about 400 messages in five days, including a request from Paducah, Kentucky, for serum.[64]

Finally, DeSoto devoted some space to the flood in the Mississippi Valley. Although it threatened extensive damage between Cairo and Memphis, the disaster did not take place. Moreover, the residents along the Father of Waters, used to floods, pretty much took it in stride. This included hams. They served at New Madrid, at Caruthersville and Cape Girardeau. On the Tennessee side of the river, Tiptonville was threatened by flood waters. Contact with a smooth-working Tennessee net was established, and communications with the Red Cross and other agencies was maintained. In Arkansas, when an ice and sleet storm pulled down wires, it was amateurs who maintained for three days links between Little Rock and St. Louis.

By mid–February the crisis was over. Clinton DeSoto concluded that the radio service in the disaster had been the best ever. First, he cited "major progress in preparedness and technique"; second, he found the 'phone stations a spectacular success; and third, he noted "the complete absence of personal traffic"—hams occupying the air with personal interests, especially trying to get into the disaster area. That more work should be done was also clear. Hams should be equipped with emergency gear, still better organization was called for, and the necessity of working through clearing and coordinating centers was apparent. But with all this, the terrible Ohio River flood was a shining hour for ham radio.[65]

It would seem that honors would surely have been bestowed upon someone involved in communications during the great flood. And indeed, the William Paley Award for 1937 did go to a brave, dedicated ham named Robert T. Anderson, W9MWC. He was involved in helping the residents of Shawneetown survive the flood.

Located in a region of small towns in southeastern Illinois, Shawneetown has a historic background. First settled on a flood plain along the Ohio River in 1800, it became important to western settlement when a land office was established there during the War of 1812.

Flooding came with the territory: the little town was inundated every year until 1884, when the residents built levees. Again and again, however, the bastions were neither high enough nor strong enough to withstand the Ohio's flood waters. In the disaster of 1937 the residents, numbering about 1500, felt safe behind their 60 foot levee, but they failed to realize that backed-up flood waters were cutting them off.[66]

Anderson lived in Harrisburg, a city of about 10,000 twenty-three miles due west of Shawneetown. He packed his portable rig and batteries into an open boat and set out for the beleaguered town. He was engulfed in swirling flood waters under blizzard conditions, with the thermometer at 12 degrees above zero. He nearly capsized more than once. Somehow on the way he rescued eleven stranded people at a tiny settlement called Junction. At a railroad crossing six miles from Shawneetown, Anderson had to give up. There, in freezing weather, he set up his rig and made first contacts with other hams, and obtained help, including food and clothing for the little community; eventually he helped in the town's evacuation. He then went on to Shawneetown, installed his equipment on a boat, and for forty hours without stopping kept in contact with relief organizations. Exhausted, he was hospitalized briefly but returned to the boat and continued manning his station. The *Post-Dispatch* wrote that his heroism would long be remem-

Robert T. Anderson won the 1937 William S. Paley Award for services in and around Shawneetown, during the Ohio River flood (from the Collections of the Harrisburg District Library, Harrisburg, Illinois).

bered along the Ohio River. At the presentation of the Paley Award at New York's Waldorf-Astoria Hotel, it was said that "he worked for four days, getting only ten hours sleep, to obtain means
of evacuation for the 1,500 inhabitants of Shawneetown which was threatened with inundation
by the Ohio River."[67]

Indeed, the terrible Ohio River Flood of 1937 may have been amateur radio's finest hour.

The last great tragedy of the decade was the New England hurricane of 1938. Only the beavers
knew it was coming. In Palisades Interstate Park north of New York City they worked all night
cutting more trees, floating them to their dams, slapping more mud where it would do the most
good, and in general reinforcing their watery homes. When the hurricane struck, the beavers'
dams held. They prevented the flooding of three arterial highways and the certain destruction
of one bridge.[68]

Humanity did not sense the impending disaster. The hurricane birthed on September 4th
once again in the eastern Atlantic near the Cape Verde Islands. It took just twelve days to cross
the ocean and then speed up the Atlantic Coast. As it advanced northward it moved faster and
faster, up to sixty miles an hour, which strengthened its force. On September 21 the eye was
above New Haven, Connecticut. Just as luck would have it, the gale made landfall at high tide.
Sustained winds were 91 miles an hour with gusts to 121 with one recorded at 125. Downed power
lines resulted in fires in Connecticut. Communications were down.[69]

Most horrifying were the experiences of recreational home dwellers along the Atlantic
beaches, most especially along the Rhode Island coast. Those who survived told of noticing a

strange color in the sky, and a rising wind. But they thought little of it; autumn had arrived and the climate was making its normal change. Quickly, however, the winds reached hurricane velocity. This was frightening enough, but the scene that had many of them temporarily transfixed was the tide. It did not come all at once as a surfer's wave, but gradually rose until, when it was too late for many of them, they realized waves on top of a surge on top of high tide raised the sea to a height of fourteen to eighteen feet—some reports say to thirty, even fifty feet. And that mass of water was racing onto shores lined with summer homes. Many buildings were reduced to kindling wood in an instant while their occupants fought to stay alive.[70]

The total impact of the storm was 564 dead, 1,700 injured; 2,600 boats destroyed, 3,000 damaged; and 8,900 homes and buildings destroyed, 15,000 damaged. Narragansett Bay was badly hit, as was Falmouth and New Bedford. Downtown Providence was submerged under 20 feet of water just as offices and stores were closing for the day. As with all hurricanes, excessive rainfall, preceded by several excessively rainy days, resulted in heavy flooding. Highways were washed out as well, as were parts of the tracks of the New York, New Haven, and Hartford Railroad. The Connecticut River rose 19.4 feet above flood stage.[71]

As during the Ohio River flood of the previous year, the amateur contribution was substantial. However, as its reporter, Clinton B. DeSoto, wrote, "ham radio generally reeled in its tracks for one ghastly round of twenty-four hours before pulling itself together." No one was prepared for the terrible wind, high tide and surge and flooding that took place (except the beavers). Even so, within twenty-four hours of the disaster hams were on the air, helping throughout New York and New England.[72]

Most counties had ham clubs which as a formality had chosen emergency coordinators (ECs) whose duty it was to accomplish as soon as possible just what their name implies. On Long Island, where 62 persons were known dead and five million dollars' damage done (in 1937 dollars), the EC was a woman, W2JCX (name not given), appointed to the position just six days before. She got her baptism by wind, flood, and tide, and did a reputable job. As the hours passed more and more ham stations throughout New York and New England got on the air to help. (New York City and Brooklyn did not experience a communications emergency.)

At Norwich, Connecticut, the worst-hit inland town, amateurs along with the mayor, the state police and the national guard commandeered from the reluctant owner a 300-watt generator. They worked communications through one station, W1EBO. At Hartford amateur relief work was coordinated through three centers, the state police barracks, Red Cross headquarters, and ARRL headquarters in West Hartford. One amateur had to lash himself to his roof while re-erecting his antenna. So beneficial was his station for the Red Cross that the agency retained it for two weeks to help with its relief work. Messages were often delivered by Boy Scouts. The station maintained by Connecticut State College at Storms was for a time the only outlet into northeastern Connecticut.[73]

At Provincetown, at the tip of Cape Cod, a ham, W1KPW, handled traffic for the entire cape, working continuously for 60 hours, catching 4 hours sleep, and then continuing his watch. At Harwich a ham built his emergency rig, including winding coils, in three hours, and was operational; at Martha's Vineyard a ham, W1JMJ, was the sole communications outlet for 32 hours. Often we find the hams working closely with the police. At Haverhill, for example, a base station was set up at city hall and three mobile rigs reported on events in a 32 mile area.

Rhode Island suffered worst from the tidal surge. Other amateurs pitched in throughout the state as flooding of the Connecticut River and other streams caused havoc in New Hampshire and Vermont. The multiplicity of their tasks is shown by the request at Morristown, Vermont. The United Farmers asked the ham operator to locate routes open to them for milk shipments to Boston—and the milk was delivered via Montreal and Sherbrooke, Quebec.

It was his work at the town of Westerly that earned Wilson E. Burgess, W1BDS, kudos as

Outstanding Amateur of the Year and the Paley Award for 1938.[74] He was manager of the appliances department at the Montgomery Ward store in Westerly. The town, population in 1938 of about 12,000, is located near the mouth of the Pawcatuck River, forming the Connecticut–Rhode Island boundary. It is on Block Island Sound, about fourteen miles east of New London, Connecticut. It was arguably the hardest hit of all Rhode Island communities. When the wind blew out the windows of the store, Burgess realized that a hurricane was on the way. Grabbing B batteries and a storage battery for his rig, he started for home. Downed trees and flying debris endangered his progress. Fortunately, he ran into another ham who offered to help. They commandeered a small truck and made it to Burgess's home, only to find the garage and his antenna gone. Not to be deterred, Burgess set out in a 65-mile-an-hour wind with a coil of wire and a pair of pliers to erect another one. Rounding a corner, the wind whipped the pliers from his hands; they were found the next day with the handle embedded deep into a tree trunk.[75]

At 8:00 p.m. Burgess sent out his first QRR and for the next 56 hours, leaving his transmitter just once for a couple hours' sleep, and aided by three other hams, his rig was Westerly's only contact with the rest of the world; for a time, W1BDS was "the only signal emanating from Rhode Island." Red Cross authorities, Boy Scouts, police, reporters, and tearful survivors invaded his home. Message after message poured out—names of Westerly's dead, calls for boats to save those marooned in their homes, pleas for bread, power, serum, planes, and caskets. For three days "the tiny radio room off the kitchen in the frame house on a hilltop was Westerly's only contact with the outside world.[76]

Burgess's achievement was recognized. He was lauded by United States attorney general Cummings, by Roosevelt's aide Harry Hopkins, and by other notables. The Western Union Telegraph Company acknowledged his contribution and presented certificates of public service to him and the three hams who worked with him.[77]

When the terrible wind and tide and floods were gone, the American Red Cross sent a congratulatory letter to ARRL headquarters in West Hartford. Colin Harris, assistant national Red Cross director, on behalf of the organization, expressed his organization's "sincere appreciation of the splendid cooperation received from the amateur radio operators.... Never before," the letter read, "have we relied so heavily on the amateur radio operators and never before has the work been carried on so effectively."[78]

Thus far the emergencies mentioned have been primarily east of the Mississippi River, but devastating disasters took place over these years in the vast American and Canadian West also. Below is a mere sampling of the many disasters in which hams offered their services. Amateurs did relief work around Kerrville, Texas, in late June 1932, when the Guadalupe River swept down the valley with a forty-five foot wall of water. In 1932 forest service radios for the first time were used exclusively in communications during a massive forest fire in Ventura and Santa Barbara counties, California; all the operators were amateurs. From December 31 to January 1, 1933-1934, floods struck in the vicinity of La Crescenta and Montrose, California (in northern Los Angeles County, at the foot of the San Gabriel Mountains). Thirteen inches of rain had fallen. Thirty-eight people died, 60 were missing, 400 homes were demolished and automobiles and bridges were washed away. A Montrose ham, W6FCE, and his XYL (wife) were almost swept away in the torrent, but made their way to another ham's residence, W6EAH, where everything was in place. They were sending their first QRR by 8:00 a.m. Other hams likewise helped, but none had skirted death so closely.[79]

Earthquakes are another disaster that has prompted hams to serve their fellow citizens. On March 10, 1933, an earthquake measuring 6.2 on the Richter scale struck Long Beach, California. A hundred fifteen people died and property damage was enormous. Ten minutes after the quake amateur Al Martin, W6BYF, was on the air informing the world of what had happened. For a few hours amateurs were the only ones in contact with the rest of the world. They hastily salvaged radio parts from damaged shacks and put together their transmitters and receivers, often with the aid of civic and military authorities. A school boy did some of the best of the early work for the authorities. He lived near the ocean where residents fled, fearing a tidal wave, but he bravely stuck to his transmitter. The Federal Radio Commission (predecessor of the FCC) complimented amateurs for their work, while a certificate of appreciation by the City of Long Beach was presented to Martin for his devotion to duty. Located in the American Legion building that was in the midst of earthquake damage, Martin stayed at his rig for eight days, handling some 3,000 messages.[80]

One more example of amateurs contributing mightily in emergencies, using their incredible ingenuity as technicians and gadgeteers, involves an assistant lighthouse keeper named Henry Jenkins, W7DIZ. His place of work was the Tillamook Rock Light Station in Oregon. It was situated twenty miles south of Cape Disappointment at the entrance to the Columbia River, a mile off Tillamook Head. In the early morning of October 21, 1934, a gale with a wind velocity of 109 miles an hour struck the lighthouse. The entire station was submerged during the gusts, 16 panes of lantern glass were smashed and the light went out; at almost the same time the telephone went down. Six feet of the west end of the rock upon which the lighthouse was located broke off; and rocks weighing up to 50 pounds tore through the windows and the roof. This was a serious situation. Ships at sea depended upon that light.

Henry Jenkins was a ham possessing great amateur ingenuity. There was not a transmitter at the facility so he set about creating one. He cannibalized an Atwater-Kent receiver for some parts and used waxed paper from bread loaves, tinfoil, parts of a telephone, and brass doorknob plates to create a workable rig. He made a hand key with a piece of spring brass. It was crude, but it did the job. With it he contacted hams who notified the coast guard of the problem; radio signals were then sent to all shipping warning that the Tillamook lighthouse was out of commission.[81]

Lesser crises were reported in almost every issue of *QST*. Every one was an unforgettable event to those involved. Many were of local importance but received little publicity outside of the area. It is difficult to mention any of these incidents as typical, for like human beings, every incident is different. But whether involved in major disasters or brief, local emergencies, amateurs' activities constitute "Exhibit A" when defenders go before congressional committees or the Federal Communications Commission, or its predecessors. It is not hyperbole they present: amateurs really did good work, and still do, in trying times.

# 14

# Amateurs on the Home Front: 1940–1945

Amateur radio entered with confidence and optimism the twenty-three months from January 1940 until America's entrance into the war on December 7, 1941. In 1940, with the war under way since the previous September, DX-ers were out of luck, but the number of Field Day participants continued to grow. It was noted that as of June 30, 1939, there were 53,558 licensed amateurs in the United States. In May 1940, QST mentioned with pride that with the European war eight months old, hams were remaining neutral. This meant that the league's voluntary neutrality rules were being honored. In this same period operations in about sixty per cent of the membership of the IARU had been banned. When, in the spring of 1940, war activities suddenly expanded, the FCC issued Order Number 72: amateur radio operators and amateur stations were forbidden to "exchange communications with operators or radio stations of any foreign government or located in any foreign country." The order did, however, allow continued communications between hams within the United States.

In May of 1940 a YL, Ethyl Smith, W7FWB, of Wenatchee, Washington, noticed a lace-bordered advertisement in QST for DeSoto's Two Hundred Meters and Down. It started her to thinking: just how many "YL key twitchers" were there? She wrote a letter to QST: she would like to know, she wrote, "how many there are, how old they are, how they got interested, whether they're key twitchers or tonsil busters, how long they've had their tickets.... Perhaps we should band ourselves together in a YLRL [Young Ladies Radio League] ... and make these woman-ignoring authors sit up and take notice."[1]

Ms. Smith received enough replies from YLs in hamdom to take steps, with other interested YLs, to form the YLRL. In October 1939, they drew up a constitution, elected a president (Ms. Smith), a vice president, and a treasurer. In November they issued their first monthly news sheet. OMs suggested it be named "Parasitic Oscillations" or "Spurious Radiations," but the girls settled upon "YL Harmonics." In a short time they had 71 members from 30 states, Alaska, Canada, Puerto Rico, and Hawaii. Membership fees were low, but the form to fill out asked so many questions one YL asked if they wanted her fingerprints.[2]

The Young Ladies ranged in age from 13 to 73 with their average age 30; the first president, vice president, and treasurer were still in their 20s. Thirty-nine percent were unmarried, attending college or teaching; two were actresses. Three-fourths used CW and most of their time on the air was spent rag chewing. They had even established their own net, on 40 and 80 meters.[3]

As war clouds gathered amateurs reflected upon their situation should war break out. It was noted that manufacturers were offering the "most interesting line of apparatus and parts that has existed in the history of radio." As for the government, it was gathering amateurs' biographies and fingerprints as an obvious necessity, but more than that, officials had indicated an awareness of the usefulness of amateurs in time of war. "They realize," it was observed, "that amateur radio is the most remarkable kind of training school, one in which ardent devotees train themselves in the intricacies of a very complex art at their own expense!"[4]

Conscription had come. That familiar letter beginning with "Greetings" was arriving at the homes of millions of eligible men. Hams were reminded that the armed forces wanted radiomen. QST wisely suggested that when arriving at the induction center, hams should have their licenses with them, "*because right there is the place for you to announce your radio qualifications and press hard for an assignment to radio work*" (italics QST's). Amateurs were urged to increase their code proficiency; to enlist in the air corps, navy, or marines, if they so wished, but by all means to let the authorities know that they were radiomen. And for those too old or ineligible for physical reasons for the service, the home front would continue to need them. Natural disasters never take time out for war. The national guard, which operated amateur stations, was being called to active duty, leaving a few amateurs at home who would have to "start from scratch." Finally, with over a million men away, radiomen still at home should arrange relays so that lonely soldiers could contact their loved ones. In short, there was plenty of work for every amateur, male or female, civilian, soldier, sailor, marine and merchant marine, when war came.[5]

Yet, in spite of a war in Europe and the feeling of inevitability that the United States would soon be a combatant, much went on as usual. Captain Irving Johnson's Dutch pilot boat, the *Yankee*, put to sea October 27, 1939, on its third round-the-world voyage, even though war had already broken out in Europe. Aboard, excited about working DX, was amateur Oakes A. Spalding, W1FTR. DX-ing was legal until Order No. 72 was issued on June 5, 1940; the only requirement until then was that the radioman practice strict neutrality. As for Spalding, his trip included perhaps more social meetings with amateurs than had the previous *Yankee* ham, Alan Eurich.[6]

Even as war clouds gathered, Antarctic explorations continued. In 1940, looking back just twenty-eight years to Sir Douglas Mawson's expedition of 1911–1914, and the primitive radio facilities he had at hand, we find ourselves in awe at the enormous progress that had been made. In June 1939, the U.S. Antarctic Service was created by act of Congress and another expedition to Antarctica planned with Admiral Byrd in charge. This was a government operation although both congressional appropriations and private contributions funded the project. The incentive was to claim land previously explored by Lincoln Ellsworth and Byrd but which was being contested by other nations.[7]

Radio communications loomed large in expedition plans. The circuits were all naval with official communications handled according to naval practice. Then there was broadcasting. This time the policy was non-commercial—that is, the broadcasts, picked up by RCA, were to be free to any and all broadcasters. Finally there was the third category—amateur radio. These circuits were to carry personal messages back and forth and, it was hoped, during the long winter night, be able to send 20 and 40 meter 'phone messages to stations in members' home towns. "If there is a member of the ice party from your city," DeSoto wrote, "you will find yourself on the preferred list when answering KS4US—CQ's!"[8]

Following the expedition's return to the States, Clay W. Bailey, communications officer for the U.S. Antarctic Service (and member of the expedition) congratulated the radiomen for their good services. He regretted that CW (code) was held to a minimum because "everyone worked 'phone." Amateur communications, he stressed, were of inestimable value "in permitting most of the men on the ice to have periodic direct 'phone contacts with their families and friends, and providing a channel for unlimited personal messages." The rapid progress in radio communications was illustrated by the fact that in 1934, from Antarctica, Boston could not be worked on 20 meters; in 1940–41, 'twas no trouble at all."[9]

Proof that amateur radio had advanced to a state of reliability that boded well for the future, be it a future of peace or war, was ably demonstrated in 1941. The ARRL teamed up with the Red Cross to handle emergencies during the presidential inaugural, January 20, 1941. Amateurs provided communications for ten first-aid stations, which were also kept in touch with the master control station at the Red Cross chapter building. The results were highly satisfactory. Two months later a nationwide *Red Cross–ARRL Facilities Test* took place. Again, the results were satisfactory in every way.[10]

In the years just prior to the outbreak of the war, ham activities during disasters were marked by improved preparations, better coordination, and among hams, a more sophisticated understanding of what should be done in the handling of communications during disasters. It all boded well for the future. In fact, American amateurs had never been so well organized, their equipment had never been so trustworthy, their dedication had never been greater than in the two or three years leading down to December 7, 1941.[11]

And the war came: December 7, 1941. Never before, and probably never again, will this nation be so united in a war effort. Time diminishes events. Of massive concern to the citizenry during those years 1941–1945, today, two generations later, the war is a fact of history diminished by more recent events: Korea, Vietnam, Desert Storm, 9/11 and the second war against Iraq. Information conveyed 24/7 via television and radio emphasizes the events of today and reduces even more the events of yesterday. It is too bad. To study World War II, to realize the enormity of the effort that went into this massive two-ocean war, is to create a sense of awe, pride in America, and great respect for the accomplishment.

It used to be said that food and munitions were the most important elements in winning a war, but in World War II there was a third element: communications. Never before had the ability to contact, coordinate, and plan operations been so dependent upon information. This was especially true in naval, army, and marine operations in the South and Central Pacific, where troops were on many islands and naval contingents were spread out at sea. The necessity of keeping in touch goes without saying; important too was radio intelligence, listening to the enemy's messages and breaking their code.

Moreover, radio had new uses besides communications. "In many ways," it was noted, "this is a war of radio gadgets." Hams were urged to continue experimenting, raid their junk boxes, and come up with new appliances. A case in point was something in 1942 called "radio locating" but soon to be announced as radar: radio detection and ranging.[12]

Radio locating was being put in use by the United States Army and Navy as rapidly as possible. It is questionable, however, whether in December 1941, officers in command had accepted it as a reliable contrivance. It was said to be capable of detecting objects, determining their position, and even identifying the nature of the blob (or blobs) on the screen. One site was on the Hawaiian Island of Oahu near Kahuku Point on the northern tip of the island, 250 feet above sea level. The importance the military gave the station may be surmised from the status of those assigned to man it: two privates. This even though the site could furnish warnings of attack on Pearl Harbor, which lay to the south on the island of Oahu.

The two privates, Joseph Larue Lockard and George E. Elliott, went on duty on December 7th at 0400. At 0700 Private Lockard began shutting down the unit because that hour marked the end of the morning's work. But suddenly the oscilloscope revealed something bigger than the two privates had ever seen. Lockard took over from Elliott, believing that something was wrong with the rig, but it proved to be in good working order. What was it? They thought it might be planes from a navy carrier, or possibly B-17s due in Hawaii at about that time. No matter: their message of what the radar indicated was slow in getting through and was discounted even as the two privates followed the blips until they disappeared in the "dead space" about twenty miles out. What they had watched on radar was the first wave of Japanese planes headed for Pearl Harbor. But the information did not get to men in command, and even if it had, from two privates operating a device of questionable reliability, would it have been accepted? Probably not. Chalk it up to a peacetime mentality, a feeling that the Japanese were inferior, that they could never bomb Pearl Harbor, that the system for conveying information from radar to command was not perfected, that the officers, both army and navy, did not yet have full confidence in the system. The fact remains, however, that radar—radio detection and ranging—had revealed the arrival of the enemy's aircraft.[13]

Lockard, who was subsequently promoted to staff sergeant, was mentioned favorably in the Roberts Report, the first official description of what happened at Pearl Harbor. It was said that he had remained on duty on his own time with a desire to improve his skill with the radio locator, and by doing so had inadvertently detected the approach of the Japanese planes. For this achievement the Veteran Wireless Operators Association at its 17th annual dinner in New York City presented Lockard, in absentia, with a Marconi Memorial Scroll of Honor; the date was February 21, 1942. This is not quite the description of the event as recorded in the two sources mentioned in footnote 26—Private Elliott appears to have been the one urging Lockard to remain at the radar and telephone the information, and Lockard appears to have agreed to do so reluctantly—and if he had telephoned "at least fifty blobs" instead of "a big blob," things might have worked out differently. But probably not.[14]

Two Honolulu hams, Frank Firth, K6QUD, and his YL, Helen, K6TCW, were enjoying breakfast when the attack came. They immediately warmed up their rigs and began doing duty for the army. Helen was employed with the 9th Signal Corps in the message center in Oahu, including weekends, Christmas, and the New Year. QRV was her motto: I AM READY.[15]

As we have already noted, amateurs were being prepared for hostilities ever since the outbreak of the European war in 1939. They had been urged to improve the accuracy and speed of their sending and receiving, master the intricacies of their rigs so that they could repair them, join the league's emergency corps, get affiliated with nets, and even sign up for service in the army—especially the signal corps—the navy, marines, merchant marine, and coast guard. After America entered the war QST printed a page titled "Radio Apparatus for War Use," in which the amateur, besides registering his or her name, address, and call letters, listed the apparatus he or she was willing to sell and at what price. Many amateurs lent their rigs to the services; most, as in Forrest Bartlett's case, never saw them again. They chalked up the loss to patriotism.[16]

In October 1942, QST printed a form titled "Registration of Personnel Availability." Besides the usual information, the amateur was asked his or her education, radio experience, and licenses held. This form was not intended to steer hams into the armed services, but to link them to civilian tasks in which hams were badly needed. These included code instructing, design and development, engineering testing, production superintending, research, theory instructing, and more. The amateur was also asked his of her preference for location and the "necessary salary."[17]

The truth was, at no time in their short history had amateur radio operators been in such demand. The armed services needed them, industry needed them, the merchant marine needed them, and, because natural disasters and emergencies do not honor war, the civilian population needed them. Officers were always glad to have an amateur among the radiomen in their units. Operators who had been force fed training rarely had the amateur's abilities. After all, the ham had mastered code on his own. He was likely to be faster and more accurate with it than a trainee. The amateur was probably technologically minded and had worked on the insides of transmitters and receivers; quite likely he had built his own rig. When radios went on the fritz in the jungles of Burma, on a destroyer-escort, or in the desert of North Africa, the call went out for "sparks." If he was known to be an amateur, he was the one to repair the rig. He rarely failed.

In many ways amateurs really were preparing for war before Pearl Harbor. They had registered with the ARRL. Membership in ham clubs meant that group discussions, which resulted in group action or plans for group action, were in place throughout the country. If amateurs were involved with a net—and increasingly nets had been created throughout the nation—then they were discussing the need for preparedness should war come. This involved plans for portable rigs installed in cars, batteries or gasoline motors to run generators and furnish power when electric power was out, and choice of coordinators operating at a central location considered safe during catastrophes.

Before war broke out there were already more than 400 USO (United Service Organization) units throughout the United States. The National Catholic Community Service group took steps to establish amateur stations at them. The first USO amateur radio station was established at New London, Connecticut, to much fanfare. These stations were to work in coordination with ARRL nets and the AARS (Army Amateur Radio System—to give way after the war to MARS—Military Amateur Radio System). Such a message service for military and naval personnel was great for morale. Some enlisted men sent messages to relatives and received replies within twenty minutes. When war came the stations were closed.[18]

In the early months following December 7th amateurs were encouraged to announce their abilities and go into employment making use of them. Instructors in radio work, including the teaching of the code, were at a premium. Often classes were organized in communities, using public schools or libraries for classrooms. So too were women encouraged to volunteer for radio work. One out of every five production workers in radio was a woman. As Clinton DeSoto reported, "Girls wind the coils, rivet the subassemblies, assemble the frail filaments and fragile grids, lay out the wiring harnesses and solder the connections to the radio equipment used by the fighting men." DeSoto urged the two thousand licensed young lady hams to make use of their skills to further the war effort. In New York City the American Women's Volunteer Service in February 1942 turned out its first class of seventeen young lady operators, who marched down to the Federal Building and passed their exams for Class B amateur tickets. They would be used as trainers or as operators at land-based stations, thus relieving male hams for sea duty or duty with the army or marines.[19]

Censorship was slapped on amateurs just as soon as the attack on Pearl Harbor took place. All transmitters were ordered silenced. However, this was considered a temporary injunction. The government realized that amateurs were of extreme importance during emergencies and disasters at home. The problem was, how to silence amateur transmitters but at the same time have some that, under certain circumstances, could be operational and contribute to relief during natural calamities. Surprisingly, several thousand amateurs stations were actually reactivated shortly after December 7 for war emergency communications, but on January 9, 1942, at the request of the Defense Communications Board, their activities were again terminated.[20]

The situation with regard to policies and regulations was understandably chaotic when war came, and it took precious time for the FCC, the FBI, the Defense Communications Board (DCB), and the Office of Civilian Defense (OCD), to say nothing of army and navy concerns, to cooperate, coordinate, and come up with a plan. An Amateur Radio Committee had been working with the DCB for well over a year prior to Pearl Harbor. The league had even chosen six widely-known amateurs, geographically spaced, as regional advisers to the DCB, but their work was not to begin until January of 1942; the war came before that.[21]

It is difficult to know whether these preliminaries helped along the steps toward a workable policy for amateurs. In its February 1942 issue, QST indicated that an official plan would soon emerge providing for the employment of uhf (ultra high frequency—10 meters or less, 28–30 megacycles). This was the "line-of-sight" communication, free from the static and fading of the lower frequency bands.[22]

Nothing had yet been worked out, but the league urged all amateurs to register with ARRL emergency coordinators in their communities; the coordinators had been asked, in turn, to act as liaisons between amateurs and the local civilian defense commander. Further details of the predicted regulations were also given.[23]

War or no war, it still took time. Not until early June of 1942 did the FCC announce new regulations; since January all work with amateurs had been at a standstill. Known as Part 15 of the FCC rules (peace time rules constituted Part 12), the complex regulations were published as a booklet. To be known as the War Emergency Radio Service, or WERS, domestic amateur radio was to consist of two kinds of stations: the civilian defense station, which would be of great interest to amateurs; and state-guard stations, which were of concern to members of the state guard. (The latter are presumed to be local national guard units controlled, it was anticipated, by the army.) In both cases the stations were to exist solely for emergencies. They were to operate on megacycles 112–116, 224–230, and 400–401 (2½ and 1¼ meters). The 5 meter band, expected to be used, was not mentioned.[24]

Nothing like WERS had existed before. It did not contemplate the operation of amateur stations as such. The system was to work as follows: a license was to be granted to a municipality or one of its agencies to be responsible for all stations engaged in emergency civilian-defense communications including fixed, mobile, and portable rigs. Individual transmitters would be given unit numbers with their call a four letter one beginning with the letter K or W followed by the unit number. The municipality held the sole WERS license.

Then bureaucracy and paranoia set it. The "instrumentality of local government" had to apply for the license, submitting a detailed plan of its operations along with a map, its intentions for carrying out its charge and proposals for monitoring, supervising, and recruiting personnel and apparatus. All of this was to be directed and supervised by the radio aide of the communications officer under the commander of the local citizens' defense corps.

For each licensed operator the amateur—or it could be a commercial operator—had to be applied for by the WERS operation in the community and the application had to include two passport photos; the permit, when issued, was to contain one of those pictures. This was in addition to certification of loyalty and proof of ability as demonstrated by the holding of an FCC license.

It was pointed out that technicalities concerning transmitting were going to be "annoyingly difficult." Yet possibly the greatest weakness of the system was this:

> The confinement of the authorized operations to emergencies attendant upon enemy activity makes it impossible for WERS to deal with those precipitated by natural disasters. It is to be deplored that the authority was not extended to cover this other essential field of civil relief.

Testing and adjusting of the stations was to be nationwide, twice a week, two hours at a time, on Wednesdays and Sundays; after November it was felt that the system was well enough under control to reduce practice periods to just two hours on Sunday afternoons. Remember that individuals were not granted licenses; the licensee was the municipality or the unit within it assigned to run WERS. A key person in the whole setup was the radio aide of the CDC (Citizens' Defense Corps) communications office; he was "the master oscillator of the hook-up." Reading between the lines, QST questioned that person's abilities when the CDC itself was not completely assembled. Moreover, if the ultimate aim was a WERS system covering a fifty square mile area, the radio tasks were clearly too complex for just one man to handle.

Amateurs were ambivalent towards WERS. After waiting nearly six months, the government had finally come up with a civilian defense emergency communications system. But it was not an amateurs' system; it did not involve the use of amateur stations as such, nor was it confined to amateurs. Yet who was to man WERS except amateurs who were too young, too old, or physically rejected from the armed services? True, they could not operate their home stations with their own calls, and their uhf bands had been taken from them, but there was a war on. So *QST* editorialized for hams to offer their services as operators, lend their uhf apparatus, train more operators, build more apparatus, and help select radio aides. In addition, it was a certainty that they "would have to build more stuff—from the parts we have on hand or can still buy, or from old BCL sets, or from tomato cans if necessary." Very important for the CDC coordinator was the choice of the radio aides. They could make all the difference between a smooth-working WERS setup and a failing one. "Your talents and your gear are both needed," *QST* editorialized.[25]

To add to amateurs' frustration, on June 19, 1942, the FCC ordered all hams to file a description of their rig, receive a certificate, and "conspicuously affix" it to their transmitter. This was to be accomplished by August 25th.[26]

So, within six months after Pearl Harbor, a system of radio relief in emergencies was finally worked out. As with the hero of the short story "A Message to Garcia," the hams did not officially complain about the complexities built into the FCC's Part 15. The league pitched in with enthusiasm, doing all in its power to bring about the quickest possible creation of WERS throughout the nation, and especially along both coasts where the possibilities of a bombing or shelling were most likely to occur. "We expect the ARRL, Emergency Corps, led by its Coordinators, and the affiliated radio clubs to become the nuclei of the War Emergency Radio Service organization," wrote the acting communications manager of the ARRL, John Huntoon. He then mapped out a model plan for hams to use in creating a WERS in their community.[27]

He began by recommending that hams get in touch with the local civilian defense coordinator, inform him of the creation by the government of WERS, and then, after the hams had met and discussed the situation, suggested submitting a name or names of the best radio aide for the contemplated WERS. (In a bulletin the OCD had recommended such a move.) He advised hams to take inventory of their 2½ meter rigs. The WERS central location, it was suggested, should be the same as the center for the air raid warning system the army had set up throughout the nation. The base once decided upon, next the amateurs for the warning district should all meet there and make plans for applying for licenses, learn of the administrative setup, and master an understanding of their duties. Huntoon then gave diagrams for the setup of a WERS unit.[28]

Amateurs had not waited for the OCD to issue its WERS plan. In Providence, Rhode Island, foreseeing the needs involved in ARP (Air Raid Precaution) services, the police radio engineer had obtained permission to use the police department radio system for civilian defense communication. A mobile radio patrol was planned using uhf equipment manned by radio amateurs in emergencies. When in January 1942 all amateur transmitters were banned, the Providence police department obtained from the FCC a permit for police-experimental communications. It was given four designated frequencies, all close to 2½ meters. The city installed equipment purchased from amateurs and swore in about fifty hams to man the rigs at designated locations. The system was operational before WERS went into effect.[29]

The State of Massachusetts was likewise ahead of the OCD and the FCC in making plans for civilian defense communications. Many towns had maintained disaster nets for years. The Massachusetts Committee on Public Safety had a radio communications department, and during the brief period when shortwave transmitters were allowed after December 7, many towns began discussions toward getting their nets operational. Then came the ban on transmitters. "Seemingly, the greatest accomplishment of these meetings was the consumption of dozens of doughnuts and gallons of coffee," wrote John Doremus, head of the radio communications department. "Yet these informal rag chews did bring about an exchange of ideas that crystallized into a 'six-point' plan that was communicated to Dean James M. Landis [head of the OCD] near the end of March."[30]

They received "a pleasant reply" informing them that plans very similar to theirs were underway and should be issued shortly. "This," wrote Doremus, "was the assurance that we needed to start the ball rolling again!" Organizational meetings were held at cities and towns throughout the state in which the role of radio in civilian defense was discussed. It was realized that a radio communications network with "exacting control" was needed, to be operated "by personnel with unquestionable loyalty to our country." By the time the government had released the WERS program the state had already been divided into nine regions, each containing one or more official warning districts. Central headquarters, district stations, and mobile stations were planned. At the time the article was published, the state had applied for WERS permits but had not yet received the necessary licenses.[31]

From the beginning of their plans, it was clear that competent operators would be in short supply. Not to be deterred, Massachusetts set up a statewide radio instruction system offering three tracks: to train operators for a restricted radio-telephone permit; for amateur licenses, and for second-class commercial radio-telephone licenses. Before the end of the summer of 1942 the state had 32 schools operating, with over 2,000 students. Over 85 percent of the students were enrolled in the amateur course and most instructors were amateurs in their own right.[32]

Such activity was not restricted to the State of Massachusetts. The St. Paul, Minnesota, Radio Club, overcoming inertia and taking note of the simple statement of "something should be done," finally got to work on the problem. It began conducting code classes nine times each week. The sessions, operating on a "graduated progressive system," brought in about 325 Twin City residents; the club also had a fifteen minute radio program "on a leading broadcast station" teaching code "in a serial story" over the air. The teachers were all members of the radio club, patiently teaching the sounds of letters in code, and bringing the students along, five letters at a time. Such energy and action was not uncommon among the hundreds of ham clubs throughout the nation.[33]

The first WERS license went to Akron, Ohio. Again hams were ahead of the government in plans for disaster aid, be it bombs or floods. Members of the Buckeye Short-Wave Radio Club, led by John A. Bailey, the club president and ARRL emergency coordinator, began assembling rigs with 2½ meter bands, prepared a plan providing for two-way communications with sixteen 2½ meter transmitters of both fixed and portable type, and noted that about 40 operators were available. In January 1942, prior to WERS, the local executive council of the Summit County Civilian Defense Corps accepted the proposal, whereupon it was submitted to the FCC. The request specified the use of 12 licensed operators operating on the 2½ meter band. All operations were to be restricted to Summit County and the City of Akron. Much correspondence ensued, and it was not until WERS had been announced that Akron received its license—the first WERS license issued.[34]

Some amateur clubs were so well prepared that quick submission according to the rules set down by the OCD and FCC were all that was needed to receive a license. The Queen City Emergency Net of Greater Cincinnati had been in existence since the terrible Ohio River flood of 1937. It quickly joined WERS, the local civil defense organization, the auxiliary communications group of the Fifth Corps Area Division of Fort Haynes, Columbus, Ohio, and the Federal Bureau of Investigation; in 1943 it also affiliated with the Red Cross.[35]

As the months went by the league emphasized that the FCC and the OCD expected WERS to be handled primarily by amateurs. "OCD has tossed the ball to ARRL on the whole job of lining up amateurs and their gear for WERS, doing the explaining, [and] helping in the appointment of good radio aides," editorialized QST. To furnish competent operators it ran articles such as "Training Civilians for Wartime Operating" and "Training Auxiliary Operators for WERS"; to help assemble rigs it ran articles such as "A Transceiver for WERS," and "Building WERS Gear from Salvaged B.C. Sets." In a two year period QST published more than 100 articles involving WERS.[36]

By January 1943, WERS licenses had been issued to 53 communities and the number of applications—and acceptances—was growing. Of 22 states with WERS licensees, 17 were coastal and the rest were inland. Of 26 states with licenses pending, 6 were in coastal areas. And of 19 inland states without WERS, 12 had applications on file; just 7 had made no attempt to fire up a WERS system. "Some of the problems have been enough to make the amateur out in the field wonder whether the regulations weren't deliberately meant to be so discouraging that there would be no WERS and hence no security worries," commented QST. Not only was there a short-age of qualified operators and equipment, but on top of that often petty politics were involved. Officials of the Air Raid Precaution Service (ARP) were jealous of power granted to WERS. The paperwork involved was tremendous, with applications being bounced back to the applicants again and again. In spite of all this, by August of 1943 there were "in the vicinity" of 200 licensees. WERS was definitely working.[37]

By November 1943, the number had increased to 223 licensed CD (civilian defense) WERS, 10 SG (state guard) WERS, and 6 CAP (civil air patrol) WERS. By March 1944, the number had increased to 261 CD-WERS, 14 CAP-WERS, and 11 SG-WERS. (The state guards were succes-sors to national guard units, all of which had been called up to active duty. The SGs took over national guard armories, blew the dust off the shortwave rigs and with a WERS license from the FCC and the OCD, became operational; they were under the control of the army, with the bless-ing of state governments.)[38]

By May 1945, when the war ended in Europe and within months would be over in the Pacific, WERS numbered 273 CD-WERS. (The number of CAP-WERS and SG-WERS is not given.) Thirty-six states and the District of Columbia had WERS protection while twelve states were as yet unlicensed; all of them but South Carolina and Delaware were inland, most of them in the far Middle West or Rocky Mountain region. All cities with a population of a half million or more, with just one exception, were protected with a coverage of 92.9 percent by WERS licensees; 60.0 per cent of cities of a quarter million or more, and 60 per cent of cities with a hundred thousand or more.[39]

The next question to be asked is, how well was the system working? The fear of bombings, shellings, or landings by the enemy in this war, while justified, proved to be groundless, but emer-gencies caused by natural disasters continued. In December 1942, the Ohio River suffered one of its worst floods in history. Because of the war effort, much of the suffering, of the damage to farms and cities and towns along the way, was not publicized. Amateurs who had participated in communication emergency work during floods were well aware of their contributions, which had saved people's lives.

In this flood, however, while civilian defense boasted of mobilizing a hundred thousand trained relief workers, and although WERS was a part of civilian defense, amateurs were not allowed to participate "at what additional cost in lives and misery we will never know," wrote K.B. Warner. "And why not?" he asked, and answered his question: "Because the confounded paperwork of making out a WERS application takes two men and a nipper several weeks of spare time to accomplish, and then it bounces back a couple of times before it is acceptable form." So civilian defense workers went without communication. Certainly, Warner suggested, the author-ities should allow the temporary assumption of amateur operations for such a restricted area, exclusively for relief. The amateurs' contributions to such emergencies had been aptly demon-strated over and over again. Even WERS amateurs using 2½ meters could have helped. But lit-tle could be done, he predicted; amateurs would just have to wish for a more realistic viewpoint in Washington.[40]

A few months later—from May 18th to 24th—the Mississippi River and some tributaries went on a rampage. It struck the central Mississippi Valley, including parts of Indiana, Illinois, Missouri, Arkansas, Oklahoma, and even northeastern Texas. Heavy rains swelled the river and many of its tributaries, breaking records going back ninety-nine years. Thirty-four people were killed and 150,000 were made homeless. By this time more WERS stations had received licenses, and radio amateurs, working as WERS operators, received much of the credit for preventing what could have been a far more disastrous situation.[41]

The WERS licensee at Granite City, Illinois, covered the surrounding area including the towns of Venice, Nameoki, Mitchell, and Madison. The first distress call came from Madison, where flood waters had backed into storm pipes and the pumping station was threatened. The Office of Civil Defense contacted WERS director Jim H. Adamson. He in turn alerted eight radio operators and placed three portable transceivers in operation. These were ordered to the emergency area, from whence they contacted the OCD offices in Madison's city hall, telephone communication being lost. From there the U.S. Army Engineer Depot was notified of conditions; it in turn sent crews of fifty men, working in four hour shifts, to man the levees and secure them.

Three WERS portables were established at strategic places: where the "Chain of Rocks" levee joins the Chouteau Slough levee; on the highway leading to the Chain of Rocks Bridge; and one near the bridge. Even with sandbags piled four feet high above the levee, it was clear by 9:00 p.m. that the levees could not withstand the pressure much longer. Warnings were given to the men and their trucks to move out. But the WERS operators stuck to their posts and when the levees broke, the six hams with their three portables were temporarily stuck on the bridge. Still other portables were assigned spots where work crews were servicing the levees. The hams sent messages to the net control station at Granite City, which in turn notified authorities of where workers and equipment were needed. At another levee work continued for many hours, with the WERS operators remaining on duty up to 48 hours without relief. Afterwards the chairman of the local civilian defense council of Granite City said that "if it were not for our radio communications, I doubt very much if we would have been able to win out in our battle to save the levees in this area."

Peoria, Illinois, fought the flood threat from the Illinois River. Whether or not it was licensed by WERS, the Peoria Amateur Radio Association furnished two of the club's generators and two privately owned ones to fill in until the Caterpillar Company could set up a more reliable source of power. The Red Cross commended the amateurs, saying that "their prompt service ... helped avert what might have been the most serious disaster to befall the community." At Anderson, Indiana, the OCD-WERS system was activated with a control center and six mobile installations in place; it handled more than 220 messages, making possible the rapid shifting of men and equipment to save the levees. At Fort Wayne, Indiana, WERS worked well, operators remaining at their stations through the night, sending 219 messages.

WERS also functioned satisfactorily when an 800 foot section of a dike on Lake Erie's southern shore, about fifteen miles from Toledo, broke and flooded a large area. With telephone communications out, WERS operators filled in and maintained communications.[42]

On a few occasions WERS made contributions more closely affiliated with the war. On February 5, 1944, an accidental discharge of a rapid-fire anti-aircraft gun on a United Nations freighter anchored in New York harbor brought 250 air-raid wardens searching for 56 unexploded shells said to have landed on Staten Island. Seven WERS portable-mobile units were assembled because of the lack of good telephone communication; headquarters was in the basement of an apartment house that was a sector of the air-warden service. Each time a shell was discovered, WERS conveyed the news to operations headquarters and army personnel were sent to remove the unexploded missile. Previously, the local WERS had carried on official drills, but when a real crisis arose, the system worked well.[43]

Possibly WERS's greatest challenge took place in September 1944, when a hurricane raced up the Atlantic coast. The fear was of an equivalent to the devastating 1938 hurricane, so WERS licensees were alerted from the Carolinas north. "Amateurs and other operators in the War Emergency Radio Service responded nobly to the call," reported Carol K. Witte, acting communications manager. Fortunately, although the big wind did harm to the New Jersey, Long Island, and Connecticut coasts, the hurricane veered eastward so there was not a repeat of the 1938 tragedy. The Roselle, New Jersey, WERS established two base units and had mobile units reporting fallen trees, hot wires, and other dangers. In Connecticut all WERS units were on active duty. Several dozen mobile units were in operation throughout the state, aiding police and firemen, warning of downed trees and hot wires, and even helping in the evacuation of 65 adults and 58 children from seaside homes.[44]

How well did WERS work? It was a compromise between the free amateurs who could pitch in when an emergency struck and a situation where all transmitters were sealed. It was tied up in the red tape of a nation at war. The need was recognized, but the fear of sabotage, of spies, and of fifth columnists using amateur radio to inform the enemy, prompted the government to create a tightly controlled system that forbade all amateurs save those fingerprinted and accepted by WERS to participate, and then only for practice drills and emergencies—and even the emergencies were at first specified. By mid–1943, however, the system was coalescing.

Then the war ended. FCC order No. 127, November 15, 1945, rescinded all WERS licenses and regulations. WERS was history. The league then announced the reconstitution of its emergency corps, "instituted before the war and kept alive through the War to help WERS." Every one of the nearly 300 WERS-licensed cities, it stressed, should continue to have provisions for emergencies; it should be a functioning organization. This was the aim of the ARRL Emergency Corps. It is still operational.[45]

# 15

# Amateurs at War and Beyond: 1941–1950

By the end of 1943 it was estimated by QST that there were at least 25,000 amateurs in the armed forces. Before Pearl Harbor Forrest Bartlett had been approached by the navy because of his knowledge and previous experience copying the Japanese Kana Morse code. He volunteered and passed the physical but was rejected because of colorblindness. After the U.S. entry into the war, he tried again, but no waivers for colorblindness were being granted. This seemed shortsighted on the navy's part since operators with Kana copying ability were in extremely short supply, and what hindrance is colorblindness to copying code? With this door closed he volunteered to join a nine member team of specialists being organized by Press Wireless for assignment to General MacArthur's headquarters in New Guinea. This group's purpose was to provide a high speed radio link to the United States, separate from the military, to serve the press corps covering the anticipated campaign to retake the Philippines.[1]

Twenty-five thousand amateurs in the armed forces! These were licensed hams. They were terribly important for the war effort, but they were a minuscule force compared to the number of radiomen needed in this all-out, two-ocean war. One set of statistics places the number of technicians trained by the armed forces at 500,000. The signal corps trained 145,000 in operations, maintenance, and repair; the army air corps 143,000; navy radio schools graduated 125,000; the marines 10,000; the coast guard 65,000; and finally the maritime service, 5,000. Not all mastered all aspects of radio—code and 'phone and the theory and technicalities involved in radio—but a surprisingly large number did. (Some were trained as electricians, not radiomen.)[2]

A favorite story among men and women in advanced training during the Second World War concerned a young soldier who dropped his pencil during a lecture. When he retrieved it he whispered to his classmate, "Did I miss anything?" To which came the classmate's whispered reply, "Only a year of college algebra." Indeed, if there was one aspect that these educational programs all had in common, it was speed. Every one, with the possible exception of the coast guard, was planned to train army, air corps, navy, marine, or maritime personnel, male or female, in the intricacies of radio as fast as was humanly possible. To a remarkable degree, they all succeeded, in spite of major differences in training techniques.

Even before the war broke out for America, classes were well underway, especially in the area of electronics including radio. There was, for example, the ESMWT program: Energy, Science, and Management War Training. More than 200 colleges and universities participated, offering more than 1,000 courses. The army signal corps inaugurated a massive civilian training program that was well under way prior to December 7, 1941, offering courses in designated fields.

Using as a basis of authority powers granted by Congress during the First World War to encourage vocational training, the signal corps inaugurated a massive civilian training program aimed at producing radiomen—operators both CW and 'phone, and repairmen. It was administered by the Office of the Chief Signal Officer with the cooperation of the U.S. Department of Education and the U.S. Civil Service. Until it was eliminated in 1943, the enlisted reserve corps constituted most of its students. As of DeSoto's writing for the March 1943 issue of QST, 27,000 trainees were attending 151 schools in 38 states; they were being trained by 1,700 instructors, many of whom held amateur licenses.[3]

Nor were women left out of the drive for radio people. By 1942 the YLRL (Young Ladies' Radio League) had grown to a membership of 250. The total number of licensed YLs was estimated at 2,000 or more. They were urged to get involved in defense activities that made use of their skills. Thousands worked making and assembling radio gear on factory assembly lines, took jobs as announcers or engineers with commercial broadcasting, and worked on-ground airline radio facilities. Some even maintained radio gear in everything from Jeeps to M3 tanks. As early as the spring of 1942 the commander of Fort Monmouth estimated that over 400 women had replaced men in radio work. Several YLs were instructors at the air corp's Scott Field, teaching code and the intricacies of radio apparatus. Some women amateurs were innovators in their own right: YL Jean Hudson even pioneered the teaching of radio at a girls' camp in New Hampshire.[4]

When the war came, all the armed services—the signal corps of the army, army air corps, navy, marines, coast guard, and maritime service—already had radio schools in place (and were still expanding). Clinton DeSoto, who visited these camps when they were in full operation, went into great detail about their teaching methods. With the exception of the coast guard school, instruction was of the hot-house variety, aimed at cramming into the minds of the young military personnel as fast as possible radio theory, CW, transmitting and receiving, 'phone, and servicing the apparatus.

Let us continue with the signal corps system. Draftees were encouraged to bring their amateur licenses to the reception center and let it be known that they were hams. If their request was honored, they would be assigned to some branch of the service needing radiomen, not necessarily the signal corps. (Of course, one of the jokes during the war was how the army assigned truck drivers to be cooks and cooks to be truck drivers, the service ignoring their civilian work. But apparently many a ham did get assigned to radio work.) If the ham had been a member of the Army Amateur Radio Service, he was also to bring proof of that to the center. With it, according to War Department orders, he was to be assigned to the signal corps forthwith.[5]

Let us assume, then, that our twenty-one-year-old ham draftee has been accepted and assigned to the Army Signal Corps Radio School at Fort Monmouth, New Jersey. He would be entering perhaps the greatest concentration of radio training in the world. In 1945 official signal corps statistics estimated that seventy-five percent of the officer staff of the radio division were amateurs. The task confronting these instructors was formidable. As DeSoto pointed out, any ham believing that a signal officer's assignment was a safe one believed so in ignorance. Often the signal corps soldier advanced ahead of the main body and was likely to find himself in hand-to-hand combat with the enemy. Whereas in the First World War he had carried for defense a .45 pistol, in this war he was equipped with Garand semi-automatics and the "lethal carbine—short, handy, its stubby snout full of quick-firing death." The signal corps, its soldiers said, is its own infantry.[6]

What DeSoto found at Fort Monmouth was this:

> We learned ... about the finely-balanced training that the officer-candidates, and the enlisted men as well—every department of the school, in fact, receive. Training designed to give them the maximum of useful practical knowledge in a minimum of time. Training stripped of every lost motion, of all superfluous minutes, designed to equip them with every item of useful knowledge within

their sphere—but not a moment wasted on cumbersome details that would slow up the time required to turn them into the efficient, competent fighting force America needs so desperately and soon.

If he was already a competent ham, the young man could advance from rookie to specialist in three months. (One third would be sent to the Wire Division, where they learned the technicalities of telephone operation; two-thirds would go to the Radio Division.) Besides categories according to student progress, the school was divided into specific training of radiomen: radio repairmen, field radio operators, fixed station radio operators, and telegraph printer operators. Our bright ham would be trained in one of the technical branches. The prevalence of hams among the students could be confirmed by the call letter carvings appearing in profusion on their wooden desk tops; call letters were there from every United States AARL district. Our bright ham did very well, applied for OCS (officer candidate school), was admitted and within fifteen months walked out of Fort Monmouth as a second lieutenant in the army signal corps.

Unlike our bright ham, most of the recruits would be sent to the Radio Operating course. Surprisingly, the corps liked to take green recruits who did not know the code. No time was expended in teaching them radio theory. They had to learn how to transcribe messages onto a mill (a typewriter) or a stick (a pencil). For clarity they were taught to print, not use cursive. Twenty wpm was about the limit of competence in doing this; fifteen wpm sending was considered satisfactory. Fixed station operators had to copy on the mill at thirty-five wpm.

Air force personnel mastering Morse code at the Sioux Falls, South Dakota, Air Force Base, March 2, 1943 (courtesy National Archives).

**Air force radioman at his position in an aircraft (courtesy National Archives).**

They were trained under what was known as the "Z-letter" system, "in which each letter is sent at a speed corresponding to 20 w.p.m., with extra spacing between the individual characters to bring the actual speed down to beginning levels of 5 w.p.m. up. An ingenious gadget called a 'clacker' accomplishes this by a multiple rotary selector and stepping relay which picks out every second—or third, or fourth—character from a fixed-speed 20 w.p.m. tape and routes it to the student's phones."

The final stage of training involved "on the air" communication, operating small transmitters "which look like small ham Field Day portables. They must tune the rig to the designated frequency and pick up the correspondent station, but must copy despite the QRM (interference), follow the drift of the other signal, and in general solve all the problems they may encounter in actual communication—minus bullets from the enemy!"

Eight thousand men were trained at Fort Monmouth. They repaired superheterodynes and complex transmitters in tanks and planes, advanced ahead of troops in the field, worked the all-important communications behind the lines, and because communications were so terribly important in World War II, made a substantial contribution to victory.

What about the army air corps? To find how it trained radiomen DeSoto traveled to Scott Field, Illinois, the parent radio school for the corps. (But note that there were also training schools at the Stevens and Congress hotels in Chicago, at Madison, Wisconsin, and Sioux Falls, South Dakota.) At one time about 30,000 personnel were undergoing training at these camps. The parent center was Scott Field. When running at full capacity it functioned twenty-four hours a day, with students actually training during the ordinary sleeping hours, and sleeping in barracks set apart and with signs "Personnel Sleeping" posted to keep the area quiet. Scott Field even trained instructors, among whom were thirty to forty women, one of whom, Mrs. Carrie Jones, was referred to by the commander as "one of my best men."

After twelve weeks the soldier-students were supposed to receive 16 wpm with heavy QRM; they were also trained in air corps procedures and networking. One of their final tests was working on rigs purposely made inoperative, but still installed in a score or more "crack-up" planes, non-flyable because of crashes but still standing.

When the training was over the air corps divided the graduates into those who would be assigned technical jobs—ground jobs servicing the planes—and those who would be sent as radiomen on the bombers. Those were also tested for their abilities to man machine guns, and if found acceptable, were sent for a few weeks to a base for training in the use of the firearms.[7]

Twice QST visited Gallups Island, a small spot in Boston Harbor that was formerly a U.S. Health Service Quarantine Station. Here the maritime service, first under the direction of the coast guard and later under the War Shipping Administration, trained radiomen for the merchant marine. These were young civilians between eighteen and twenty-three years of age, with at least two years of high school behind them. At peak enrollment the Gallups Island facility had more than 1,000 enrollees, with a staff of 280 of whom 34 were instructors. Again, a hothouse training schedule was in operation: 29 weeks down from an initial 40. Courses were organized under five headings: code, procedures, theory, lab, and seamanship. Students were expected to pick up a word a week in code until by the end of 25 weeks they could do 23 wpm. At the end of the 29th week the students paraded in review and were then transported to the Boston Inspection Office of the FCC, where they were expected to—and almost always did—pass the test for their 2nd class radio telegraph license.[8]

Both the first QST visitor, Clark Rodiman, and the second visitor, Clinton DeSoto, were impressed by the training at Gallups Island. They liked the drawer each candidate had at his desk, filled with most of the tools he would ever need to repair a rig. They liked the training— two hours of code and an hour of touch typing—and after four weeks, correlating the two. They approved of the lab, which included all the rigs most likely to be on a merchant marine vessel, and the way the enrollees were taught to service them. They were also impressed by the training the enrollees received in the care and servicing of batteries, which would be used at sea.

DeSoto visited Camp Hood in Texas, where tank destroyer instructions were taught. Here radio was a must: every tank was two-way equipped and every man in the outfit could double as a radio operator. Each battalion had six radio electricians attached to the unit to keep the radios in operation. Training lasted just eight weeks, DeSoto reported, "the shortest and perhaps most intensive course of its kind given in any of the military services." Three basic sets were in use, and only three. The soldier was expected to be able to take just 8 wpm of code "under hazards of battle in a moving vehicle at night by the light of a dial lamp, under machine gun fire or with the sound of artillery paralyzing his ear drums."[9]

Even the branch complimenting itself as being the toughest of the tough, the United States Marines, made use of radio. DeSoto visited Camp Le Jeune, North Carolina. There he found radiomen trained in a twelve week, no-frills course, after which they were immediately assigned to combat areas. When he was there, 800 men were enrolled in the radio school. The marine aim was for the radioman to learn as quickly as possible how to use the radio while in the field under fire. Of the twelve weeks training, just eight were in the classroom; the other four were in the field under conditions identical to actual warfare save that no enemy bullets were whizzing by. When in the field the trainee had to put up with intense QRM—realistic interference effects— including recordings of Japanese voices, exploding artillery shells, machine gun fire, and bombs,

Marines repairing radios on Guadalcanal (Marine Corps Photo).

conditions that might be experienced in the Solomon Islands. Marines copied with the stick, typewriters not being common in foxholes.[10]

And the radiomen really were in the field when in actual combat. "As fighters," wrote De Soto, "these radio Marines must carry the radio gear (usually as two- or three-man crews) in addition to their regular packs and a complete complement of weapons. They're trained to set up stations and start brasspounding on a minute's notice wherever their outfit stops—and on even shorter notice they're prepared to drop 'phones and key and grab an automatic rifle in case of attack."[11]

While the army tank destroyers, the marines and the army air corps concentrated on quick, no-frills instruction in radio, the coast guard, war or no war, stuck with its usual training policies. According to DeSoto, the CG maintained the longest and most intensive radio training of any of the military services. Billets were in the Hotel Morton on Virginia Avenue in Atlantic City, and a half block away was the school in the old Elks Building. In the two-story-high code room were charts of typewriter keys and other symbols of importance, and clocks, all of them on Greenwich time, because that is the time the mariners use. *Semper paratus*—always ready—is the coast guard's motto, and to make sure the radiomen were always ready, their course was of

*Opposite top:* Marines operating radio at the "nerve center" of Tarawa (Marine Corps Photo). *Opposite bottom:* Marine Jeep with transmitter and receiver installed (Marine Corps Photo).

six months' duration. Accuracy and dependability were stressed. Code was the heart of their instruction.[12]

Training was based on the reality that an operator might be called for active duty prior to completing the course. For this reason the coast guard radiomen were taught in parallel the first three categories into which their instruction was divided: code, procedure, and watch standing; theory and material came on the fourteenth week. On his very first day the student sat down in the big code room with headphones on and listened while the entire code alphabet was sent to him at 4 wpm. On that same day he became acquainted with a typewriter and began learning touch typing. After four weeks, taking code and typing coalesced. But every day he was also instructed in procedure and watch standing—meaning that he was learning the way of sailing in the coast guard. Calisthenics were part of the daily routine also, but were designed so as not to tighten up the muscles used for typing and sending code.[13]

The coast guard stressed accuracy. "A single error might lose an entire convoy," one of the officers told DeSoto. Students learned to take code without any QRM, the coast guard believing that learning pure code was tough enough without added interference; if the students mastered the code, they would conquer the QRM. Sending code was also stressed. Students were taught "how to hold a hand key for maximum control and minimum fatigue" and they learned the basic rhythm of the code and the spacial relations between dots and dashes. One side of their split headphones would be keyed to their own sending; the other side to ideal sending of the same code letters. In this way the sender so mastered code sending that by the end of the course it was difficult to tell the automatic sender from the student's. Yet receiving was stressed even more. For ten weeks the young man listened to transmissions, keeping the log and other details of the operation as if he were aboard a coast guard vessel. Moreover, he was being taught theory and learning the hardware, how to care for it and how to repair it.[14]

Most of the young men—the ideal ages were nineteen and twenty—finished the course. Those that did not were likely to be washed out after the first month. They were treated well, an officer explaining to them that radio was simply not the cup of tea for some men, and this did not mean that they were not intelligent—they just weren't keyed to radio.

The services recognized that there were women amateurs—about 2,000 with licenses—and that they could be of extreme value as radio operators. The Women's Auxiliary Army Corps—the WAACS—recognized the value of women with technical abilities and opened the way to radio instruction by way of an aptitude test. If accepted they were sent to a civilian radio school in Kansas City, Missouri, under signal corps supervision. Those passing the course would replace army air force enlisted men. As for the Women Appointed for Voluntary Emergency Service—the WAVES—they would be sent to the radio school at Madison, Wisconsin. The coast guard also had its women's auxiliary, the SPARS, so named for the coast guard motto, s(emper) par(atus)—always prepared. Amateurs could join but it is not clear whether or not the service had provisions for women radio operators.[15]

What about the navy? That branch of the service had always had great interest in radio. This was to be expected: after all, radio's first valuable use was on board ships. In 1940, almost two years before the United States entered the war, the navy was already encouraging hams to join up. In the navy radio was still not considered the only means of communication. The navy encouraged amateurs to apply as signalmen, men who "carried on with searchlights, yardarm blinkers, blinker guns, semaphores, flag, and sound signals." The pitch for enlisting seemed to consider amateurs as potential experts in all forms of communication, not just radio.[16]

When the great need arose for radiomen, the navy divided its schools between those training in electrical engineering and radio maintenance (EE and RM) and those teaching code and 'phone operators. The seven EE and RM schools were scattered at college campuses around the nation. The enrollees, many of them hams or radio technicians, were put through a twelve week course. They were trained in DC theory, mathematics, AC theory, and radio; they used text-books and the slide rule, and practiced mechanical drawing, and conducted experiments on all manner of electrical gadgetry. In the first three weeks of their third and last month they had lectures four hours a day and four hours devoted to radio experiments. Each man built a uhf super-heterodyne as a part of his training. Superior students were likely to be sent to still another, more advanced training school.[17]

Of far more interest was the navy's experience in teaching radio operators. Surprisingly, the service appears to have had no previous experience in this discipline. In September 1940, Captain William Baggaley and a few reserve officers were ordered to set up a naval radio training station. The site was the old Fitch Home for Veterans in the town of Noroton Heights, Connecticut. They were given a month to renovate the seventy-five-year-old buildings and bring in everything needed for a boarding school—bunks, bedding, dishes, silverware, code tables, typewriters, radio equipment, even new kitchen equipment. (As soon as they opened the doors the cork insulation fell out of the old iceboxes.) When the month was up the job was done—they thought. Just before the first class arrived a check-off revealed that the silverware had not arrived. "They scoured the neighborhood; they drove to New York; and before the day was over they had begged and borrowed enough silverware to equip the class when it arrived."[18]

Then there was the matter of the training. "There was no blueprint ... no standard course

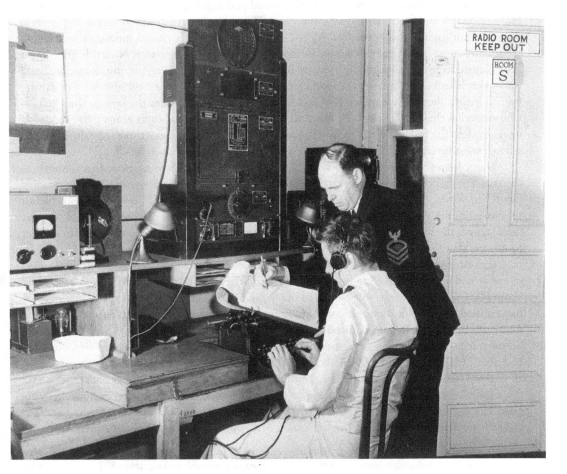

The radio room at Noroton Training School where radio messages from other parts of the country were received (courtesy National Archives).

of study. Much of the equipment ... had to be improvised—and that was where the junk-boxes in the shacks of the hams on the staff did their part!" Gradually the course took form. Each student who completed the course was ordered to fill out a form a month after going on active duty that, by the answers, revealed strengths and weaknesses in the program. "Gradually the course took form," wrote DeSoto. "It was done by ham methods, by cut-and-try, by constant experimentation, by a persistent dissatisfaction with the best thus far accomplished—and therefore it worked."

Fortunately most members of the first class were hams. So much conversation involved radio that sometimes it seemed like a "perpetual ham convention." One of the officers was made aware of this one night when he inadvertently flashed CQ on his flashlight, only to find replies from flashlights from darkened dormitory rooms "like a hundred busy fireflies."

Even after several months had elapsed and new enrollees had arrived, half the teaching personnel of the station were hams. The Noroton station turned out radiomen in four months. When they graduated they were also trained in the other methods of communication used by the navy. The school was operated in strict naval fashion, so that graduates had a clear concept of "the Navy way." And Noroton was a tough school: about twenty-five percent of the candidates washed out. Noroton's system was considered successful. It was adopted by other navy radio schools, which included the RCA Radio Institutes and radio schools at Cornell and Harvard.

The Ferrying Division of the Air Transport Command operated its own training program. Experience had demonstrated that there was a world of difference between the abilities of ground radio operators and flight radio operators. The ATC's schools were designed specifically to train radiomen for flight radio operation. All the students had already passed through a basic radio school and some had already flown on foreign flights. Two schools, one at Nashville and one at Long Beach, were consolidated into one at Reno, Nevada. The training was dominated by hams who were described as "among the world's finest." Among them were a number of WACS who found "little difficulty in learning CW and have been an inspiration to the boys at the school—who marvel at their code speed." It was a six week intensive course, eight hours a day six days a week, the first thirty days on the ground followed by fifty hours in "one of three specially equipped aircraft working out radio problems in the air." There were many washouts.[19]

Major James W. Hunt, W5TG-W5CCU, was sent by the ATS to a desert area somewhere in Asia. He described the disaster at the small field where he was assigned. "The wind died down," he wrote. "I have never before seen such a mess. The sand had been so thick in the air that we had built the entire airport 50 feet above the ground and, when that wind died and the sand settled, the whole mess fell 50 feet to the ground. This just shows you some of our smaller problems."[20]

It is estimated that at least a half million and possibly three-quarters of a million members of the armed forces mastered radio during the war. When the war began the United States had about 60,000 licensed hams; it is estimated that by war's end between 25,000 and 30,000 of them had enlisted in the army, navy, air corps, marines, coast guard, and merchant marine. According to the director of Naval Communications, Rear Admiral Joseph R. Redman, at least 6,400 of the licensed hams served in the navy, marines, and coast guard. As of December 7, 1941, the navy had about 650 officers and 3,000 radiomen in its Naval Communications Reserve, and almost all of them were either hams or had begun their careers in radio as hams. They constituted, said the admiral, "the first and most valuable source from which the needs of expansion were supplied." To estimate the contribution, it must be realized that no one of our enemies had as many as 1,000 amateurs at hand when the war came. The Admiral added this about American hams:

> Literally, they did everything. They learned quickly, and taught others well. They designed, serviced, operated and supervised in the fields of communications and electronics. They served with ability, enthusiasm and unfailing devotion to duty. Their individual accomplishments have won praise on every ship, on every beachhead, and in every laboratory, radio station, or communica-

**Students practicing receiving International Morse code on radio sets at Noroton Training School, Noroton Heights, Connecticut (courtesy National Archives).**

tion office in which they have served.... To amateur radio as a whole I would like to tender an added word of appreciation by the Navy Department, for performance "in the highest traditions of the Naval Service."[21]

The commander added a few touches to the navy's experience with hams. Rag chewers, he noted, had a harder time learning the navy's needs than hams who engaged in relaying. The rag chewers carried the information in their heads, and found it difficult to set it down on type-writer or in pencil (on the mill or on the stick). "Amateurs as trainees had the edge on other students," he wrote, "because of the inherent enthusiasm born of their ham radio background."

What is to be said of these concentrated instruction schools? Save for at most 30,000 licensed hams, the half to three-quarters of a million radiomen—which also means radar men and electrical engineers—received such training. They served their country successfully, proven by the outcome of the war. All the descriptions save that of the coast guard program seem to empha-size a speed of instruction placing the young students under constant stress. Even though they were usually free from noon Saturday until reporting in on Sunday evening, even though there was recreation and physical education including not only calisthenics but games, the five days each week of eight, ten, and twelve hours, lasting for twelve to nineteen weeks—or twenty-four with the coast guard—seem almost inhumane and beyond the capacity of ordinary people. Yet thousands attended, and the great majority made it through the courses and so on to the war.

But let us remember that most of these candidates were young men or women, most of them under age twenty-five. They had undergone physical examinations and been accepted into the services in good health; then they had attended boot camp for six or eight weeks, learning the army, navy, air corps, coast guard, or maritime ways of doing things. Physically, many left boot camp in the finest physical condition they would ever enjoy. And they had learned to take orders and handle themselves as soldiers or sailors or marines.

When they entered the radio schools they were immediately subjected to peer pressure—to keep up with the buddies whom they met and who often became life-long friends. They helped each other. They ate, slept, played, and above all worked and learned together. Most had few outside problems: their families, friends, and sweethearts were proud of them. Stars hung in the windows of their homes.

Moreover, the stringency, the stress, was possibly overdone in the QST articles. Of course the rookie who couldn't "cut the mustard" was under stress, sometimes because he or she just wasn't mentally created for radio. But for most, when twenty or thirty students were at work on the identical phase of instruction, they learned from each other as well as from the instructor. And the instructor had to be a reasonable individual—if he antagonized too many, failed too many, denigrated too many, that instructor was going to be called on the carpet. A camaraderie was created that made the tough regimen easier. Those valiant youngsters are in their eighties now, and are leaving us at the rate of a thousand or more a day. But if you know one of them, ask how he or she remember the weeks of instruction. Don't be surprised if the reply is that it was *fun!*

Their training over, the radiomen went into action. QST requested stories of ham experi-ences in the war, but received relatively few. It even offered a $25.00 war bond for any story accepted, and the Hallicrafters Company offered a hundred dollars. The paucity of entries was attributed to the time lag between battles and letters describing them, and secondly, to the "mod-esty and congenital inability of the ham, soldier, or a combination of both—to sense the drama of his experience." Some stories came directly from the ham, editors put some together from newspaper clippings, and a few came from friends or relatives who brought the editors letters relating such experiences. For several months during the war QST ran a section titled "Hams in Combat." But to attempt even an encapsulation of radiomen's war experiences, whether culled from QST, newspapers and magazines, or even books would be ludicrous. There are thousands of stories out there, of hams who improvised parts from old receivers, or used empty tomato cans that could be cut up and used to make a transmitter work; found new ways to string up an antenna, or discovered power sources where they were least expected to exist. Of signal corps men with tank companies who salvaged gear from disabled vehicles and thus created that won-derful ham source of parts, a junk box. Of brave men on torpedoed ships, both maritime and naval. Of marines advancing into enemy terrain with the new walkie-talkies. We are acutely aware of their accomplishments; they were substantial, and contributed measurably to the winning of the war.[22]

One ham who wrote of his experiences as a radioman-gunner on a B-25 was technical sergeant Charles W. Tinsley, W8HGC. He described his first combat mission, in which he had to determine their position, which he did correctly, landing safely with just five minutes of gas left. While in the service, he wrote, "I never lost sight of the fact that, at heart, I was still a ham. I sincerely believe that the bond of fraternity built up among amateurs has done much to aid radio communications in this war. Wherever I went, in the States, on the way to the fronts, I constantly came in contact with amateurs."[23]

Some of their efforts were not specifically war related, but they demonstrate the ingenuity of hams. A licensed ham stationed on the Anzio beachhead in Italy described how several radio men rigged a foxhole version of a crystal set using a razor blade and safety pin. As described, "a station was found by moving the point of the safety pin, anchored at the other end, over the opposite end of the blade from where it is connected to the coil and antenna. The 'phones," the writer said, "were inserted between the pin and the ground side of the coil." Reporting that reception "was good," the correspondent added that they were "rewarded by reception of a jive program (along with German propaganda) aimed at the American forces from an Axis station in Rome."[24]

Let us not forget the part played by amateurs in maintaining morale. There were the Armed Forces Radio Stations, such as the "Mosquito Network" in the South Pacific and the "Bridge to Victory" in the Aleutians, managed and programmed and announced by radiomen, many of them hams. In 1944 the army and navy agreed to run them on a fifty-fifty basis, with a governing board consisting of a representative of Navy Welfare and the other of Army Special Services.[25]

And speaking of morale, hams on duty in remote theaters of war made major contributions totally apart from their assigned duties. An example was given by an army operator on the Burma Road, who wrote of his contribution:

> Many a homesick soldier has talked directly to his father, mother, brother or friend thousands of miles from the Burma jungles, and believe me, the Army has no form of morale uplift to compare with these direct contacts. I have seen soldiers enter "the shack" at a prearranged time, to talk to a brother or sister in Hawaii, Japan, or ETO [Eastern Theater of Operations] and upon hearing their voice cry for half an hour unable to talk to them. But believe me, that release of pent-up emotion is something to behold, and I am proud to say I was able to help, small as my part may have been. And the operation of this station does not in any way differ from hundreds of other American stations throughout the occupied areas. They are filling a gap no other Army activity can meet.[26]

And they did it all with humor. Take Major J.W. Hunt, assigned to a remote post in Alaska: he complained of the blue termites which ate up the steel antenna tower after it had crystallized and fallen into the snow. The termites ate their remote control lines too. Then he "encountered the terrible Kee bird." Flying back one night from Eskimo Joe's tavern, on the other side of the Pole, they heard this terrible noise. "It was the wail of the dreaded Kee bird," Major Hunt reported. "It sat on the ice and shrieked Kee! Kee! Ke-rist it's cold up here!"[27]

What remains to be told is the social history of amateur radio during the war. Hams constituted a brotherhood—even a sisterhood, for women were also prominent as radio people—that gave them entry to homes in other countries where radio shacks or ham antennas existed. When amateurs in a theater of war became aware of each other, they worked to get together, to have a hamfest. Briefly, while at the meeting, officers and enlisted men were equals. Whether the radiomen created in jig time by the concentrated training necessitated by the war participated also is a moot point. It was the *amateurs*, so aware of their hobby, who worked to bring themselves together whenever and wherever possible, war or no war. Their success was remark-able. Here are a few of the hamfests, and their locations, and how they came about.

First, we should mention hamfests at the training stations. Whether all the stations had meetings of amateurs is not known—although they almost certainly did—but the radio school at Fort Monmouth had at least two hamfests mentioned in QST, and quite possibly, several more. The first one, deemed highly successful, was held in September 1942. Hams attached QSL-size cards to their uniforms. The meeting room had district placards hung around the walls. Clinton DeSoto emphasized the differences from a meeting in "Podunk," the soldiers standing straight and their uniforms immaculate, saying "sir" before speaking. There was an army air about the meeting. But as minutes went by, informality set in, and when they began singing, it became more of a typical civilian hamfest:

> Hail, hail, the hams are here
> Lemme hear your call sign,
> Then I'm gonna send you mine.
>
> Dit dit dah—dit dit dit dah—
> This is our way of Talking.
> Dit dit dit dah—dit dit dit dah—
> That's why you hear us squawking.
>
> We're in the code-room now
> With sweat upon our brow.
> We're going to send until the end—
> This is our Army now.[28]

At least two Anglo-American hamfests were held during the war. The first took place in September 1944 at the behest of the Radio Society of Great Britain. It was held in the Mostyn Red Cross Club near the Marble Arch in London. Seventy hams attended, including one Canadian, twenty-one Britishers, and the rest Americans with every American radio call district represented. Postwar plans and activities were discussed; refreshments and rag chewing ended the meeting. The second get together, held on October 28, 1944, was attended by 150 British and American hams.[29]

A hamfest was held in North Africa, probably in late 1943. Neither the exact location or the month are given, but it came about when a routine inspection of radio facilities by U.S. Signal Corps officers discovered their guide was a well-known Scottish DX fan. The upshot was an informal hamfest in which all but two U.S. districts were represented. Much of the talk was about amateur radio after the war. Not a hamfest but of considerable importance was a United Nations Amateur Radio Convention held in Cairo on May 5, 1944. The Cairo Amateur Radio Club extended an invitation to all the participants and another U.N. convention was scheduled for November 24, 1944.[30]

A mid–Pacific chapter of the Society of Amateur Radio Operators congregated in Hawaii for three meetings in early 1945. Corporal Jack Walker, a native of Chicago, was on a South Pacific Island on which the signal corps had erected an armed services broadcast station. He thought he recognized the voice of the announcer, looked him up, and found that he was a fellow Chicagoan with whom he had had many a rag chew in prewar days. Their meeting, 8,000 miles from home, resulted in other hams congregating there and plans made for a hamfest. It may have taken place but the corporal was shipped out before the meeting.[31]

The conviviality of amateurs was apparent in Australia and New Zealand, South Africa and the British Isles. Amateurs in these countries wrote to QST to have their names, addresses, and call letters listed, emphasizing that visiting ham GIs were welcome. "Like ham music," commented the editors, "ham language is universal and any American or Canadian who finds himself at loose ends for companionship and conversation is overlooking a good bet if he doesn't take advantage of the invitations."[32]

Amateur Captain William H. Graham met "some of the lads down in Australia." He wrote how Australian amateur Wal Ryan, VK2TI, corralled him "and used up his entire month's gas ration—taking me home for the day, showing me some sights.... He had quite a bunch of the Sidney amateurs in for the evening, too, and honestly ... I never saw such hospitality. They made me feel like I was the great white warrior come from America to save their country single-handed. I was prouder, I believe, than at any time in my life I was an amateur!"[33]

Before the war ended hams were already concerned over the condition of their hobby after the Axis powers were defeated. Determined to keep their wonderful hobby alive, Chicago amateurs held a "Hamboree" on May 13th, 1944. The event was sponsored by the Chicago Area Radio Club Council, with donations from local dealers and manufacturers; attendance was free. The meeting was held in the Sherman Hotel with the latest army signal corps equipment on display. More than five hundred "hams, ex hams, and would-be hams" were represented from every call area.[34]

VE Day was proclaimed on April 8, 1945; Japan agreed to surrender on August 14, 1945, and formally surrendered on September 2. The terrible Second World War was over. The day after President Truman announced the end of the war with Japan, the ARRL moved for amateur reopening. On the 21st the FCC put amateurs back on the air on the single 112-Mc band. This was great! However, granting amateurs additional bands was held up because the armed forces still needed those they had requisitioned for the war. All presently licensed hams were allowed on the air, their otherwise expired licenses having been extended both during the war and immediately after, until operations returned to normal. On November 15th, 1945, hams were allowed back on every frequency above 28 Mc. They could operate at 10 meters and even 5 meters but had to wait for their favorite 80, 40, and 20 meters. It was a far cry from the battle to regain their privileges amateurs they had had to wage after the First World War. On July 1, 1946, the FCC returned half of the 7 and 14 mc bands (half of 40 and 20 meters) to hams. "Thus we are back in the DX business again," the editors commented. Another minor change was the increase of call areas—what had been inspection districts—to ten, later expanded to twenty.[35]

It was predicted that there would be "plenty of cussing" before transmitters were brought up to operational level. "The look at the rig. The trip to the cellar. The cobwebs.... The lack of wire. The lousy solder.... The changing of coils. The connecting up of the oscillator. The closing of the switch.... The open grid leak. The dirty contacts.... But boy, it's worth it when you think of the DX ... and your old pals ready to sked you."[36]

The war had not been over long before it became clear that thousands of radio parts, even entire rigs, were available as war surplus. Hams made great use of this opportunity, so much so that in its 1947 index QST ran a section titled "Surplus Conversion" and in 1948 "Surplus Corner." Several articles appeared instructing radiomen in the how-to of converting this equipment to ham use. A brief perusal of those articles drives home to the layman just how knowledgeable and sophisticated, how curious and how determined, the amateurs were. For example, the article "Operating the BC-645 on 420 Mc" began by reminding amateurs that the rig had been "in the supersecret class." So secret, in fact, that instruction manuals were not available, nor was information "regarding the control and indicator accessories and connections to them from the various plugs and connectors on the BC-645." Never mind. Start by "tracing the schematic diagram through a maze of cabling and terminal boards used in mounting resistors and capacitators." And so on. True amateurs approached such challenges with the zest of a hungry child sitting down to a Thanksgiving dinner.[37]

In hamdom, "back to normal" advanced rapidly. At league headquarters applications for membership in the RCC—the Rag Chewer's Club—were again coming in. The first post-war Field Day was held June 22–23. "It is *the* ham outing of the year," stressed QST. Emergency corps members would find it the ideal time to test their old equipment and try the new. Military affairs were not completely forgotten: twenty-seven amateurs were aboard the USS *Avery Island*, the electronic control ship for Operation Crossroads, the designation for the experimental dropping of two atom bombs on Bikini Atoll.[38]

The end of the war did not put a halt to natural disasters. Amateurs were present to give aid. In April 1946, an earthquake followed by a tidal wave with a crest of 60 feet hit Ikatan, a very small village on Uninik Island in Southwest Alaska. In November 1946, the worst snowstorm to hit northern Idaho in 44 years disrupted all communications. Amateurs provided contact for the Northern Pacific Railroad with their trains out of Sandpoint, Idaho. Relays were established between Spokane, Washington, and Moscow, Idaho.[39]

The sophistication born of both peacetime and wartime experiences was revealed when a hurricane bearing 125 miles an hour winds struck Palm Beach, Florida, on September 17, 1947. The Florida Emergency 'Phone Net with the coast guard station provided the entire state of Florida with emergency coverage. Amateurs had become well managed with plans made long in advance, and they worked very well.[40]

The real test of amateur emergency capabilities came in the period between April 9 and midnight, April 17th, 1947. Two calamities, one natural, one an accident of commerce, struck the American Southwest. The first, the natural one, was a tornado that leap-frogged from one place to another over a span of about 185 miles. It happened in a region sparsely populated in 1947 and just about as thinly settled even today. It started at the tiny farm town of White Deer, about 40 miles northeast of Amarillo and advanced to the city of Woodward, Oklahoma. There the twister tore up 100 city blocks.[41]

Sparsely settled the region was, but people did live there. Estimates were 150 fatalities, 1,800 injured, 3,000 homeless, and $8,000,000 damage, in 1947 dollars. And here, in "the first major post-war communications emergencies," amateur radio chalked up its first overwhelming success. Meeting the crisis was not a simple task. Distances between towns there are measured in fifties and hundreds of miles. The weather hardly cooperated: rain, sleet, and snow pommeled the area for nearly a week. Yet, as soon as news spread of the disaster, hams were on the move. Five days after the catastrophe, officials declared the situation well in hand. Hams had the satisfaction of knowing that amateur radio had played a major, and successful, part in returning the region to normal conditions.[42]

Within forty hours some of them found themselves glued to their rigs again, helping in a man-made emergency. Possibly the most devastating non-natural emergency until the horrors of September 11, 2001, took place on April 16th and 17th at Texas City, Texas. Located about six miles north-northwest of Galveston, it is a deep water port connecting with Houston, Galveston, and the Gulf of Mexico. Monsanto Chemicals is located there as are other petrochemical industries. The city had expanded rapidly during the Second World War, with a population in 1947 of about 16,000. A French freighter, the *Grandcamp*, was in the harbor taking on 51,000 sacks of ammonium nitrate fertilizer. A fire broke out in its hold and most of the fire department was there trying to snuff it out. A crowd, including many children, had gathered at the waterfront to watch the activities. Intermingled with the black smoke was a fascinating bright orange color. A tug boat had been ordered to take the boat out of the harbor but it had not arrived. Suddenly—the word hardly fills the need—a little after 9:00 a.m. the *Grandcamp* exploded. A column of smoke rose 2,000 feet in the air and was followed about ten seconds later by a second shock wave. "Hot chunks of steel that were once her hull rained down on homes and buildings and pierced oil tanks, touching them off in a holocaust of fire. The Monsanto Chemical Company's new $19 million dollar styrene plant experienced three explosions and was immediately in flames; buildings collapsed, windows smashed and sent shards of cutting glass upon people; a miniature tidal wave rushed 150 yards onto shore, destroying everything in its wake. Rescue went on throughout the day and into the night. Then, at 1:10 a.m. on April 17th, the *High Flyer*, another freighter loaded with sulfur and ammonium nitrate, exploded with the most violent of all the blasts. It even destroyed a nearby ship, the *Wilson B. Keene*. Nearby Baytown awoke to rattling windows and "a fine mist of black oil rained on Galveston."[43]

By noon, just three hours after the initial disaster, amateur portable and mobile rigs were in Texas City and on the air. One of the hams was a neurosurgeon who in his spare moments contributed his amateur status to communications. Army, navy, coast guard, U.S. Engineers, the FBI, and the local and state police also made use of their radio facilities. Nearly two dozen amateurs remained in the "gas-filled, explosion-torn city." They handled requests for additional hospitalization, ambulances, plasma, oxygen, and gas masks. One of the first men on the scene handled more than 200 messages. At the Texas City message center in city hall over 300 messages were handled for army, Red Cross, Salvation Army, and telephone officials. FCC men arrived from Houston to supervise and coordinate activities.

The amateurs had many narrow escapes. When the *High Flyer* blew up with an explosion greater than that of the *Grandcamp*, two hams who were on the air and had remained in the city braved the second explosion, sending messages throughout the night. A naval communications unit was just a hundred yards from the *High Flyer*, but fortunately the operators survived with minor cuts and bruises.[44]

And so hams in peacetime returned to their vital tasks of furnishing communications in emergencies. Now there were more hams as a result of the massive training in radio that had taken place during the war. They helped work the emergency nets, participated in the annual Field Day which really prepared them for disasters, and were anxious to contribute their expertise. Western Union reinstituted its cooperative program with the ARRL whereby its agents in 3500 communities were ordered to be in touch with local amateurs so that cooperation could be immediately forthcoming in case of an emergency. The disasters continued; they are a part of life, and they continue to the present. Writing about them becomes repetitious, but those involved never forget the experience.[45]

The return to pre-war activities was inclusive. Commander Donald B. MacMillan, the Arctic explorer, set out in the *Bowdoin* on June 29, 1946, on his twenty-fourth expedition into the Arctic. The FCC had granted the expedition a license for two-way amateur work. Operator Bill Matchett, W1KKS, of Manchester, Connecticut, was the radioman. His sophisticated rig, with an antenna looming seventy feet above the deck, was a far cry from the crude sets Commander MacMillan had taken along decades before.[46]

Commander Finn Ronne, USNR, commanded an Antarctic expedition to Palmer Island, 1,000 miles south of Cape Horn. During the expected eighteen months' duration, amateur communications were allowed above 4,000 kc. Lawrence DeWolfe Kelsey, W3LYK, was the operator. In late November 1947, the Gatti-Hallicrafters Expedition left New York City for Africa, where it proposed to explore the Mountains of the Moon and carry out shortwave radio research. Two radio operators, William D. Snyder, W0LHS, and Robert E. Leo, W6PBV, accompanied the explorers.[47]

Captain Irving Johnson and his sailing vessel, the *Yankee*, in early November 1947 started from Gloucestor, Massachusetts, on another round-the-world cruise. It was a different vessel, however. The captain had obtained a 92 foot brigantine, an ex–German pilot ship taken by the British as a war prize. Alan Eurich, W7HFZ, who had been along before, went as far as Panama, then returned to the states; the *Yankee* was to pick up another operator in Hawaii. The vessel's call letters were WEXO.[48]

Receiving greater publicity was the voyage of Thor Heyerdahl's balsa-log raft, *Kon-Tiki*. A Norwegian ethnologist, Heyerdahl was convinced that Polynesia had been settled from America rather than from Asia. To prove his point—or try to—he built a forty-five-foot raft of primitive construction, assembled a crew of five which included two radiomen and radio call letters L12B, and set out from Callao, Peru (seven miles west of Lima), on April 28, 1947. The wind blew them west; on August 7, one hundred and one days after leaving Calao, they landed on a reef on Ralora Island. It is one of the Tuamotu Islands, a French-controlled archipelago about 300 miles south of the Marquesas.[49]

The ARRL, when approached by the Norwegians for cooperation, cheerfully accepted. *Kon-Tiki* would have two radio transmitters, one of normal maritime channels and the other, using the 7-, 14-, 28-, and 50-mc bands, for amateurs. The league singled out a number of hams to assist. The real problem for Heyerdahl concerned keeping transmitters and receivers operational in the humidity of the sea on a wooden raft that floated very close to the water. The rigs were designed and built by an American company to carefully planned specifications. Some were based upon a *QST* article that had appeared in the July 1941 issue giving instructions for building "A Versatile Portable-Emergency Transmitter." Batteries, it was decided, would be dry cells that could be charged with a hand generator. Although they had serious difficulties with batteries and the rigs, primarily due to the excess humidity, *Kon-Tiki's* radiomen contacted hams on schedule, and many others. After a few weeks "a smoothly-working long-haul network of amateur stations developed ... in North America, Canal Zone, and Norway ... handling the flow of information from the raft." One of the hams, although not mentioned in the article, was Gene Sykes, W400, of West Palm Beach, Florida.[50]

The war had not been over for many months before hams were again aware of their precarious position in the world at large. At home the FCC appeared to be friendly to the hobby, but as had been evident through all the years of radio, other nations took a dim view of hamdom. Some countries maintained a monopoly over communications. They had always looked upon amateur radio as if it were a competitor, and they wanted its power restricted. Probably even more nations were paranoid, believing that amateur radio expedited disorder and revolution. Being aware of this hostility, amateurs were alert to any international meetings involving radio. Every one was a potential threat to their rights.

The first post-war international meeting was a preliminary one, with no signed documents or formal agreements. It was held in Moscow between September 30th and October 21st, 1946. China, France, the USSR, the United Kingdom and the United States sent delegates. Of the nineteen American representatives (plus accompanying secretary, stenographers, a fiscal officer and an interpreter), five were nongovernmental and one of those was the league's assistant secretary, A.L. Budlong. The purpose of the conference was for the nations to come to preliminary understandings prior to the world telecommunications conference, due to meet in early 1947.[51]

The Moscow meeting was, according to Budlong, congenial, with delegates meeting half way on many subjects. However, he added, "to those new to the game (as well as many old timers) the aggregate of the proposals of the other countries at Moscow must seem to paint a pretty black picture for the future of amateur radio, particularly as concerns our DX bands, at the next world conference." Other nations wanted to restrict frequencies desired by the amateurs. American hams were staunchly defended, however, by the chief of the FCC's frequency service, Captain Paul D. Miles, USN (Ret.).[52]

The Atlantic City International Telecommunication and International Radio Conferences met from June through October, 1947. It was attended by nearly 1,000 delegates at a cost of about $12 million, with a total of around 75 committee meetings a week. It really divided into three conferences: the first was to revise the Cairo regulations; the second, the plenipotentiary meeting, which began in July, set out to revise the work of the Madrid Conference; and in August an International High-Frequency Broadcasting Conference took place. The league was well represented because as usual, there were fears of other nations cutting amateur operations almost to death. Official representatives for the United States, men from the FCC, the State Department, and other agencies, were solid, ardent defenders of hams. The result was that this tremendous international conference left the amateurs with about the same rights and band allocations they had previously enjoyed.[53]

In accordance with the provisions agreed upon at this meeting (which divided the world into three regions), the Fourth Inter-American Region 2 Radio Conference was held April 25 to July 11, 1949, in Washington, D.C. Ten American republics, including the USA, plus colonies of European states in the Western Hemisphere, were represented by about 200 delegates. The purpose was to revise regulations set up in 1940 at the Santiago Radio Convention, including possible changes in the Santiago Regional Allocation Table.[54]

The ham presentation at the conference began with a statement as valid today as it was in 1949 and would have been in 1920, of the merits of amateur radio, now described as the Amateur Radio Service. It emphasized its "value to the public as a voluntary noncommittal communication service"; its continuation of the amateurs' "ability to contribute to the advancement of the radio art"; its encouragement "of a program which provides for advancing skills in both the communications and technical phases of the art"; its "establishment of a reservoir of trained operators, technicians, and electronics experts"; and finally, "the amateurs' unique ability to enhance international goodwill." These contributions have been continued throughout amateur radio history right down to the present. Amateur rights emerged from the conference essentially unscathed.[55]

Also, in the fall of 1949, the FCC reviewed American amateur regulations. International meetings, as mentioned above, over the past eighteen months had resulted in drastic changes involving radio communications—not just amateur but embracing all of radio as it was then practiced. Again, the hams emerged as strong as ever. It was emphasized that this was due to the unified efforts of the league and other groups that were determined to protect amateurs' rights. The league issued a 200 page statement presenting its rationale for the rules and regulations under discussion, and walked away with success. Writers of amateur history, suggested the editors of QST, "may well find [it was on] the date of October 10th ... that the amateur body presented a striking demonstration of unity such as perhaps never before existed in our ranks."[56]

And so the tempestuous decade of the 1940s continued. The navy continued its strong interest in the amateur. At the ARRL National Convention in 1948, Rear Admiral Earl E. Stone praised hams, their part in the winning of the war, and urged their continued enthusiasm. He stressed that the navy still considered "CW operation to be the fundamental requirement of a radio operator." He noted that, if a ham had the money, he could purchase ready-made rigs, "but," he added, "the young amateur of today may fail to learn much that is fundamental if he passes up the opportunity of assembling some of his own equipment." He should learn "by cutting and trying"; it was the "know-how" acquired through the trial and error of building one's own equipment that made the amateur so vitally important in emergencies. The navy continued its Navy Day activities and in other ways encouraged hams to continue with their hobby.[57]

The league continued to encourage CW also, "just as we have urged that the first transmitter and receiver be home-built." It showed concern because so many newcomers to hamdom were purchasing their equipment ready made and using 'phone exclusively. Why 'phone? The editors speculated that it was because of fear of being criticized as part of the learning process. "Don't shy ... simply because your code speed is a bit on the shaky side. Give us a call and request QRS [help]. We'll be more than glad to comply." In the same essay the editors urged old timers to be kind and patient with newcomers.[58]

As for the army, in mid–December 1948, the secretary of defense announced creation of the Military Amateur Radio System, or MARS. Both the army and the air force were involved. To be a member of MARS an amateur had to be a veteran or affiliated with one of their civilian components, such as the ROTC or the national guard. Eventually it allowed qualified civilians into its ranks. Incentives to join included, for veterans, one retirement credit for each three hours of MARS network participation; the acquiring of surplus parts which were obsolete for the army or air force "but are just right for the 'junk box'"; instruction in standard operating procedures; the use of military frequencies and call signs; network operations; and service integration, i.e., the coordination of all radio facilities in times of emergencies. MARS still thrives.[59]

Technological progress continued. The coming of television created new problems for hams, who were accused of interfering with television reception. It was not long before hams themselves began experimenting with their own private televisions, and this has continued down to the present. In October 1948, Technical Editor Byron Goodman ran a short essay titled "The 'Transistor'—an Amplifying Crystal." Time was, he commented, when such devices were classed with "sky hooks, left-handed monkey wrenches and striped paint." The device was then described quite accurately, but the writer added, "It doesn't appear that there will be much use made of Transistors in amateur work, unless it is in portable and/or compact radio amplifiers." The noise figure was said to be poorer than with vacuum tubes, which could reduce its usefulness for amateurs. However, Mr. Goodman added this thought: "These clever little devices are well worth keeping an eye on."[60]

Time passes on, and the old guard gives way to newcomers. In November 1948 death came to Kenneth Bryant Warner, who for nearly thirty years was secretary and general manager of the ARRL. "If it was Maxim who conceived our League, it was Warner who breathed into it life and energy and vitality, whose balanced judgment and clear vision ensured its growth and success," reads his obituary. "With his passing we suffer the loss of a great leader, an untiring servant in the cause of amateur radio." On April 27, 1949, death came to Clinton B. DeSoto, a former editor of QST. His Two Hundred Meters and Down has been the basic history of amateur radio down to 1936, and his short non-fiction book, Calling CQ, published just before the Second World War, is a delightful account of a number of amateur exploits. Both Warner and DeSoto were outstanding, dedicated men. I am writing this today at age eighty-six; it comes as a shock, the early demise of these fine men, Warner at age of fifty-four, DeSoto at just thirty-seven, both from heart ailments.[61]

By 1950 there were an estimated ninety thousand licensed amateurs in the United States. Their contributions to humanity in times of emergencies and disasters continued, their clubs and associations and meetings carried on, their DX-ing was as vigorous as ever. Yet in many ways the real social story of amateur radio is described adequately in its first fifty years. In that half century amateur radio advanced in tandem with automobiles and airplanes, medical knowledge, population growth, world wars, and peace-time activities. Looking back at those fifty years we find ourselves amazed at the progress of technology. As the great improvements for humanity proceeded with breathtaking speed, from the Wright brothers' short flight to routine passenger traffic over the oceans and around the world, as the polar regions and the jungles previously unexplored were crossed, and as radio became more reliable and sophisticated, we have to compliment and give thanks to the hams. Curious, hard working, technologically inclined, they advanced communications more than any research lab. In communications, they brought us to where we are today.

Their experimentation continues, with satellites and television and computers entering their orbit of interest. So too does their social history. Their continuing enthusiasms are significant to the hobby, but compared with the first fifty years in which radio advanced from the incredibly crude to the incredibly sophisticated, the story loses much of its fascination. Those first fifty years constitute the truly exciting, adventurous, developmental era of amateur radio. But be assured, it is just the first phase in the history of this fascinating avocation.

# Epilogue

Why end this book as of the year 1950? It is because the story of ham radio's development essentially takes place in the first fifty years of the twentieth century. Having been created, accepted, regulated, and achieved permanent status by 1950, the story after that becomes one primarily of repetition. The one great exception is in the area of technology, and, save for minimal descriptions necessary to the story, that has not been our concern.

It was in those years, *circa* 1900–1950, that a scientific marvel grew from an experiment that worked, to the art that exists today. This was Marconi's sending messages by wireless at his Italian villa followed in time by the incredible possibilities held out by the three dots received at St. Johns, Newfoundland, from his station in Wales across the sea. The concept was simple enough, and demanded such a small monetary investment, that technically-minded gadgeteers discovered they could build transmitters and receivers and carry on experiments and contact others with like interests (later known as hams). They were especially active in the United States, where there were no government restrictions until 1912, and even after that the restrictions were limited. Moreover, the American standard of living made it possible for the ordinary person to indulge the hobby. Ham radio was flourishing before commercial radio using AM technology exploded into reality in the early 1920s. Experimentation, expansion and achievement continued through the era.

The first fifty years of the twentieth century were years of incredible scientific and technological development, and they brought about changes that touched much of human activity. At the turn of the century automobiles were still put-putting in the "get-a-horse" stage. Aircraft evolved from wood and linen contraptions in 1903 to jet aircraft carrying thousands across continents and oceans. In 1900 much of the Arctic and Antarctic remained unexplored, as well as parts of the Amazon, Africa, and Borneo. Ham achievements advanced in tandem with the improvements in automobiles, in heavier-than-air and lighter-than-air developments, in the expansion of the uses of electricity, and with explorations. Amateur contributions in times of disaster became more sophisticated and efficient. When World War II came along, the ham operator made contributions in the instruction room, on the battlefield, and in the realm of invention.

The first fifty years were the period of initial experimentation and development. They were exciting years.

In April 1949, the Fourth Inter-American and Region 2 Radio Conference, with twenty-five countries represented, convened in Washington. No major changes in amateur radio affairs took place. The House of Representatives passed a resolution commending amateurs. And in the December issue of QST, the ARRL editors proclaimed October 10, 1949, as important to amateurs as were Marconi's three dits sent across the Atlantic on December 12, 1901. As a result

231

of the unity of the amateur position, on the latter date the FCC had gone along with almost every item desired by the hams. United States allocations of kilocycles were decided upon, by and large to the satisfaction of the amateurs.[1]

Effective January 1, 1951, the FCC handed down new rules for licensing. It created a Novice class license. This was an entry level grade, examination for which required answers to basic questions and demanding just five words per minute in a code test. The license was good for just two years and was not renewable. To maintain license status the novice was required to pass the examination for one of the higher license grades prior to the expiration of the Novice license. In the 1951 restructuring a Technician Class was created, as well as an Advanced Class (replacing Class A) and finally an Amateur Extra Class. These two classes were granted additional frequencies.[2]

The old BCL (broadcast listener) problem appeared again because transmitters interfered with television reception. Caused by harmonic radiation, the only sure solution was a transmitter that was enclosed in a metal housing that confined the residual harmonic energy within the enclosure. Moreover, all connections leaving the housing had to be filtered. The problem resulted in manufacturers quickly producing what were known as TVI Proof (television interference proof) transmitters. Nevertheless TVI must be blamed for a general slowdown in ham activity, particularly on the popular hf bands.

Under the guise of esthetics but with the days of TVI remembered, communities began adapting ordinances that severely limited outside antennas. The result was an FCC ruling stating such regulatory action was within federal government jurisdiction. This prevented communities from banning antennas completely but it could not touch homeowners associations and CC&R (covenants, conditions and restrictions) rules. The problem continues today.

As stated above, the major changes in amateur radio since 1950 have been due mainly to technology. "To cover them in detail," writes W6OWP, "would require a book-length manuscript." Fascinating as the progress in communication technology is, it has not been the primary subject of this book. Nevertheless a brief overview is called for.

More and more, advancements in technology have seen the days of home construction declining. Today's average amateur station features an array of manufactured equipment, the central player of which is the transceiver, transmitter and receiver in a single package. Except for high power amplifiers, transistor development has done away with vacuum tubes. Coexistent with this development has come improved circuitry. Smaller, more compact equipment is not a manufacturer's move for a competitive edge, it's just the practical application of new construction techniques. These techniques would be difficult if not impossible for the "home brew" amateur of today to duplicate.

Still, the experimenters, tinkerers and those anxious to try new techniques don't give up easily. New ideas to improve circuit design, reliability and operation efficiency are always showing up in QST and contemporary radio and electronic publications. The long time standard AM voice transmission has given way to the more efficient Single Sideband (SSB) system for voice communication. FM (Frequency Modulation) is now the standard for communication in the vhf/uhf regions. And a small group of amateurs began using teletype machines to mechanize sending and receiving. The FCC took notice, making legal frequency shift keying (FSK)—a system where the carrier frequency changes with keying impulses rather than being interrupted as in CW transmission. FSK was adopted by commercial services in the 1940s and became one of the building blocks for what became known as digital communications.

The "amplifying crystal"—what we know as the transistor—began getting attention with an article in the March 1953 QST by George Rose, K2AH. This was just the beginning. Transistors were expensive, but hams began acquiring them and discovering their potentialities. By the mid–60s what had become known as Solid State had amateurs the world over experimenting.

Especially intriguing were the multiple function "chips," less than an inch long and ⅜ of an inch wide. The steps from the chip to the computer were complex, but steady.

In 1957 came Russia's launching of Sputnik. Its radio made a beep-beep signal that amateurs could pick up during orbital passes favorable to their location. Four months later the army put *Explorer I* into orbit, and development progressed steadily to the first moon landing. The potential of satellites was so obvious to amateurs that in December 1961 they launched their own satellite, dubbed OSCAR (Orbital Satellite Carrying Amateur Radio). Since then, dozens of satellites for amateurs have been placed into space by American and foreign amateur groups. A practical means by which launch vehicles accommodate carrying an amateur satellite is to use it in place of ballast, thus adding useful cargo to the craft's primary payload. The "birds" use transponder technology to accomplish their space mission. A signal to the satellite (uplink) triggers an on board transmitter (downlink) to automatically repeat the received signal back to earth. Depending on the orbital characteristics of the individual satellite, its access time and geographic coverage can be calculated, thus enhancing its use for communicating over a wide area.

Nor should changes for amateurs wrought by the computer be overlooked. Soon communications programs pairing the computer with the amateurs' transceiver came on the market. The bulky and noisy Teletype machine could now be replaced with the quiet computer keyboard. The same keyboard could be used for sending Morse code. Received signals could be viewed on the computer's monitor screen.

All the above hardly scratches the surface of the experimentation and development that has continued to the present. The technically minded, curious radio amateur is still with us, and he or she is busy. Where those in earlier years were experimenting with new circuits or antennae of their own design, the young recruits of today are trying their hand at computer programming, designing interface units to integrate their transceivers with computers or becoming involved with the design and construction of low power (QRP) equipment using the latest in miniature component technology developed by today's manufacturers.

But hams still have an incredibly important part to play in modern society. For example: September 11, 2001:

After the terrorist attacks the New York–Long Island section emergency coordinator, Tom Carrubba, KA2D, reported that hams were supporting emergency officials and American Red Cross relief, helping at Red Cross shelters and New York City American Red Cross headquarters as well as giving aid to the New York City Office of Emergency Management. Lower Manhattan's telephone system was operating problematically because of overloading. "The Amateur Radio ops (operators) are doing a great job under very difficult and strange conditions," said Carrubba, "but this is what they have trained for; they are doing it well." Local clubs volunteered gear, repeaters and operators. Amateurs were working 12-hour shifts per day, "plus or minus three or four hours, mostly plus." Simultaneously an "upbeat" crew of two dozen amateurs were staffing six amateur radio stations in the immediate vicinity of the Pentagon. And in Pennsylvania Kevin Custer, W3KKC, arranged communications with the Red Cross, the Salvation Army, the Pennsylvania State Police, the FBI, and other federal agencies on the scene.

After September 11 more than 300 hams volunteered in excess of 5,000 work hours during the aftermath of the disaster. Many initially did the "iron man act," working thirty or forty hours, running on adrenaline. After that they received some rest and were able to unwind, then back they went on minimum twelve-hour shifts.

The response may have been Amateur Radio's finest hours (since the 1937 Ohio River flood).[3]

December 25, 2004: an underwater earthquake at the violence of 9 on the Richter scale brought on a Tsunami that killed at least a hundred thousand persons; eighteen aftershocks were reported, some registering up to 6 on the Richter scale. Immediately amateurs in India began handling "hundreds of pieces of health and welfare traffic regarding people missing and from

**Forrest Bartlett, age 88, fixing his antenna. Hams remain hams until they become "silent keys"!**

relatives of those living in the Andaman and Nicobar islands, close to the earthquake's epicen-ter." A ham physician named Sarath, 4S7SW in Maratha, Sri Lanka, was soon on the air request-ing food, clothing and medicines for relief. The Calcutta Amateur Radio Society set up a control station in Calcutta. The Indian army assisted amateurs in setting up stations close to the epi-center, and therefore the worst hit area.[4]

Worldwide contributions of amateurs are needed when disasters occur, and disasters to come are as certain to occur as disasters that have happened before.

So too will the amateurs still be with us, and for that we should be grateful.

And should the reader have a son or daughter who shows curiosity about how radio, the television, or the computer works, do encourage the child. When the instrument is ready for the trash heap, give it to that curious kid. Let him or her take it apart. Find a basic pamphlet or write the ARRL and obtain one of their instructional books. Encourage the curiosity, help the boy or girl step by step into the wonderful world of amateur radio.

You will be doing not only your son or daughter a favor, but, indirectly, the world at large.

# Appendix A

# Morse Code Alphabet

| | | |
|---|---|---|
| A .− | N −. | 0 −−−−− |
| B −... | O −−− | 1 .−−−− |
| C −.−. | P .−−. | 2 ..−−− |
| D −.. | Q −−.− | 3 ...−− |
| E . | R .−. | 4 ....− |
| F ..−. | S ... | 5 ..... |
| G −−. | T − | 6 −.... |
| H .... | U ..− | 7 −−... |
| I .. | V ...− | 8 −−−.. |
| J .−−− | W .−− | 9 −−−−. |
| K −.− | X −..− | Fullstop .−.−.− |
| L .−.. | Y −.−− | Comma −−..−− |
| M −− | Z −−.. | Query ..−−.. |

# Appendix B

# Q Signals and Other Abbreviations

## Q Signals

A Q signal followed by a "?" asks a question. A Q signal without the "?" answers the question affirmatively, unless otherwise indicated.

QRA—What is the name of your station?
QRG—What's my exact frequency?
QRH—Does my frequency vary?
QRI—How is my tone? (1–3)
QRK—What is my signal intelligibility? (1–5)
QRL—Are you busy?
QRM—Is my transmission being interfered with?
QRN—Are you troubled by static?
QRO—Shall I increase transmitter power?
QRP—Shall I decrease transmitter power?
QRQ—Shall I send faster?
QRS—Shall I send slower?
QRT—Shall I stop sending?
QRU—Have you anything for me? (Answer is negative)
QRV—Are you ready?
QRW—Shall I tell _____ you're calling him?
QRX—When will you call again?
QRZ—Who is calling me?
QSA—What is my signal strength? (1–5)
QSB—Are my signals fading?
QSD—Is my keying defective?
QSG—Shall I send _____ messages at a time?
QSK—Can you work break in?
QSL—Can you acknowledge receipt?
QSM—Shall I repeat the last message sent?
QSO—Can you communicate with _____ direct?

QSP—Will you replay to _____?
QSV—Shall I send a series of Vs?
QSW—Will you transmit on _____?
QSX—Will you listen for _____ on _____?
QSY—Shall I change frequency?
QSZ—Shall I send each word/group more than once? (Answer, send twice or _____)
QTA—Shall I cancel number _____?
QTB—Do you agree with my word count? (Answer negative)
QTC—How many messages have you to send?
QTH—What is your location?
QTR—What is your time?
QTV—Shall I stand guard for you _____?
QTX—Will you keep your station open for further communication with me?
QUA—Have you news of _____?

## Abbreviations for Morse Code Transmissions: Prosigns and Other Procedural Signals

Prosigns, which appear with a bar over the letters, are symbols formed by running together two characters into one without the inter character space. This creates an abbreviation for the most common procedural signals.

CW—Phone (meaning or purpose)
AA—(Separation between parts of address or signature.)

AA—All after (use to get fills).

AB—An before (used to get fills).

ADEE—Addressee (name of person to whom message addressed).

$\overline{\text{ADR}}$—Address (second part of message).

AR—End of message (end of record copy).

ARL—Used with "check," indicated use of AARL numbered message in text.

$\overline{\text{AS}}$—Stand by; wait.

B—More (another message to follow).

BK—Break; break me; break-in (interrupt transmission on CW; quick check on phone).

$\overline{\text{BT}}$—Separation (break) between address and text; between text and signature.

C—Correct; yes.

CFM—Confirm. (Check me on this).

CK—Check.

DE—From; this is (preceding identification).

$\overline{\text{HH}}$—Error in sending. Transmission continues with last word correctly sent.

HX—Handling instructions. Optional part of preamble. Initial(s). Single letter(s) to follow.

IMI—Repeat; I say again. (Difficult or unusual words or groups.)

K—Go ahead; over; reply expected. (Invitation to transmit.)

KN—Specific station answer only.

N—Negative, incorrect, no more. (No more message to follow.)

NR—Number. (Message follows.)

PBL—Preamble (first part of message).

N/A—Read back. (Repeat as received.)

R—Roger; point. (Received; decimal point.)

SIG—Signed; signature. (Last part of message.)

$\overline{\text{SK}}$—Out; clear. (End of communications, no reply expected.)

TU—Thank you.

WA—Word after (used to get fills).

WB—Word before (used to get fills).

N/A—Speak slower.

N/A—Speak faster.

N/A—There is no CW equivalent.

# Appendix C

# Common Abbreviations in Amateur Morse Code

| | | | |
|---|---|---|---|
| ABT | about | SED | said |
| AGN | again | SEZ | says |
| ANI | any | SKED | schedule |
| BCNU | be seeing you | TKS or TNX | thanks |
| BTU | back to you | TT | that |
| BK | break | TU | thank you |
| BTR | better | TDA | today |
| CRD | card | VY | very |
| CUD | could | WX | weather |
| CUL | see you later | XMTR | transmitter |
| DX | distance | YF | wife |
| ES and U UR | you, your | YL | young lady |
| FB | fine business | 73 | best regards |
| GA | go ahead | 88 | love and kisses |
| GB | good-bye | 33 | love you too |
| GDA | good day | XYL | ex-young lady (wife) |
| GE | good evening | R | received OK |
| GN | good night | Rgr(Roger) | also sometimes used as above |
| GUD | good | HNY | Happy New Year |
| HAM | amateur operator | MX | Merry Christmas |
| HI | laughter | Fone | phone |
| HR | here | CWCode | (his CW sending is good) |
| HRD | heard | TTY | teletype transmission |
| HV | have | MTR Meter | (the output meter reads normal) |
| HW | how | Freq Frequency | (his frequency is correct) |
| NIL | nothing | Kc | kilocycle |
| NW | now | Mc | megacycle |
| OB | old boy | Sat(s) | satellite(s) |
| OM | old man | Ant | antenna |
| OP or OPR | operator | Pblm | problem |
| OW | old woman | LL | landline |
| PSE | please | Msg | message |
| RCVR | receiver | | |

# Chapter Notes

## Introduction

1. A detailed defense of amateur radio is in Susan J. Douglas, *Listening In* (Baltimore: Johns Hopkins University Press, 1999), pp. 328–346.

## Chapter 1

1. Experimenters had been delving into the concept of wireless for many years, and choosing Marconi's achievement as a beginning date of amateur activity is done only for convenience.

2. December 15, 1901, pp. 1–2; W.B. Jolly, *Marconi* (New York: Stein and Day, 1972), p. 106; Gavin Weightman, *Signor Marconi's Magic Box* (New York: Da Capo Press, 2003), pp. 88–107.

3. Orin E. Dunlap, Jr., *Radio's 100 Men of Science: Biographical Narratives of Pathfinders in Electronics and Television* (Freeport, New York: Books for Libraries Press, 1944, reprinted 1977), pp. 11, 67.

4. Ibid., p. 10. A drawing of Hertz's experiment is in William Maver, Jr., "Wireless Telegraphy—Its Past and Present Status and Its Prospects," *Annual Report of the Board of Regents of the Smithsonian Institution for the Year Ending June 30, 1902* (Washington, D.C.: Government Printing Office, 1903), p. 263.

5. Dunlap, p. 37; Joseph B. Lebo, "The Man Before Marconi," *QST*, August 1948, pp. 42–44; J.S.V. Allen, "Amateur Radio in 1882," *QST*, December 1940, p. 28; Weightman, pp. 62, 109, 117; Clinton B. DeSoto, *Two Hundred Meters and Down: The Story of Amateur Radio* (No Place: The American Radio Relay League, Inc., 1925), p. 11; Margaret Cheney, *Tesla: Man Out of Time* (Englewood Cliffs, N.J.: Prentice-Hall, Inc., 1981), pp. 74–178; "Marconi Wireless Telegraph Co. of America v. United States," *United States Reports: Cases Adjudged in the Supreme Court of the United States*, vol. 320, June 21, 1943, pp. 1–80.

6. DeSoto, pp. 16–17.

7. Headquarters Staff of the ARRL (American Radio Relay League), *The Radio Amateur's Handbook*, 9th edition (West Hartford, Conn.: The ARRL, Inc., 1932), p. 1.

8. Paul Schubert, *The Electric Word: The Rise of Radio* (New York: The Arno Press of the *New York Times*, 1921, 1971), pp. 19–21.

9. G.G. Blake, "How to Make a Wireless Telegraph," *The Boys Own Paper* (London, England), vol. 27, Saturday, November 2, 1904, pp. 110–111; an earlier issue gave instructions on building a dry battery: vol. 19, Friday, September 4, 1897, pp. 779–780. As the art improved, so did the sophistication. See A. Frederick Collins, "Design and Construction for a 100 Mile Wireless Telegraph Set," *Scientific American Supplement*, no. 1605 (October 6, 1908) pp. 15712–15714.

10. In 1905 Hugo Gernsback, an immigrant from Germany, advertised the first wireless kit, the Telimco Wireless Telegraph Outfit. Macy's sold it for $7.50 (Weightman, *Signor Marconi's Magic Box*, p. 202); Amateurs chose their own call letters then. Often it was their initials. The San Lorenzo Valley Amateur Club, *Interlink* 2, no. 4 (September–October 1997): 3.

11. Winthrop Packard, "The Work of a Wireless Telegraph Man," *The World's Work* 7, no. 4 (February 1904): 4467–4470; Eugene P. Lyle, Jr., "The Advance of Wireless," *The World's Work* 9, no. 4 (February 1905): 5842–f5848. Page 5845 has a small reproduction of a page from the *Cunard Daily Bulletin*.

12. Weightman, 230–235; W.P. Jolly, pp. 178–179. Erick Larson, *Thunderstruck* (New York: Random House, 2006) is a recent book about the Crippen murder with additional information on Marconi.

13. Lawrence Perry, "Commercial Wireless Telegraphy," *The World's Work* 5, no. 5 (March 1903): 3195; Packard, p. 4467.

14. Schubert, *The Electric Word*, pp. 70–72; Susan J. Douglas, *Inventing American Broadcasting, 1899–1922* (Baltimore: The Johns Hopkins University Press, 1987), p. 95.

15. Lawrence Perry, "Commercial Wireless Telegraphy," p. 3195.

16. Lyle, "Advance Wireless," p. 5847; Kenneth Wimmel, *Theodore Roosevelt and the Great White Fleet: American Sea Power Comes of Age* (Washington, D.C.: Brassey's, 1998), p. 155. The worldwide interest in wireless may be surmised from the names of the systems tried by the U.S. Navy: Ducretet, Rochefort, Slaby-Arco, Braun, DeForest, Fessenden, Bull, and Telefunken (Lyle, p. 5846).

17. The *New York Times*, February 13, 1910, lists fourteen "Notable Instances of Rescue Made by Wireless." Among them is the foundering off the Alaskan coast on August 17, 1909, of the Steamship Ohio. The Marconiman was George E. Eccles, who was one of just five who lost their lives; 150 were saved. Eccles is commemorated on an

obelisk in Battery Park, New York City, dedicated to wireless operators who lost their lives while on duty in sinking ships. He had previously been employed as dispatcher for the Canadian Northern Railway. There had been a collision causing excessive damage but no loss of life, for which he felt responsible. He quit his job and without informing his friends, went to Alaska and became the *Ohio's* radio operator. The Seattle *Daily Times*, August 27, 1909.

18. "A Wireless Victory: How the Use of the Marconi Telegraph Averted a Great Maritime Disaster," *Harper's Weekly* 53, no. 2719 (January 30, 1909): 7.

19. *New York Times*, January 24, 1909, p. 1.

20. Ibid., pp. 24, 27.

21. *New York Times*, January 27, 1909; Allan Chapman, *The Radio Boys' First Wireless* (New York: Grosset and Dunlap, Publishers, 1922), pp. 1–2.

22. Many books have been written about the *Titanic* disaster. Sources consulted include Richard Garrett, *Atlantic Disaster: The Titanic and Other Victims of the North Atlantic* (London: Buchan and Publishers, 1986), pp. 151–157; the *Times* of London, April 14–28, 1912; *New York Times*, April 14–22, 1912; W.P. Jolly, *Marconi*, p. 179; 181–184; Susan Douglas, *Inventing American Broadcasting*, pp. 216 ff.; Robin Gardiner and Dan Van Der Vat, *Titanic Conspiracy: Cover-Ups and Mysteries of the World's Most Famous Sea Disaster* (New York: Citadel Press, 1998, first published in England in 1995) pp. 8, 52–53, 80–82, 85–86, 205, 219; Charles Pellegrino, *Her Name Titanic* (New York: Avon Books, 1990, first published in England, 1988), pp. 237–328.

23. Gardiner and Van Der Vat, pp. 52–53; 80.

24. This incident is described in greater detail in Jolly, *Marconi*, pp. 182–183. Harold Bride went back to sea, and in 1922 was radio operator on the *Cross-Channel Ferry*. Then he disappeared, covering his tracks so well that, when he became a traveling salesman, no one even knows what he sold. He married and had three children. Bride was finally identified after his death on April 19, 1956, of cancer, in a Glasgow hospital. It was said that after the *Titanic* disaster his hair turned white. We can speculate that the notoriety he earned by his dedication during the tragedy bothered him to the point that he had himself disappear. Yet he is believed to have retained his amateur status, working hams all over the world. See Charles Pellegrino, pp. 237–238.

25. Gardiner and Van Der Vat, p. 139.

26. March 1, 1912.

27. Mention of the amateurs is in "Wireless Telegraphy and Wireless Telephony," Report to Accompany House Resolution 182, 61st Congress, 2nd session, March 29, 1910. The "danger" referred to stemmed from the October 1906 decision of the San Francisco school board to segregate Orientals in San Francisco schools. Although President Roosevelt settled the affair amicably, Japan was incensed and "rattled the saber" in protest.

28. Clippings from the *Examiner* are included in Hearings under House Joint Resolution 95, "A Bill to Regulate and Control the Use of Wireless Telegraphy and Wireless Telephony," Committee on Naval Affairs, Subcommittee on Naval Law, 61st Congress, 2nd session, January 8, 1910, pp. 110–112. The *Examiner* article is included in the testimony of H.S. Gearing, a naval equipment officer, writing to the committee on November 4, 1907.

29. The Senate of the United States Committee on Commerce, "Hearings on the Bill (S. 7243) to Regulate Radio Communication," 61st Congress, 2nd session, April 18, 1910.

30. Sixtieth Congress, 1st session, Senate Committee on Foreign Relations, *Exhibit A*. The correct title of the Berlin document, as submitted by President Roosevelt to the Senate, is "International Wireless Telegraph Convention ... with Service Regulations Annexed Thereto, a Supplementary Agreement, and a Final Protocol, All Signed at Berlin on November 3, 1906, By Delegates of the United States and those of Several Other Powers."

31. January 28, 1909. The incident was further discussed by the *Bremen* operator before the Roberts Committee on Naval Affairs, Report to Accompany House Resolution 182, 61st Congress, House, 1st session, March 29, 1910.

32. Robert A. Morton, "The American Wireless Operator," *Outlook*, January 1910, pp. 131–135.

33. Ibid.

34. House of Representatives, *Joint Resolution 95*, "A Bill to Regulate and Control the Use of Wireless, Telegraph, and Wireless Telephony," 61st Congress, 2nd session, 1909–1910.

35. The Senate of the United States Committee on Commerce, "Hearings on the Bill (S. 7243) to Regulate Radio Communication," 61st Congress, 2nd session, April 18, 1910, pp. 12–19.

36. Susan Douglas, p. 225.

37. The Senate of the United States Subcommittee on Commerce," Hearing on the Bills (S.3620) and (S. 5334) to Regulate Radio Communication, 62nd Congress, 2nd session, March 1, 1912.

38. Public Law 264, U.S. Statutes at Large, "An Act to Regulate Radio Communication," August 12, 1912; Taft signed on August 17th.

# Chapter 2

1. Correspondence with Forrest Bartlett. Details of the initiation are in "Hamdom," *QST*, January 1934, p. 33.

2. Obituary in the *Times* of London, November 25, 1916, p. 10; Iain McCallum, *Blood Brothers: Hiram and Hudson Maxim—Pioneers of Modern Warfare* (London: Chatham Publishing, 1999). See also Hiram Percy Maxim, *A Genius in the Family: Hiram Stevens Maxim Through His Son's Eyes* (New York: Harper and Brothers, 1926).

3. Obituary in the *New York Times*, February 18, 1936, section L-23; see also Alice Clink Schumacher, *Hiram Percy Maxim* (Cortez, Colorado: Electric Radio Press, Inc., 1998).

4. Hiram Percy Maxim, *Horseless Carriage Days* (New York: Harper and Brothers, 1936, 1937).

5. Obituary, *New York Times*, February 17, 1936.

6. Schumacher, pp. 53–54.

7. Ibid.

8. Clinton B. Desoto, *Two Hundred Meters and Down* (Hartford, Connecticut: The American Radio Relay League, 1936), pp. 34–35.

9. Sixty-first Congress, 2nd session, Hearings on S. 7243, Testimony of W.E.D. Stokes.

10. DeSoto, p. 38.

11. Ibid., pp. 38–43.

12. Ibid.

13. Ibid.

14. Ibid., p. 41; *QST*, no. 1 (December 1915): 3–5.

15. Ibid. By the mid–1920s the navy had become a defender of amateur radio.

16. *QST*, February 1916, p. 20.

17. *QST*, January 1916, p. 5.

18. *QST*, March 1916, pp. 57–58; *QST*, April 1916, p. 65.

19. William H. Kirwan, "Washington's Birthday Amateur Relay Message," *QST*, April 1913, pp. 65–66; *Scientific American*, February 19, 1916, p. 16. Descriptions of the preparations differ somewhat in the two periodicals.

20. *QST*, April 1916; *New York Times*, February 22, 1916, p. 16.

21. *Literary Digest* 51 (July 3, 1916): 15; *QST*, June, July, and August 1916.

22. For examples see *QST*, June 1916, "Long Distance Work in May," p. 140, and a letter from a Cleveland, Ohio, amateur boasting of working stations up to a thousand miles away (p. 143).

23. *QST*, February 1917, pp. 34–35. For Pershing's foray into Mexico, see Herbert Molloy Mason, Jr., *The Great Pursuit* (New York: Smithmark, 1970).

24. *QST*, February 1917, p. 35.

25. "ARRL First Transcontinental Radio Relay," *QST*, February 1917, p. 21.

26. "QRM," *QST*, January 1917, p. 26.

27. *QST*, pp. 35–36. Mrs. Candler may have been responsible for "OW" (for Old Woman) being used as the feminine of "OM" (Old Man) in amateur parlance (*QST*, October 1916, p. 303).

28. A.C. Campbell, "First Trans-Continental Relay Fails," *QST*, February 1917, pp. 40–43.

29. "Trans-Continental Traffic Begins," *QST*, April 1917, p. 18.

30. Ibid.

31. "The Trans-Continental Record," *QST*, April 1917, p. 17.

32. *New York Times*, March 8, 1917, p. 8.

33. *QST*, May 1917, pp. 18; 24.

34. "Where Are We Bound?" *QST*, February 1917, p. 36.

# Chapter 3

1. *Literary Digest*, January 16, 1915, p. 93; *New York Times*, November 21, 1915.

2. *New York Times*, January 14; March 6, 1916.

3. *QST*, January 1917, pp. 25, 28, 32–34; July 1917, p. 19.

4. *New York Times*, August 10, 1914, section IV.

5. *New York Times*, January 29, 1914.

6. *New York Times*, July 16, 1914.

7. *New York Times*, August 6, 1914.

8. Ibid.

9. Kenneth Macksey, *The Searchers: Radio Intercept in Two World Wars* (London: Cassell, 2003), p. 35. See also Paul Schubert, *The Electric Word: The Rise of Radio* (New York: Arno Press and the New York Times, 1971 reprint of 1928 edition), pp. 128–129; Diana Preston, *Lusitania: An Epic Tragedy* (New York: Walker and Company, 2002), pp. 159–163; *New York Times*, August 7, 1914.

10. Keith Yates, *Graf Spee's Raiders: Challenge to the Royal Navy, 1914–1915* (Annapolis, Maryland: Naval Institute Press, 1995), pp. 119–143; Barrie Pitt, *Coronel and Falkland* (London: Cassell and Co., 1960, paperback 2002). The wireless operator on the HMS *Glasgow* heard "almost continuous, high-singing Telefunken signal notes" (Pitt, p. 1; see also p. 65).

11. *New York Times*, August 20, 1914.

12. *New York Times*, August 30; September 1, 3, 4, 1914.

13. *New York Times*, September 5, 1914.

14. *New York Times*, September 10, 13, 15, 1914; Schubert, *The Electric Word*, p. 104.

15. *New York Times*, September 10, 13, 15, 1914.

16. *New York Times*, September 25, October 27, 1914.

17. *New York Times*, April 4, 8, 1917.

18. *New York Times*, November 7, 1914.

19. *New York Times*, November 7, 8, 25, 1914.

20. *New York Times*, November 11, 15, 1914. If there was such a station, details of its discovery and its closing are not available. Another Mexican station and its closure are discussed below.

21. *New York Times*, November 18, 1914.

22. *New York Times*, November 13, 15, 1914.

23. *New York Times*, November 23, 1914.

24. Ibid.

25. *New York Times*, December 16, 1914.

26. Daniel Allen Butler, *The Lusitania: The Life, Loss, and Legacy of an Ocean Legend* (Mechanicsburg, Pa.: Stackpole Books, 2000), p. 56. According to articles in the *Times* of London (March 30, 31, 1915) the *Falaba* was not bound for Liverpool, as Butler writes, but was outward bound for the West African Coast. Taylor, the Marconiman, is listed as one of the survivors. Dr. Bernard Dernburg, ex-colonial minister of Germany, in an interview to the *New York Times*, April 1, 1915, excused the torpedoing. "She [the *Falaba*] was working her wireless apparatus and called for help. That is probably why they sank her."

27. Butler, pp. 152, 162, 168; Adolph A. Hoehling and Mary Hoehling, *The Last Voyage of the Lusitania* (New York: Henry Holt and Co., 1956), pp. 74, 85, 103, 115, 154, 171. The Hoehlings list as references relatives of survivors, among them S. Alex Leith. This may be the source of their information on Robert Leith. See also Diana Preston, *Lusitania: An Epic Tragedy* (New York: Walker and Co., 2002), p. 119.

28. Thomas S. Bailey and Paul B. Ryan, *The Lusitania Disaster: An Epic in Modern Warfare and Diplomacy* (New York: The Free Press, 1975), p. 119.

29. Ibid. See also David Ramsay, *Lusitania: Saga and Myth* (New York: Norton, 2001) and Butler, *The Lusitania* for up-to-date analyses of the tragedy.

30. *New York Times*, April 23, 1915.

31. *New York Times*, May 10, 1915.

32. *New York Times*, May 11, 1914.

33. *New York Times*, June 12, 1915; *QST*, September 1917, p. 18. The *Times* consistently misspelled his name (Zemmick rather than Zennick).

34. *Literary Digest*, September 11, 1915; quoted from the *Electrical Experimenter*, September 1915.

35. *New York Times*, July 1, 1915. The *Journal* article was reproduced in the *Times*.

36. *New York Times*, July 22, 1915.

37. Harden Pratt, "Sixty Years of Wireless and Radio Experience" (n.p., n.d.; typescript in the possession of Forrest A. Bartlett).

38. Ibid.

39. Ibid.

40. Macksey, *The Searchers*, pp. 46–47.

41. *QST*, April, 1917 "Our Country Calls Us," p. 25; "If We Are Closed Up," p. 26; "Department of Defense," p. 32.

42. *QST*, May 1917, p. 3.

43. "War," *QST*, pp. 4, 13.

44. The Old Man, "Rotten!!," *QST*, June 1917, pp. 7–9.

45. Ibid.

46. "Reports of Division Managers," *QST*, July 1917, p. 21; "Now the Army," *QST*, 7 September 1917, p. 15.

47. The Old Man, "Something Rotten Somewhere," *QST*, pp. 7–10.

48. "Concerning Phantom Antennas," *QST*, p. 2. The words "Antenna" and "Aerial" are used interchangeably.

49. "Another Season Opens, But—," *QST*, p. 16.

50. Ibid., p. 27.

51. *New York Times*, April 10, 18, 1917.

52. *New York Times*, May 19, November 8, 1917; March 5, 8, and July 30, 1918.

53. *New York Times*, May 19, 1917.

54. The Old Man, "Something Rotten Somewhere," *QST*, September 1917, pp. 7–10. Zenneck was interned first at Ellis Island and then at Fort Oglethorpe, Georgia. After the war he returned to Germany, remained prominent in radio experimentation, and in the fall of 1928 returned to the United States to receive the Institute of Radio Engineers' Medal of Honor. *Proceedings*, Institute of Radio Engineers, vol. 16 (August 1928): 1025.

55. Navy Publicity Bureau, "What the Naval Reserve Offers Men of the ARRL," *QST*, August 1917, pp. 3–4; Bill Woods, "in the Service," *QST*, p. 17.

## Chapter 4

1. Jan Davis Perkins, N6AW, *Don C. Wallace, W6AM: Amateur Radio's Pioneer* (Vestal, New York: The Vestal Press, Ltd., 1991), pp. 1–3, 26–34, 237. Unless otherwise noted, the information on Don Wallace and his service on the *George Washington* is based upon this book.

2. Ibid., pp. 26–34.

3. Ibid. A brief biography of Fred Schnell is in *QST*, November 1919, pp. 29–30.

4. House of Representatives, 64th Congress, 2nd session.

5. Ibid., pp. 13–36.

6. Ibid., p. 147.

7. Ibid., pp. 180, 195, 209, 215–221. Sarnoff's criticism could have had something to do with his rejection for an officer's position with the navy, although the service's anti-Semitism could also have been a factor.

8. Ibid., 321.

9. Ibid., p. 222. The reference to decrement refers to regulations pertaining to spark transmitters. The term had to do with the wave formation produced by the electrical discharge across the transmitter spark gap. It minimized the interference caused by spark transmitters.

10. Ibid., pp. 222–225.

11. *Scientific American*, February 3, 1917, p. 116; March 17, 1917, p. 276.

12. The Senate of the United States Committee on Commerce, "Hearings on the Bill (S. 7243) to Regulate Radio Communication," 61st Congress, 2nd session, April 18, 1910, p 407.

13. House of Representatives, 65th Congress, 3rd session, *Hearings Before the Committee on Merchant Marine and Fisheries on A Bill to Further Regulate Radio Communications*, December 12, 13, 17, 18, and 19, 1918, pp. 5–24.

14. Ibid., pp. 39–40.

15. Ibid. Maxim's statements extend from p. 238 to 272, indicating that the committee considered his testimony of considerable importance.

16. Ibid., pp. 272–273.

17. Ibid., p. 314.

18. But an entrepreneur named David Sarnoff *had* envisioned the revolution in commercial radio. He anticipated profits from the sale of receivers.

19. The latter three words are in his essay "A Few Random Remarks," *QST*, February 1926, p. 26.

20. *New York Times*, July 15, 1919.

21. *QST*, September 1919, "The Amateur Situation," pp. 5–6.

22. "Operating Department," *QST*, August 1919, p. 16.

23. *New York Times*, September 28, 1919, section II, p. 3.

24. "Daniels Only Knows," *QST*, October 1919, pp. 11–12.

25. "Ban Off: The Job Is Done, and the ARRL Did It," *QST*, supplement for October 1919 issue.

26. Jan David Perkins, *Don Wallace*, p. 36.

27. Quoted in Perkins, *Don Wallace*, p. 36.

28. *QST*, December 1920, pp. 5–6.

29. Sixty-sixth Congress, Senate, 1st session, "Hearings Before a Subcommittee of the Committee on Naval Affairs," part 2, October 1919.

30. Gerald Carson, *The Roguish World of Dr. Brinkley* (New York: Rinehart and Co., 1960), pp. 199–202; R. Alton Lee, *The Bizarre Careers of John R. Brinkley* (Lexington, Ky.: The University Press of Kentucky, 2000), pp. 164–167. Brinkley used the Toggenberg breed because it did not smell. Milford, a town of about 500, is just west of Manhattan and north of Junction City, Kansas.

31. Jonathon Daniels, *Shirt-Sleeve Diplomat* (Chapel Hill, N.C.: University of North Carolina Press, 1947), pp. 373–374.

## Chapter 5

1. Sources for the 1920s are Frederick Lewis Allen, *Only Yesterday* (New York: Harper and Brothers, 1931); Preston William Slosson, *The Great Crusade and After, 1914–28*, The History of American Life Series (New York: Macmillan, 1931); and Mark Sullivan, *Our Times*, vol. 4, "The Twenties" (New York: Charles Scribner's Sons, 1971).

2. Allen, *Only Yesterday*, pp. 164–167.

3. Many sources are available for information about the beginnings of commercial radio. I have used Sidney W. Head, *Broadcasting in America*, 3rd ed. (Boston: Houghton Mifflin Co., 1972); Peter M. Lewis and Jerry Booth, *The Invisible Medium* (New York: Macmillan, 1989); Susan Smuylan, *Selling Radio* (Washington, D.C.: Smithsonian Institution Press, 1994); and Susan Douglas, *Inventing American Broadcasting, 1899–1922* (Baltimore: Johns Hopkins University Press, 1987).

4. "With Our Radiophone Listeners: Getting Started Listening," *QST*, pp. 47–48, 51.

5. Marvin R. Bensman, *The Beginning of Broadcast Regulation in the Twentieth Century* (Jefferson, N.C.: McFarland, 2000), pp. 49–51.

6. Ibid.

7. *QST*, February 1923, p. 58.

8. "Strays," *QST*, April 1923 p. 63.

9. Editorial, "The 'Phones and Amateur Radio," *QST*, March 1922, pp. 29–33.

10. Among many sources for the history of radio since 1920 are Susan J. Douglas, *Listening In* (New York: Times Books by Random House, 1999) and Susan Smuylan, *Selling Radio*.

11. The FCC has always been heavily influenced by politics with the general trend being toward more and more monopoly control of both radio and television. See Sidney Head and Christopher Sterling, *Broadcasting in America; A Survey of Electronic Media*, 5th ed. (Boston: Houghton Mifflin Co., 1987).

12. Quoted in Head and Sterling, p. 153.

13. "Rotten Broadcasts," *QST*, May 1922, pp. 12–15.

14. *QST*, pp. 51–54.

15. "With Our Radio Listeners," *QST*, July 1922, p. 54; S.R. Winters, "Radiophone Broadcasting of Weather and Market Reports," *Radio News*, February 1922, pp. 717 ff; S.R. Winters, "Government and Amateurs Join," *Radio News*, May 1923, pp. 768 ff.

16. Gernsback, "Results of the $500 Prize Contest," *Radio News*, February 1922, p. 1450.

17. "A Word to the Novice," *QST.*, April 1922, pp. 38–39.

18. Ibid.

19. Editorial, "Unscrambling the Eggs," *QST*, April 1923, p. 31. Rochester, New York, had first tried the plan.

20. "Radio Market News Service," *QST*, July 1921, p. 24; "Wireless Market Reports Used by Many Agencies," *QST*, March 1922, pp. 48–49. The reports had been extended to North Platte, Nebraska, Rock Springs, Wyoming, and Reno and Elko, Nevada.

21. *QST.*, July 1921, p. 25.

22. Herbert H. Foster, "Radio in the National Guard," *Radio News*, January 1923, p. 1259; Editorial, "Onward," *QST*, October 1925, p. 7; K.B. Warner, "The Army Links Up with the Amateur," *QST*, pp. 22–24.

23. Editorial, "The Army-ARRL Affiliation," *QST*, December 1925, p. 7; "Army-Amateur Notes," *QST*, May 1926, p. 100.

24. Tom C. Rives, captain, signal corps, "Signal Corps Training in Citizen's Military Training Camp," *QST*, April 1926, p. 47.

25. Captain A.C. Stanford, USA, liaison agent, "The Purpose of Army Amateur Affiliation," *QST*, April 1927, pp. 33–34.

26. "Schnell Returns," *QST*, November 1925, p. 25; "Recognition for Our Traffic Manager," *QST*, January 1926, p. 54.

27. Hiram Percy Maxim, ed., "Another Chance to Put One Over," *QST*, February 1925, p. 2.

28. K.B. Warner, untitled editorial, *QST*, August 1929, p. 7; R.G.H. Mathews, "The Amateur and the Naval Reserve," *QST*, pp. 17–19.

29. Editorial, "Friends of Ours," *QST*, June 1921, p. 27; J.H. Dillinger, PhD, chief of Radio Laboratory, Bureau of Standards, "The Radio Work of the Department of Commerce," *QST*, pp. 18–21.

30. Editorial, "Help the Cops," *QST*, April 1921, p. 25; "Amateur Radio Recovers a Stolen Auto," *QST*, May 1921, p. 16.

31. Porter T. "Rip" Bennett, 51P, "The Land of Blue Lightning," *QST*, December 1923, pp. 22–23.

32. "Dern the Amateur," *QST*, June 1923, p. 36; "The League's Radio Information Service," *QST*, August 1923, p. 36.

33. "City Ordinances," *QST*, April 1923, pp. 32–33.

34. Editorial, "Bugaboo Nr. 1,234,567,890," *QST*, May 1923, p. 52.

35. "Radio Interference Ordinances Cannot Limit Transmitting Stations," *QST*, June 1927, p. 43, 45; Paul M. Segal, "Municipal Ordinances on Radio Transmitters," *QST*, September 1927, pp. 25–26. Segal finds five groupings of ordinances, and gives constitutional reasons why they are illegal. See also A.R. Budlong, "Municipal Ordinances on Radio Transmission Unlawful," *QST*, January 1928, pp. 28, 70. Budlong was assistant to the secretary of the league.

36. K.B. Warner, "The Washington Radio Conference," *QST*, April 1922, pp. 7–11. The following paragraphs are from this report. Hoover's philosophy towards radio is spelled out in S.R. Winters, "An Interview with Secretary H.C. Hoover," *Radio News*, October 1924, pp. 472, 568.

37. K.B. Warner, "The Radio Telephony Conference," *QST*, June 1922, pp. 15–17. An excellent compilation of the history of radio regulation is Marvin R. Bensman, *The Beginning of Broadcast Regulation in the Twentieth Century* (Jefferson, N.C.: McFarland, 2000).

38. K.B. Warner, "The New Radio Bill," *QST*, July 1922, p. 32, text on p. 56; "The White Bill," *QST*, January 1923, pp. 40–41.

39. Editorial, "Unscrambling the Eggs," *QST*, April 1923, p. 31.

40. Editorial, "The State of the Amateur," *QST*, July 1923, pp. 33–35.

41. Editorial, "The New Short Waves," *QST*, September 1924, p. 7.

42. Editorial, "The New Problems," *QST*, December 1924, p. 7; K.B. Warner, "The Third National Radio Conference," p. 16; A.R. Kennelley, "The Conference in Relation to Amateur Activities," p. 8, Ibid. The conference also called for an end to spark transmitting, a suggestion with which the ARRL heartily approved.

43. "The New Problems, *QST*, December 1924, p. 7; Bensman, *Beginning of Broadcast Regulation*, p. 113.

44. Bensman, *Beginning of Broadcast Regulations*, p. 117.

45. Editorial, "Going Up," *QST*, January 1926, p. 8.

46. K.B. Warner, "The Fourth National Radio Conference," *QST*, pp. 33–36.

47. Ibid.

48. "Radio Legislation Pending," *QST*, March 1926, p. 44; editorial, K.B. Warner, "The Problem of Regulation," *QST*, June 1926 P. 7; editorial, K.B. Warner, "Loyalty," *QST*, September 1926, p. 7. See also Bensman, *Beginning of Broadcast Legislation*, pp. 153–206. This presents the controversy in detail.

49. Warner, "Loyalty."

50. Ibid.

51. Editorial, *QST*, February 1927, p. 8.

52. Editorial of *QST*, April 1927, pp. 7–8. The text of the Radio Bill of 1927 is on pp. 39–44.

53. K.B. Warner, "Radio Regulation Returns," Ibid (May 1927), pp. 15–17, 26. Warner's statement to the Commission is included.

54. K.B. Warner, Editorial, Ibid., (May 1925), p. 7.

55. Editorial, K.B. Warner, "Interference," *QST*, June 1925, pp. 7–8.

56. F.H. Schnell, "ARRL Vigilance Committees," *QST*, May 1925, p. 37.

57. "Sending Licenses Suspended," *QST*, May 1925, p. 37.

# Chapter 6

1. J.O. Smith, "The Operating Department," QST, September 1919, pp. 16–21.

2. QST, July 1919, p. 20; "Affiliating the Clubs," QST, August 1919, pp. 6–7; "The Operating Department," QST, September 1923, p. 36.

3. F.H. Schnell, "Ham Traffic in Any Old Shack," QST, September 1923, pp. 31–34.

4. Ibid., p. 32.

5. The description of the relay is entirely from Schnell (with some clarification by Forrest Bartlett) and its use is hereby acknowledged. Ibid.

6. "Daylight Communication Wins!" QST, March 1925, pp. 8–9. The ARRL attempted several "transcons." One series was on July 2, 4, and 6, 1922, another on September 23, 1923. None were successful. See F.H. Schnell, "Daylight Transcons Fail," QST, September 1922, p. 26; "Report on the Transcoms of September 23rd," QST, November 1923, p. 27. But by 1925 they were succeeding. Incidentally, Forrest Bartlett believes W6TS, with whom he talked on the air years ago, was 6TS, a pioneer radioman named Ed Willis.

7. Editorial, "The Five Point System," QST, November 1926, p. 7.

8. Ibid.

9. W.K. Kirwan, "Results of the Washington's Birthday Relay," QST, June 1921, pp. 21–23.

10. Ibid.

11. Ibid. Call letters of the winners were not listed in the article.

12. K.B. Warner, "Governors-President Relay," QST, February 1922, p. 34; QST, May 1922, pp. 17–21.

13. F.H. Schnell, "Police Chief's Relay," QST, May 1922, pp. 21–22; Boyd Phelps, "The Police Chief's Relay," QST, July 1922, p. 11.

14. T.O.M., "Rotten Hours," QST, March 1920, p. 7–8.

15. T.O.M, "Rotten Air," QST, June 1920, pp. 12–13, 23.

16. QRU, "The First Epistle from the Young Squirt to the Old Man," QST, September 1920, pp. 17–18.

17. Correspondence with Beth Keuneke, adult service coordinator, St. Mary's Public Library, February 19, 2003, including information from the Auglaze County Historical Society; QST, December 1916, p. 33; QST, July 1921, p. 21; "Anyone Who Hasn't Heard 8NH?" QST, May 1917, p. 33 (this has a photograph of Mrs. Candler's rig); QST, July 1921, p. 37. According to the Historical Society, the Candlers moved from St. Mary's to Marietta, Ohio, in 1930. (The Historical Society may have been wrong in reporting that she had earned a second class commercial license.) Charles Candler's ham activities are mentioned in QST, November 1920, p. 42, along with a portrait; his wife is pictured on page 303 of the October 1916 QST.

18. Editorial, "The Ladies Are Coming," QST, August 1917, p. 19; e-mail from Forrest Bartlett, January 2, 2000; editorial, "The Radio Ladies," QST, March 1921, p. 30, 64.

19. The Old Woman, "Beginning of the End," QST, September 1920, pp. 10, 14.

20. "The Story of the Transcons," QST, March 1921, pp. 5–12; 47.

21. Pierre H. Boucheron, "The Amateur Transatlantic Feat," Radio News, December 1920, pp. 353, 416.

22. The editor (K.B. Warner), "The Story of the Transatlantics," QST, February 1922, pp. 7–11.

23. Ibid., pp. 7–11; "Godfrey in England to Copy Transatlantics," QST, October 1921, pp. 29–32.

24. Paper presented by George F. Burghard at meeting of the Radio Club of America, "Station IBCG," QST, February 1922, pp. 30–32. See also Pierre Boucheron, "Amateurs Span the Atlantic," Radio News, February 1922, p. 697 ff. Several photographs accompany the article.

25. "Godfrey in England," p. 31.

26. Ibid.

27. Godley's achievement is based upon two articles in the February 1922 QST: The editor (K.B. Warner), "The Transatlantics," pp. 12–13; and Paul F. Godley, "Official Report of the Second Transmission Tests," pp. 14–28, 36–40, 46.

28. According to the editor of QST, it was on December 12th that 1BCG "sent a coherent message on broadcast, at 3 a.m. GMT" (Greenwich Mean Time). This transatlantic message read as follows: "NR 1 NY ck 12 to Paul Godley, Androssan, Scotland." In non–QST English the message said: "Hearty Congratulations. Burhard, Inman, Grinan, Armstrong, Amy, Cronkhite."

Godley cabled back to 1BCG that he had received their message. He also gave the results of each day's test to Marconi's powerful commercial station, which transmitted them by slow Morse code to WII, the American RCA station in New Brunswick, in order to let amateurs everywhere copy direct.

29. "More About the Transatlantics," QST, April 1922, pp. 35–36. Two articles in the New York Times, December 14 and 17, 1922, are at variance with the QST articles. They say that 99 American radio operators' signals were received by amateurs in England, France, and Switzerland "last Tuesday, Wednesday, and Thursday, the first three days of the amateur international test." They also state that the tests were run from December 12 until 31. They actually ran from December 7 to 16. See also A.E. Greenslade, "How I Received the American Amateur Stations in England," Radio News, February 1922, pp. 700 ff.

30. K.B. Warner, "The Onward March of Transocean Communication," QST, March 1924, pp. 19–21; "New World's Relay Records," QST, January 1924, pp. 18–19; K.B. Warner, "Transatlantic Amateur Communication Accomplished!," QST, January 1924, pp. 9–12. Laurence S. Lees, "Another Historic Event in Amateur Radio," Radio News, March 1924, pp. 1236, 1290, 1292, 1294, 1296. This essay concerns Leon Deloy, the French amateur.

31. "International Intermediates Expanded," QST, February 1925, p. 22.

32. Howard S. Pyle, 8FT, "The Amateur DX Card: A Study of the International Fad that 8UX Started," QST, September 1924, pp. 36–38.

33. "More About the Transatlantics," QST, April 1922, pp. 35–36, 39, 54; quotations on page 36.

34. Oscar C. Roos, "Get Ready for 'IL,' Work with Foreign Amateurs," QST, February 1924, pp. 21–22.

35. Ibid.; also, Oscar C. Roos, "A Radio Auxiliary Language for Trans-Oceanic Work," Radio News, May 1924, pp. 1565, 1674.

36. Henry W. Wetzel, "The Language of International Radio," QST, July 1924, pp. 42–45; James D. Sayers, "Esperanto Radio World Language," Radio News, August 1924, pp. 169, 204–206. See also Oscar C. Roos, "Esperanto or ILO–Which?" Radio News, October 1924, pp. 471, 546–549. On the first page of each of the above two articles are paragraphs in ILO and English, or Esperanto and English. ILO claims to be an improvement over Esperanto.

For information on ordering beer in over forty languages, including Esperanto, see http://www.esperanto-usa.org.

37. K.B. Warner, "ARRL Endorses Esperanto," *QST*, September 1924, p. 40–41.

38. E-mail from Forrest Bartlett, August 9, 2001.

39. Hiram Maxim, "The International Radio Union," *QST*, May 1924, pp. 16–17.

40. Ibid.

41. K.B. Warner, "International Amateur Radio Union Formed," *QST*, June 1925, pp. 9–13.

42. Ibid.

43. IARU News, *QST*, August 1928, pp. 60, 72, 74, 76; "IARU News," *QST*, December 1928, pp. 51, 64, 66. The latter includes the amended constitution.

44. Hiram Maxim, "Big Dividends," *QST*, December 1927, pp. 9–11.

45. "The Royal Order of Brasspounders," *QST*, July 1924, pp. 36–37.

46. K.B. Warner, editorial, "Washington," *QST*, December 1927, pp. 9–11.

47. K.B. Warner, editorial, no title, *QST*, January 1928, pp. 9–11.

48. *QST*, December 1927, p.9.

49. K.B. Warner, "The Amateur and the Radiotelegraph Conference," *QST*, January 1928, pp. 15–22; *New York Times*, October 20, 1927, p. 26.

50. Ibid.

51. Ibid., p. 22.

52. K.B. Warner, editorial, *QST*, May 1929, pp. 9–10.

53. Editorial, "The Hoover Cup," *QST*, January 1922, pp. 25–26; "The Department of Commerce Cup," *QST*, pp. 20–22.

54. E-mail from Forrest Bartlett, September 28, 1999.

55. The Old Man, "Rotten S.O.L.," *QST*, February 1921, pp. 9–15, 19, 23. All material on the St. Louis convention is from this article.

56. His name was John K. Haddaway. He went to California and in Mono Lakes ran a factory that manufactured "tiny pumps." He died in the Northern Inyo Hospital on January 17, 1990. This information is from Ms. Patricia Gatz and William Korn of the St. Louis Genealogical Society; Los Angeles Westside Genealogical Society; and Barbara Moss of the Bishop, California, Museum and Historical Society. Reference to "tiny pumps" is in Susan R. Schrepfer, "Establishing Administrative 'Standing': The Sierra Club and the Forest Service, 1897–1956," in *American Forests: Nature, Culture, and Politics*, ed. Char Miller (Topeka: University Press of Kansas, 1997), p. 132.

57. "Our First National Convention," *QST*, October 1921, pp. 7–21. See also Roscoe Smith, "The Chicago Radio Convention," *Radio News*, October 1921, p. 282. Material on the convention is from these two sources.

58. Correspondence from Forrest Bartlett, October 15, 1998.

59. "A New International Brasspounders Club," *QST*, April 1926, p. 54.

60. "The Rag-Chewers Club," *QST*, June 1925, p. 29. The QST English in the article is flawed; with Forrest's help, I have corrected it. Forrest adds that "it is unlikely there was any formal cancellation of an R.C.C. membership. The statement was a warning to members to avoid that sort of operating."

61. "The ROWH," *QST*, December 1923, pp. 8–9; F.D. Fallain, "The Story of the Royal Order of the Wouff Hong," *QST*, April 1924, pp. 22–26.

62. Fallain, "Story of the Royal Order."

63. Ibid.

64. K.B. Warner, editorial, *QST*, July 1927, p. 7.

## Chapter 7

1. Howard F. Mason (7BK), "The Radio Lizz," *QST*, December 1922, pp. 26, 28. Two photographs of the car plus a schematic are included in the article.

It should be pointed out that wireless and automobiles were linked as far back as 1903. In February of that year Lee De Forest equipped an electric automobile, calling it Wireless Auto No. 1. He exhibited it along Wall Street in New York and later at the St. Louis Exposition. See Michael Brian Schieffer, *Taking Charge: The Electric Automobile in America* (Washington D.C.: Smithsonian Institution Press, 1994), p. 85. In 1910 the owners of a Chalmers "30" used three stations, one mounted on an automobile, to keep in touch on the Glidden Tour. "Wireless Telegraph Apparatus for Contestants of the Glidden Tour," *Scientific American*, May 14, 1910, p. 395. The Chalmers won.

2. Oliver Wright, "Loops and Fords," *QST*, July 1925, pp. 33–36. Four photographs and schematics accompany the article.

3. Ibid.

4. F. Johnson Elser, "Following the Sun with a Radio Flivver," *QST*, September 1927, pp. 9–10.

5. Some examples are "Have you heard KFUH?" *QST*, February 1925, p. 20; "The Mysterious WJS," *QST*, August 1925, p. 20; T.S. McCaleb, "WJS," *QST*, August 1925; "Contact with the Expeditions," *QST*, September 1926, pp. 99–100; *QST*, October 1926, pp. 99–100; "Expeditions," *QST*, January 1930, p. 101; "Gatti African Expedition," OQ5LZ, "Bowdoin-Kent Expedition, VEILN," and "Archbold New Guinea Expedition," PK6XX, which are all in *QST*, August 1938, p. 47; and Edward A. Ruth III, W2GYL, "OQ5ZZ Calling 'CQ USA,'" *QST*, April 1939, p. 29–30. See also T.S. McCaleb, "Radio with the Rice Amazon Expedition," *Radio News*, November 1925, pp. 588, 747, 749.

6. Harry Wells, W3DZ, "The Story of PMZ," *QST*, August 1930, pp. 9–14. Unless otherwise noted this narration is based upon Wells's article.

7. "Borneo": Gail Saari, "The Penan: the Forest People of Borneo." Earth Island Institute, "The Borneo Project": www.borneoproject.org

8. *Chicago Tribune*, August 15, 1929, p. 7.

9. W6BBY apparently was not part of the expedition's scheduled contacts. How his CQ was answered by PMZ gives one the impression that W6BYY was just looking for someone to talk to and it was Wells's good fortune to make the contact. Theodore Seelmann mentions Paul Holbrook, KA1AF of Fort Mills, P.I.; Commander S.M, Mathes, KA1CY of Manila, P.I.; L.R. Potter, W6AKW of Lancaster, California; and Colonel Clair Foster, W6HM of Carmel, California, as the operators who gave Wells the greatest support. Letter written by Steelman in "Correspondence," *QST* (June, 1930) p. 52.

10. "Two Chicagoans and Radio Man Lost in Borneo," *Chicago Tribune*, August 15, 1929, p. 7. The story related how a message from Wells told of a Dyak tribesman being found "slain with a spear thrust through his neck." The

murderer was found and taken to the nearest outpost for trial, leaving the three Americans unguarded, and they had not been heard from since. This is good newspaper material but although Wells tells of the capture of the Malayan murderer, he says nothing about the party being left alone, unguarded.

11. "All-American Mohawk Malaysian Expedition," *QST*, June 1930, p. 52. The University of Chicago had no additional information on this expedition.

12. "Pan-American Highway," *The Columbia Encyclopedia*, 5th ed. (New York: Columbia University Press, 1993), p. 2060; *New York Times*, January 16, 1930.

13. Bertram Sandham, "With IPH in Mexico," *QST*, September 1930, pp. 17–20, 72, 74; *Los Angeles Times*, March 18, 1930.

14. *QST*, September 1930, pp. 17–18.

15. Harry Carr, "Links Two Americas," *Los Angeles Times*, March 1, 1930, pp. 1–2 (a map of the proposed route is included); March 16, 1930; *QST*, September 1930.

16. Carr, March 18, 1930.

17. Sandham, "With IPH in Mexico"; Carr, March 21, 1930, p. 1.

18. Sandham, "With IPH in Mexico."

19. Sandham, "With IPH in Mexico."

20. Carr, March 27, 28, 29, 30; April 1, 6, 12, 14, 20, 27; May 1. Not all of Carr's dispatches were sent by amateur radio. His feature articles were sent to the *Times* by mail or telegraph. The articles listed constitute a delightful travelogue of Mexico in 1930.

21. Sandham, "With IPH in Mexico."

22. Sandham, "With IPH in Mexico."

23. Anonymous, "Blazing a Trail Through Mexico," *Popular Mechanics*, September 1931, pp. 404–408; Bertram Sandham, W6EQF, "In the Field with IPH," *QST*, December 1931, pp. 33–35; Clinton B. DeSoto, "IARU News," *QST*, p. 60; *Los Angeles Times*, January 17–18, 1931.

24. Sandham, "In the Field with IPH."

25. Ibid.; DeSoto, "IARU News," *QST*, December 1931, p. 60.

26. Sandham, "In the Field with IPH."

27. *Los Angeles Times*, February 13, 21; April 12, 1931.

28. Alfred Aloysius Horn, *Trader Horn*, edited by Etherelds Lewis (New York: Simon and Schuster, 1927). The copy I used was the nineteenth printing, with 174,000 copies already sold–a real best-seller in the 1920s.

29. Clyde DeVinna, "Hamming with a Portable in Africa," *QST*, July 1930, pp. 27–32. Unless otherwise noted, this essay is based upon the DeVinna article.

30. "Have you Heard KFUH?," *QST*, February 1925, p. 20. The article details their contributions on a 1925 expedition in the Pacific.

31. *New York Times*, February 4, 1931, p. 21; February 8, p. 5. More on Trader Horn is in David Killingray, review of Tim Couzens, "Tramp Royal: The True Story of Trader Horn," *Southern African Review of Books* Issue 27 (September/October , 1993) http://www.uniulm.de/~rturrell/antho2html/Killingray.html.

32. Clinton B. DeSoto, *Calling CQ: Adventures of Short-Wave Radio Operators* (New York: Doubleday and Co., 1941), pp. 125–133. This book is a delightful read. It is not documented and the author has created conversation, but it is clearly accurate and thorough. DeSoto was affiliated with the ARRL and was contemporary with the incidents he relates. Long out of print, copies of the book are hard to come by.

33. Georges-Marie Haardt, "The Trans-Asiatic Expedition Starts," *National Geographic* 59, no. 6 (June 1931): 776–782.

34. "Finding the Expeditions: Motor Car Station," *QST*, July 1931, p. 50; *QST*, August 1931, p. 48; *National Geographic*, June 1931, pp. 776 ff.; Maynard Owen Williams, "The Citroen-Haardt Trans-Asiatic Expedition Reaches Kashmir," *National Geographic* 60, no. 4 (October 1931): 387–443.

35. *National Geographic*, June 1931, pp. 776 ff.

36. O.W. Hungerford, "VP3THE, the Terry-Holden Expedition," *Radio*, December 1938, pp. 17–21; 76–78. Save for the historical paragraphs, which are common knowledge, the narration is based entirely on this article.

## Chapter 8

1. Editorial, *QST*, May 1938, p. 7; *QST*, January 1935, p. 25; *QST*, May 1939, p. 11; *QST*, January 1940, p. 10.

2. E. Lowell Kelly, "Personality over the Air," *QST*, pp. 42–43, 56–57.

3. William H. Graham, "Crashing Page One," *QST*, January 1934, pp. 32, 56, 58.

4. "The History of the Olympics: 1932—Los Angeles, U.S." http://www.history1900s.about.com/library/weekly; W.A. Lippman, Jr., "W6USA–Amateur Radio at the Olympics," *QST*, August 1932, pp. 27–28.

5. W.A. Lippman, Jr., "W6USA–The World Was Its Oyster," *QST*, October 1932, pp. 10–12, 90.

6. Ibid.

7. "Amateur Radio at the Century of Progress," *QST*, August 1933, pp. 28–29; Wallace F. Wiley, "The World's Fair Radio Amateur Exhibit," *QST*, December 1933, pp. 29–31.

8. "World's Fair A.R.R.L. Convention," *QST*, July 1933, p. 8.

9. "World's Fair Amateur Radio Convention," *QST*, October 1933, pp. 23–24; I.S. Coggeshall, "When the World's Radio Speed Title Changed Hands," *QST*, November 1933, pp. 39–40. The speed contest was conducted as follows: It started at 8 words per minute (WPM). At five minute intervals the speed was advanced first to 12 WPM, then 15, 20, 25, 30, 35, and up to a speed where no one made acceptable copy. Jean Hudson made acceptable copy at the 20 WPM section and was the winner in that speed range. The overall winner was Joseph Chaplin from Press Wireless, who beat out Ted McElroy with a speed of 57.3 WPM (e-mail from Forrest Bartlett, January 27, 2004).

10. *QST*, August 1934, p. 57.

11. "Amateur Radio—A Century of Progress—1934," *QST*, June 1934, p. 34; "Amateur Radio at the Fair," *QST*, August 1934, p. 39; "World's Fair Amateur Exhibit," *QST*, p. 57.

12. E-mail from Forrest Bartlett dated October 22, 1999, and June 1, 2003.

13. "Amateur Radio at the Fairs," *QST*, June 1939, p. 23.

14. *QST*; "An Animated Radio Diagram," *QST*, July 1939, p. 26. This is a page of photographs of the diagram.

15. "Timing Ski Races via 56 mc," *QST*, July 1935, p. 57; "546 Mc. at Outboard Races," *QST*, February 1936, p. 66; *QST*, January 1935, p. 54; *QST*, March 1938, p. 38; "Radio Rifle Matches," *QST*, December 1929, p. 103.

16. Garland C. Black, "C.C.C. and the Amateur," *QST*, November 1933, pp. 36, 82; H.O. Bixby, "Third Corps Area Asks Amateur Help," *QST*, October 1934, pp. 16–17. Reference to "radio conferences" referred to government meetings where amateur radio needed defenders.

17. William Haight, "The C.C.C. Takes the Air," *QST*, January 1938, pp. 41, 86.

18. "Index to 1937 QST," pp. 130–136.

19. "The Governors-President Relay," *QST*, July 1933, p. 30; E.H. Williams, "The 1937 Governor's-President Relay," *QST*, March 1937, pp. 45, 86. "ARRL Affiliated Club Directory," *QST*, pp. 33–35.

20. "Announcing the ARRL Twentieth Anniversary Relay," *QST*, April 1934, p. 19; "Some Anniversary Greetings," *QST*, May 10, 1934, p. 110.

21. E.D. Miller, "56 Mc Rolls Up Its Sleeves: Solid Work at 5 Meters at the National Glider Meet," *QST*, September 1932, p. 29. Amateur Don Wallace conducted many experiments with five meters including transmitting from the summit of Mt. Whitney. See Perkins, *Don C. Wallace* (Vestal, New York: The Vestal Press, Ltd., 1991), p. 111.

22. "First Annual Field Day Report," September 1933, p. 35; "August '36 Field Day," *QST*, January 1937, pp. 28–30; "1937 ARRL Field Day Results," *QST*, November 1937, pp. 11–13.

23. Correspondence from Forrest Bartlett, December 21, 1994; e-mail from Bartlett dated June 10, 2001, in collection of Richard A. Bartlett.

24. Mitchell V. Charmley, *News by Radio* (New York: The Macmillan Company, 1948), p. 2. According to Charmley 8MK had been broadcasting since August 20.

25. Ibid., pp. 2–5.

26. Ibid., pp. 6–10.

27. T.R. Carskadon, "The Press-Radio War," *The New Republic*, March 1936, pp. 132–135.

28. Ibid., p. 133.

29. Ibid.

30. Ibid.

31. Ibid.; Bice Clemow, "U.P.-I.N.S. Take Lead in Effort to Retain Control of Radio News," *Editor and Publisher*, May 4, 1935, pp. 304.

32. Isabelle Keating, "Radio Invades Journalism," *The Nation*, June 12, 1935, pp. 677–678. I have not found a physical description of Moore. In one of those horrific happenings in one's life, Forrest Bartlett was reading the obituaries in the local Paradise, California, paper one day and there was Herbert Moore's name. Both men had lived for some time in the same small town, but neither knew the other was there. Had Forrest known it—or Moore of Forrest—surely they would have got together for long talks. Forrest contacted Moore's widow and was allowed to read a manuscript history of Trans Radio Moore had written. He returned it, of course; Mrs. Moore left, and attempts to contact her have failed (letter from Forrest Bartlett to author, August 19, 2001).

33. Clemow, p. 4.

34. Clemow, "Trans Radio Files Suit for Million," *Editor and Publisher*, May 25, 1935, pp. 9, 44; Ervin Lewis, "More News After This," in Herb Moore's unpublished manuscript, courtesy of Forrest Bartlett.

35. Ervin Lewis, "More News After This."

36. "Coast Press-Radio Bureau Folds; Publishers Withdraw Support," *Editor and Publisher*, July 20, 1933, p. 8; *Editor and Publisher*, July 27, 1935, pp. 10, 30; Charmley, *News by Radio*, pp. 19–22; Carskadon, "The Press-Radio War," p. 134.

37. Ervin Lewis, "More News After This." No date is given for this, but it must have been after Trans Radio ceased to exist.

38. Press Wireless was a shortwave communications system created by the press associations to receive news worldwide. It was established in 1929 and faded from the scene in the 1960s.

39. Editorial, "It Seems to Us," *QST*, January 1939, p. 7; "Hams over Chicago," *QST*, November 1938, pp. 48–51, 116, 118, 120.

40. Editorial, "It Seems to Us," *QST*, January 1939, p. 7.

41. K.B. Warner, editorial, "The Editor's Mill," *QST*, April 1938, pp. 7–15; see also Alice Clink Schumacher, *Hiram Percy Maxim* (Cortez, Colorado: Electric Radio Press, 1998).

42. Editorial, "It Seems to Us," *QST*, November 1938, pp. 7.

43. A.L. Budlong, "Cairo," part I, *QST*, January 1938, pp. 1113, 66, 68, 70, 72, 74, 76, 78. This article presents in detail the international regulation of radio from 1906 until the time of the Cairo Conference in 1938.

44. K.B. Warner, "The Madrid Conference," *QST*, February 1933, pp. 9–17.

45. Ibid.

46. K.B. Warner, "The American Regional Conference," *QST*, November 1933, pp. 19, 46.

47. "What the League Is Doing," *QST*, August 1936, pp. 21–22.

48. "What the League Is Doing," *QST*, October 1936, p. 27, 78, 80, 82.

49. "What the League Is Doing," January 1937, p. 21; *QST*, February 1937, p. 21; *QST*, March 1937, p. 32.

50. James J. Lamb and John C. Stadler, "The Fourth C.C.I.R. at Bucharest Paves the Way for Cairo," *QST*, September 1937, pp. 8–11, 102.

51. *QST*, July 1937, pp. 21–22, 50, 52, 54, and 56.

52. "What the League Is Doing," *QST*, July 1937, p. 21; *QST*, December 1937, p. 22; K.B. Warner, "The First Interamerican Radio Conference," *QST*, February 1938, pp. 9–11, 68, 70, 72, 74.

53. Warner, "The First International Radio Conference," p. 10; see also "What the League Is Doing," *QST*, January 1938, p. 24.

54. A.L. Budlong, "Cairo," part I, *QST*, January 1938, pp. 11–13, 66, 68, 70, 72, 74, 76, 78; *QST*, part II, February 1938, pp. 32–33, 88, 90, 92, 94, 96, 98, 100, 102.

55. Kenneth B. Warner and Paul M. Segel, "The Battle of Cairo," *QST*, July 1938, pp. 9–12, 45–46, 104, 106.

56. *QST*; "What the League Is Doing," April 1938, p. 18; *QST*, May 1938, p. 22.

57. "How's DX?" *QST*, p. 68.

## Chapter 9

1. The navy amateur was Fred Schnell. "Schnell Returns," *QST* November 1925, p. 25; "Recognition for Our Traffic Manager," *QST*, January 1926, p. 54.

2. Allen Villiers wrote a trilogy about the last sailing ships: *By Way of Cape Horn* (London: Hodder, 1930); *Falmouth for Orders* (Garden City: Garden City Publishing Co., 1929); and *Grain Race* (New York: Charles Scribner's Sons, 1933).

3. Villiers, *Falmouth for Orders*, p. 174.

4. M.B. Anderson, "In IARU News" (International Amateur Radio Union), *QST*, January 1928, pp. 52, 72, 74.

5. Villiers, *Falmouth for Orders*, p. 173. The *Grebe*, which rescued the *E.R. Sterling*, was commissioned as a minesweeper in 1919. According to "Historical Notes: U.S.S. Grebe," on "October 14, 1927-the Barkentine 'E.R. Sterling,' one of eighteen [seventeen?] sailing ships to leave South Australia with a load of wheat in Spring of 1927 was reported in distress seventy miles from St. Thomas *Grebe* went to her assistance and towed in to harbor of St. Thomas. 'E.R. Sterling' was dismasted through storms encountered while rounding Cape Horn. She had been at sea one hundred and eighty days" (Department of the Navy, Naval Historical Center, Washington Navy Yard, Washington, D.C.). The information appears to be wrong about the vessel being in a storm "while rounding Cape Horn."

6. These names are taken at random from the "Expeditions" section of 1927–1930 QSTs.

7. Alan R. Eurich, WCFT-W8IGO, "CQ WCFT: Further Adventures Aboard the 'Yankee,'" *QST*, March 1938, pp. 11–13, 84, 86, 88. Unless otherwise noted, the experiences of Mr. Eurich are from this article.

8. Captain and Mrs. Irving Johnson, *Sailing to See: Picture Cruise in the Schooner Yankee* (New York : W.W. Norton and Co., 1939), pp. 29–30; see also Captain and Mrs. Irving Johnson, *Westward Bound in the Schooner Yankee* (New York: W.W. Norton and Co., 1936).

9. IARU News, *QST*, April 1927, p. 68.

10. Roy E. Abbott, VK2YK, "The World's Loneliest Radio," *QST*, August 1932, p. 59. VK2YK's residence was Park Street, Dorrigo, New South Wales, Australia; the names of the amateurs on Willis Island at the time of the author's writing are not given.

11. "Pitcairn Communications History," http://www.lareau.org/pitccomm.html.

12. Alan Eurich, "CQ PITC," *QST*, August 1937, pp. 9–10, 70, 72. Unless otherwise noted, information about Pitcairn Island is from this article. A boxed explanatory statement preceding Eurich's article relates Ross Hull's experience with Pitcairn.

13. Lew Bellem, "The New PITC: Modern Radio Equipment for Pitcairn Island," *QST*, January 1938, pp. 19–20, 78, 80, 82, 90, 92. The following paragraphs involving Pitcairn radio are from this article.

14. Ian M. Dall, *Pitcairn: Children of Mutiny* (Boston: Little, Brown and Co., 1973), pp. 274–280. When Eurich was at Pitcairn there were about 250 residents; today there are fewer than 50. Many have left for the United States, where their church, the Seventh Day Adventist, welcomes them. For information on the telephone, see Peter Barnes, "Think It's Tough to Make a Call to a Tiny Pacific Island? Not Atoll," *Wall Street Journal*, July 10, 1985. In 1939, according to Irving and Electa Johnson in "Westward Bound in the Yankee" (*National Geographic*, January 1942, pp. 1 ff), the *Yankee* in Panama took on board the Pitcairn radio, which had been sent there for repairs, and returned it to the Pitcairn Islanders.

15. Michael McGiar, M.D. (K9AJ), "VP6DI: The Story of the 2002 Ducie Island Expedition," *CQ: Amateur Radio Communications and Technology* 59, no. 2 (February 2003): pp. 11–13, 16.

16. Guy Townsend "Review of *Richard Halliburton, the Forgotten Myth*," http://www.memphismagazine.com/back issues/april2001/coverstory2.htm. In writing of Halliburton's junk, they mistakenly call it the *Green Dragon*.

17. Ibid.; Rex Purcell as told to George W. Polk, "The Cruise of the Pang Jin," *QST*, October 1939, pp. 18–20.

18. Purcell, "Cruise of the *Pang Jin*."

19. "I.A.R.U. News," *QST*, August 1932, p. 53; October 1932, p. 46.

20. Philip B. Peterson "The Shenandoah is Lost" *Information Age* http://www.infoage.org/p-79Shendandoah. April 7, 1991; *New York Times*, January 17, 1924. See also Richard K. Smith, *The Airships Akron and Macon* (Annapolis: U.S. Naval Institute, 1965).

21. *New York Times*, January 17, 1924, October 20, 1924, and September 25, 1925.

22. *New York Times*, June 28, 1931, July 10, 1931, August 8, 1931, September 25, 1931. According to Forrest Bartlett W60WP, "certain broadcasting station signals were, or could be used, for navigational purposes. The 'calibrating' could involve the station frequency, its geographical location, hours of operation and the general area within the survey region that specific stations were recorded with the most reliable signals" (e-mail from Bartlett, January 4, 2004).

23. "ZRS4 and ZRS5: Akron and Macon," *The Zeppelin Library*, http:// www.history.navy.mil/photo/ac-usn; "Macon Airship and Akron Airship," *New York Times*, April 4, 1933.

24. *New York Times*, May 31, 1935, August 29, 1935.

25. The Communications Department, "The Springfield Air Races, *QST*, August 1930, p. 51.

26. The Communications Department, *QST*, October 1931, p. 44; Miami *Herald*, January 7–10, 1928. The newspaper makes no mention of such a crash, but does report a crack-up from which the pilot walked way unscathed "outside the airport."

27. Ross A. Hull, "Amateur Radio in a New Field," *QST*, November 1931, pp. 32, 74.

28. E.D. Miller, "Solid Traffic Work on 5 Meters at the National Glider Meet," *QST*, September 1932, pp. 29, 88; Ross A. Hull, "Amateur Radio at the National Soaring Meet," *QST*, September 1933, pp. 32, 64.

29. "Strays," *QST*, June 1932, p. 24.

## Chapter 10

1. Richard K. Smith, *First Across: The U.S. Navy's Transatlantic Flight of 1919*, (Annapolis, Md.: Naval Institute Press, 1973), pp. 159, 191, 86–89. Information on the flight of the "Nancys" is primarily from this thoroughly documented, definitive book. The *New York Times* has seventy-three articles on the flights, but they reveal little that is not in the Smith book. (A blimp is a non-rigid airship; a dirigible has a metal skeletal structure.)

2. Smith, *First Across*, p. 25, 29. The NC-4 may be seen to day at the Navy Aviation Museum in Pensacola, Florida. "N" stood for navy and "C "stood for Curtis, the manufacturer who built the aircraft.

3. Ibid., p. 46.

4. Ibid., p. 113.

5. Ibid., p. 206, 221–236.

6. Admiral Richard E. Byrd, *Skyward: Man's Mastery of the Air* (New York: G.P. Putnam's Sons, 1928, paperback Jeremy P. Tarcher/Putnam, 2000). The essay on Byrd is primarily from the paperback edition.

7. Ibid., pp. 206–207.

8. Byrd, *Skyward*, pp. 237, 329. The back-up pilot, Bernt Balchen, was the only one trained to do instrument flying. For a darker side of Byrd's career, see Carroll V. Glines, *Bernt Balchen: Polar Explorer* (Washington, D.C.: Smithsonian Institution Press, 1999). See also Marc Dierikx, *Fokker* (Washington, D.C.: Smithsonian Institution Press, 1997), p. 105.

9. Ibid., p. 250. Ver-sur-Mer was the sight of British landings in the Second World War.

10. Richard E. Byrd and his crew were in the *America* on their way to France at the time of the Maitland and Hegenberger flight; Byrd radioed them good luck from the plane. Byrd, *Skyward*, p. 232. Maitland and Hegenberger had used the radio beacons and reported them satisfactory, but they had trouble with their receivers. *New York Times*, August 17, 1927, p. 1.

11. An excellent book on the Dole Contest is Robert H. Scheppler, *Pacific Air Race* (Washington, D.C.: Smithsonian Institution Press, 1988). Descriptions and illustrations of several of the planes are in "Descriptions of the Dole Derby Planes," *Aviation*, August 22, 1927, pp. 414–423.

12. Letter from Forrest A. Bartlett, W6OWP, May 17, 2002; *New York Times*, August 17, 1927, p. 2.

13. According to Kenneth Meek of the Woolaroc Museum, the *Woolaroc*'s radio was one the used by Ernie Smith and Emory Bronte in their June 1927 flight to Hawaii. E-mail from Kenneth Meek, January 18, 2005.

14. *New York Times*, August 18, 19, 1927. Or possibly just two and one-half hours (Scheppler, pp. 90–92).

15. J. Walter Frates and A.L. Budlong, "Amateur Radio and the Pacific Flights," *QST*, November 1927, pp. 40–41, 80.

16. Ibid.

17. Ibid.

18. Ibid.

19. Articles in the *New York Times*, August 19–22, 1927. Because she was not yet twenty-one, Eichwadt's young wife was denied permission; Mildred Doran was twenty-two. According to Scheppler, p. 40, it was not really a Swallow airplane, but had been built by two former employees of Swallow and contained many Swallow parts.

20. Ibid.

21. Ibid.

22. Ibid.

23. Ibid.

24. Ibid.

25. *New York Times*, August 21; Frates and Budlong. Eichwaldt's complete log is included in Scheppler, pp. 108–110.

26. Frates and Budlong.

27. Emelie and Rexford Matlack, W3CFC, "The Paper, the Station, and the Man—A Brief History of the *New York Times* Radio Stations," *73 Magazine*, February 1980, pp. 54–59.

28. An excellent brief biography of Kingsford-Smith is in Norman Macmillan, *Great Airmen* (London: G. Bell and Sons, Ltd., 1955), pp. 181 ff. See also *New York Times*, June 6, 1928.

29. *New York Times*, June 4, 1928, p. 3.

30. Ibid.

31. J. Walter Frates, "Following the 'Southern Cross' to Brisbane," *QST*, August 1928, pp. 21–22.

32. Ibid.

33. *New York Times*, June 8, 1927.

34. *New York Times*, June 9, 10, 1928. Both the *San Francisco Examiner* and the *New York Times* had their amateur licenses revoked by the Federal Radio Commission for using their rigs for commercial purposes (*QST*, December 1928, p. 22).

35. *QST*, August 1928, p. 22. In Australia there is an air museum with the *Southern Cross* on display and a replica of the Fokker F.VIIb-3 has been flown. "Smithy and the Southern Cross," State Library of New South Wales: http://www.atmitchel.com.

36. *Chicago Tribune*, July 6, 1929. On page 2 of the July 25, 1929, issue appear the headlines "'Untin' Bowler's Name Gives You Chance at Prize." The story began, "What is the Meaning of 'Untin Bowler' and how did the name originate?" Winners were to receive prizes of $100, $50, and ten prizes of $10 each. I scanned later issues but found no mention of the contest.

37. F.E. Handy, communications manager, "KHEJ and the 'Untin Bowler' Awards," *QST*, October 1929, pp. 21–22; 76. First prize went to J.R. Miller of Hammond, Indiana, W9CP; second prize to F.H. Schnell of Chicago, W9UZ; and third prize to Irving Strauss, Chicago, W9AAS. In lieu of the failed flight, the ARRL expressed a willingness to forget the prizes, but the *Tribune* stated that the newspaper was impressed with the part amateurs had played, and so it gave three prizes, although originally five prizes were to be awarded.

38. F.E. Handy, Communications Manager, "ARRL Cooperates with the 'Arctic Patrol' in Mid-Winter Maneuvers," *QST*, May 1930, pp. 29–38.

39. Ibid.

40. Ibid.

41. *New York Times*, January 6, 11, 14, 27, February 3, 1930.

42. Handy, p. 29.

43. Anne Morrow Lindbergh, *North to the Orient* (New York: Harcourt Brace and Company, 1935), pp. 30–32. "The third class commercial group were restricted operator permits. The Morse Code requirement for the restricted operator permit was ability to copy fifteen five-letter groups per minute. The written exam covered basic laws, theory and radiotelegraph operation. For most anyone starting from scratch the fifteen code groups per minute would certainly have required weeks of study. Of course, Anne and her husband were talented people" (e-mail from Forrest Bartlett, March 31, 2004).

44. A.M. Lindbergh, *North to the Orient*, pp. 34–35.

45. Ibid., pp. 42–45. For a description of the rig see Dorothy Hermann, *Anne Morrow Lindbergh: A Gift of Life* (New York: Ticknor and Fields, 1993), p. 77. It is quoted from Robert Dailey's biography of Juan Trippe, head of Pan American Airways.

46. A.M. Lindbergh, *North to the Orient*, p. 54; Hermann, p. 77.

47. A.M. Lindbergh, *North to the Orient*, p. 57.

48. Anne Morrow Lindbergh, *Hour of Gold, Hour of Lead: Diaries and Letters of Anne Morrow Lindbergh* (New York: Harcourt Brace Jovanovich, 1973), p. 172.

49. Anne Morrow Lindbergh, *Locked Rooms and Open Doors: Diaries and Letters of Anne Morrow Lindbergh: 1933–1935* (New York: Harcourt Brace and Co., 1974), pp. 34, 41.

50. Anne Morrow Lindbergh, with a foreword by Charles A. Lindbergh, "Flying Around the North Atlantic," *National Geographic* 67, no. 3 (September 1934): 259–337; A.M. Lindbergh, *Locked Rooms*, pp. 52, 77, 78.

51. A.M. Lindbergh, "Flying Around the North Atlantic," p. 307.

52. Hermann, p. 128; Anne Morrow Lindbergh, *Listen! The Wind* (New York: Harcourt Brace and Company, 1938), pp. 242–243; "Flying Around the North Atlantic," p. 329. QST, "Finding the Expeditions," August 1931 p. 48, lists the call letters of sixteen groups save for one: the Lindberghs' call letters are not listed. We can speculate that with their notoriety it was requested that their call letters not be listed to avoid a deluge of hams trying to contact them. QST has nothing on the Lindberghs' 1933 survey.

53. A.M. Lindbergh, *Listen! The Wind*, pp. 225–229. In chapter 8, "My Little Room," she describes in detail the equipment in her cockpit on the *Sirius*. The plane, renamed the *Tingmissartoq*, Eskimo for "one who flies like a big bird," is today in custody of the National Air and Space Museum (Hermann, p. 133).

54. *New York Times*, December 29, 1934.

55. This narration is based upon articles in the *New York Times*, December 29, 30, and 31, 1934, and January 1, 1935; also, QST, February 1935, pp. 12, 78, and 80.

56. Amateurs constituting the radio group included G.M. Brown, W2CVV, E.H. Fritschel, W2DC, G.W. Fyler, W2HLM, R.A. Lash, W2CBO, W.J. Purcell, ex-8JS, R.W. Williamson, ex-9AHH, and H.E. Hotaling, W8DKK, all working under call letters GE1000 (QST, February 1935, pp. 12, 78, and 80).

## Chapter 11

1. Kieran Mulvaney, *At the Ends of the Earth: A History of the Polar Regions* (Washington, D.C.: Island Press/Shearwater Books, 2001), pp. 203–204.

2. Frederick A. Cook, introduction by Edward Hoagland, *Through the First Antarctic Night, 1898–1899* (London: William Heinemann, 1900), x. For his claim to climbing Mt. McKinley, see Robert Dunn, *The Shameless Diary of an Explorer* (New York: The Modern Library, 2001).

3. For more on the Cook-Peary question, see Charles Officer and Jake Page, *A Fabulous Kingdom* (New York: Oxford University Press, 2001), pp. 162–173; and Clive Holland, ed., *Farthest North* (New York: Carroll and Graf, 1999), pp. 197–221.

4. Cook, *Through the First Antarctic Night*, 290 ff.

5. Ibid., p. 310. There were fourteen expeditions to Antarctica between 1897 and 1917. See Alan Gurney, *The Race to the White Continent* (New York: Norton, 2000), p. 282.

6. Gurney, *Race to the White Continent*, pp. 393–400.

7. Roland Huntford, *Shackleton* (New York: Carroll and Graf, 2002), p. 419.

8. Sir Ernest Shackleton, *South: Journals of His Last Expedition to Antarctica* (New York: Konecky and Konecky, no date), pp. 315–16, 320, 325, 327–28, 330, 334–35, and 339. The most puzzling entry (p. 319) refers to Hooke trying to contact men at "the Hut" on Cape Evans: "If the people at the hut have rigged the set which was left there, they will hear 'All well' from the *Aurora*." They did not receive the message. F.A. Worsley, *Endurance: An Epic of Polar Adventure* (New York: W.W. Norton and Co., 1931, 1999); *New York Times*, May 14, 1916, Wireless Press story. In bringing supplies from the *Aurora* to be cached along

Shackleton's proposed route, the *Aurora* lost three men and a fourth went insane. For more about the *Aurora* being saved by a radio message, see Lennard Bickel, *Shackleton's Forgotten Men* (New York: Thunders' Mouth Press and Fitzgerald Inc., 2001), p. 220.

9. Lawrence P. Kirwan, *A History of Polar Exploration* (New York: W.W. Norton and Co., 1960), p. 265; Roald Amundsen, *The South Pole: An Account of the Norwegian Antarctic Expedition in the 'Fram,' 1910–1912* (New York: Barnes and Noble, 1976), pp. 54–89. Amundsen lists a few scientists along, but nowhere is there mention of wireless. It should be noted that when Peary reached Hamilton Inlet, where Smoky Tickles's wireless antenna was in sight, he radioed a telegram: "Stars and Stripes nailed to the pole." See also Donald Baxter MacMillan, *Etah and Beyond* (Boston: Houghton Mifflin Co., 1927), p. 35.

10. Philip Ayres, *Mawson: A Life* (Victoria, Australia: Melbourne University Press, 1999), pp. 66, 90.

11. Douglas Mawson, foreword by Ranulph Fiennes, *Home of the Blizzard: A True Story of Antarctic Survival* (New York: St. Martin's Press, 1998), p. 22, photograph of the antenna following p. 62; Philip Ayres, *Mawson: A Life*, p. 60. A meteorologist with the expedition, Morton Moyes, recorded in his diary that Hannam was an "awful liar and a skate."

12. Mawson, p. 262.

13. Ibid., p. 105. This indicates that they worked during the Antarctic night.

14. Ibid., pp. 105, 106; Ayres, p. 67.

15. Ibid., pp. 200–203; Ayres, pp. 82–83.

16. Mawson, 83; Ayres, pp. 316–317.

17. Mawson, pp. 316–319.

18. Ibid., pp. 328–319.

19. Ayres, p. 90.

20. Ibid., pp. 91–92.

21. Ibid, pp. 94–95.

22. Donald Baxter MacMillan, *Etah and Beyond or Life Within Twelve Degrees of the Pole* (Boston: Houghton Mifflin Co., 1927), foreword by Gilbert Grosvenor, pp. xvii–xix. For a somewhat different story of MacMillan's early life, see John H. Bryant and Harold N. Cones, *Dangerous Crossings*, pp. 8–10. (Incidentally, I am in no way related to Captain Bob Bartlett.)

23. Information on MacMillan is from *Who's Who in America*, vol. 14 (1926–1927), p. 1246, and from *The National Cyclopedia of American Biography* Current Volume E (1937–1938), pp. 42–43. Donald Baxter MacMillan, *Four Years in the White North* (New York: Harper and Brothers, 1918. See also the biography of MacMillan in Everett S. Allen, *Arctic Odyssey: The Life of Rear Admiral Donald B. MacMillan* (New York: Dodd, Mead and Company, 1963).

24. Information on the Crocker Land Expedition is based upon MacMillan, *Four Years in the White North*.

25. Ibid., p. 23.

26. Ibid., pp. 34–36; MacMillan, *Etah and Beyond*, pp. 121–122.

27. MacMillan, *Etah and Beyond*, pp. 13, 32.

28. Ibid., pp. 109–110.

29. Allen, *Arctic Odyssey*, p. 167–68; 206, 212.

30. Ibid., p. 313.

31. Ibid., pp. 321–322.

32. QST, July 1923, p. 9; *New York Times*, June 24, 1923, p. 8; MacMillan, *Etah and Beyond*, pp. 15–16; Obituary, *New York Times*, September 8, 1970. The *Bowdoin* is today in custody of the Maine Maritime Academy at Castine, Maine.

33. J.K. Bolles, "Arctic Explorer to Communicate with Amateurs," *QST*, June 1923, pp. 9–10, 69, 73.

34. Ibid, p. 10.

35. Ibid.

36. *QST*, July 1923, pp. 7–14; MacMillan, *Etah and Beyond*, pp. 8–9.

37. *QST*, July 1923, p. 8.

38. K.B. Warner, "West Coast Working 'Bowdoin' WNP," *QST*, November 1923, pp. 21–24; *New York Times*, July 29, 1923, p. 1.

39. MacMillan, *Etah and Beyond*, pp. 87–91, 115.

40. F.H. Schnell, "White Silence of Arctic Broken," *QST*, October 1923, pp. 10–12.

41. MacMillan, *Etah and Beyond*, p. 115.

42. Ibid., p. 105; *QST*, November 1923, p. 21–24.; *The Literary Digest*, November 24, 1923, p. 25.

43. *The Literary Digest*, November 24, 1923, p. 25. The September 1, 1923, earthquake in Yokohama, Japan, caused 200,000 deaths; on September 14, 1924, Jack Dempsey KO'd Luis Firpo in New York City.

44. K.B. Warner, "9BP Still Chief Contact with MacMillan," *QST*, December 1923, p. 23–24. A "Log of A.R.R.L. Stations with WNP" is included.

45. K.B. Warner, "Coolidge's Holiday Greetings to Macmillan Travel Via Amateur Radio," *QST*, February 1924, pp. 29–30.

46. "The 'Bowdoin' Returns," *QST*, November 1924, pp. 16–17; Donald H. Mix, 1TS, WNP, "My Radio Experience in the Far North," *QST*, November 1924, pp. 17–23.

47. K.B. Warner, "7DJ Works the 'Bowdoin' with One-Five-Watter," *QST*, June 1924, pp. 27–28.

48. "The 'Bowdoin' Returns," and Donald Mix, "My Radio Experiences in the Far North," *QST*, pp. 17–23.

49. "The 'Bowdoin' Returns," p. 16.

50. In scanning the *Atlanta Constitution*, listed as a member, I failed to find a single Alliance distributed article on the *Bowdoin*.

51. *Atlanta Constitution*, December 9, 1923, p. 4; C.P. Edwards, "The 'Arctic' Sails," *QST*, July 1924, pp. 12–13.

52. Shackleton had tried a motor car in 1908 and in 1914 motorized sleds, none of which worked well (Huntford, *Shackleton*, p. 237, 239–40, 376).

# Chapter 12

1. Edwin P. Hoyt, *The Last Explorer: The Adventures of Admiral Byrd* (New York: The John Day Company, 1968), pp. 68–94; Raimund E. Goerler, *To the Pole: The Diary and Notebook of Richard E. Byrd, 1925-1927* (Columbus, Ohio: Ohio State University Press, 1998), pp. 17–40. Byrd's *Skyward* details the expedition with MacMillan, including their flights, but makes no mention of conflict save for the statement that MacMillan forbade him to make a final flight (p. 146). Bryant and Cones in *Dangerous Crossings* (p. 29) state that dated letters prove that Byrd was dishonest in asserting that MacMillan checkmated plans Byrd had already made; MacMillan's plans had come first.

2. Bryant and Cones in *Dangerous Crossings* consulted the papers of several of the participants, yet their book, which is essential reading for anyone interested in the expedition, includes some statements that will raise eyebrows.

3. Hoyt, *The Last Explorer*, pp. 80–81; Goerler, *To the*

*Pole*, p. 23. Hoyt (p. 85) states that the navy's demands "were minimal and sensible. The Navy long wave radio was to be used for local traffic, when the planes were in the air and the shortwave sets were subject to propeller noise and other interference." If the planes were not flying, the McDonald shortwave sets, which worked better over long distances, would be used. The navy demand had been concerned primarily with maximum safety. After all, the long wave radios were tried and tested; the shortwave radios for the airplanes constituted experiments.

4. K.B. Warner, "Shortwave Communication with WNP," *QST*, July 1925, p. 20.

5. *QST*, July 1925, p. 20; *New York Times*, August 4, p. 3, August 12, p. 12, 1925.

6. *New York Times*, September 6, 1925.

7. *New York Times*, August 7, 1925, p. 2, August 16, 1925, p. 2.

8. "Scientific Aspects of the MacMillan Expedition," *National Geographic* 47 (September 1925): pp. 349–354; *New York Times*, August 7, 1925, p. 2, August 16, 1925, p. 2; Richard E. Byrd, "Flying Over the Arctic," pp. 519–521; *Times*, August 3, 1925, p. 6; August 16, 1925, p. 2.

9. Bryant and Cones, *Dangerous Crossings*, pp. 54, 123–126, 141–142, 161–163. The authors present the case thoroughly, and are strictly impartial.

10. Donald MacMillan, "MacMillan in the Field," *National Geographic* 48 (October 1925): 475–476.

11. Hoyt, *The Last Explorer*, pp. 97–106; Byrd, *Skyward*, pp. 149–186.

12. Hoyt, *Last Explorer*, 104–105; Goerler, *To the Pole*, pp. 63–71; *New York Times*, May 11, 1926, p. 2.

13. Goerler, *To the Pole*, pp. 77–79.; Hoyt, *The Last Explorer*, p. 114. K.B. Warner, "Byrd Expedition Sails," *QST*, May 1926, 32. The instruction to a radioman (or was it to Bennett?) to notify Spitzbergen of their progress is puzzling. QST states, "We have no reference to the employment of Hanson's 50-watt crystal set aboard the *Josephine Ford*, KNN, the Byrd plane, during her polar flight, and apparently it was not used" (July 1925, p. 18).

14. *New York Times*, May 11, 1926, p. 2.

15. Ibid., p. 2.

16. Amundsen's narration quoted in Holland, *Farthest North*, p. 251–252.

17. Lincoln Ellsworth, *Search* (New York: Brewer, Warren, and Putnam, 1932), 135, 142 ff. *New York Times*, May 12, 1926; K.B. Warner, "More Arctic Adventures," *QST*, July 1926, pp. 19.

18. *QST*, July 1926, p. 19.

19. Captain George H. Wilkins, *Flying the Arctic* (New York: G.P. Putnam's Sons, 1928), pp. 3–26.

20. Ibid, pp. 27–48.

21. Howard F. Mason, "An Arctic Adventure," *QST*, October 1927, pp. 9–14.

22. F.H. Schnell, "Amateur Radio to the North Pole Again," *QST*, March 1926, p. 33–36. Very little is said about these rigs, Hanson's being the ones that are mentioned.

23. Wilkins, *Flying the Arctic*, pp. 27–48 ; K.B. Warner, "Progress of the Wilkins Expedition," *QST*, May 1926, p. 38.

24. Warner, "Progress of the Wilkins Expedition," *QST*, May 1926, p. 38.

25. Ibid.; Wilkins, p. 76–77.

26. Wilkins, pp. 55–59; article by Frederick Lewis Earp for the North American Newspaper Alliance, in the *Atlanta Constitution*, April 3, 1926, p. 7; and , April 8, 1926,

p. 4.

27. Wilkins, 96–97. Mason ("An Arctic Adventure," p. 11) says it was the DN-2; Wilkins seems to indicate it was the DN-1 (Wilkins, p. 169).

28. Wilkins, p. 140 ff. Wilkins goes into interesting detail about the supplies taken along.

29. Ibid., 145–151. The "sounded" refers to determining depth of the ocean when they had landed.

30. Mason, "Arctic Adventure," p. 10. While "OK" meant just that, "KO" meant motor trouble, forced landing, out of gas.

31. Ibid.

32. Wilkins, p. 189; Mason, "Arctic Adventure," p. 10. Beechey Point is about 185 miles east southeast of Point Barrow.

33. Mason, p. 12; Wilkins, p. 189.

34. Wilkins, pp. 197–201.

35. Wilkins, pp. 212–214.

36. The ARRL had requested amateurs to stay off of Wilkins's wave band of 33.1 meters so that if he was able to use his radio he would be heard. Atlanta Constitution, April 22, 1928, p. 2.

37. Wilkins, 246; 320–326.

38. In "Polar Exploration by Airplane," in W.L.G. Joerg, ed., Problems of Polar Research: A Series of Papers by Thirty-one Authors, American Geographical Society, Special Publication no. 7 (New York: American Geographical Society, 1928), pp. 307–409.

39. Atlanta Constitution, April 22, 1928, p. 1. Wilkins shared headlines with the fate of two Germans and one Irishman who tried to fly in a Junkers monoplane, the Bremen, across the Atlantic from east to west, and made a forced landing on a remote island off the coast of Labrador. They did not have a radio but landed where there was one. Floyd Bennett and Bert Balchen tried to fly to their rescue but Bennett contracted pneumonia and was hospitalized. Lindbergh flew a serum to Canada for Bennett but he died anyway. A Canadian ace named "Duke" Schiller finally rescued the three men. New York Times and other newspapers, issues of April 20–30, 1928.

40. "More Arctic Adventure," QST, July 1926, p. 17. Bartlett would be into the Arctic many more times in the years to come. See his obituary in the New York Times, April 29, 1946.

41. A report that the Morrissey had nearly struck an iceberg appeared in the Times on Sunday, July 25, with Putnam as the author and the additional statement, "(By Wireless to NYT): ... have received a wireless from Hobbs party sent on their low power shortwave portable outfit from Ickertiok Fiord, behind Holstenberg.... Their message is believed to be the first communication north of the Arctic Circle between low power shortwave sets." The report in the Times for August 3, 1926, p. 7, was by Putnam; it was received by the Times from amateur operator J.R. Miller, 9CP, of Hammond, Indiana. Putnam sent seventeen articles by wireless to the Times, all of them received by J.R. Miller. Manley, the Morrissey's radio operator, was also from Ohio, which may explain why an Ohioan received the messages.

42. Clark C. Rodiman, "MacMillan and Party in Labrador," QST, February 1928, pp. 15–17.

43. Holland, Farthest North, pp. 266–273; Wilbur Cross, Disaster at the Pole (New York: Lyons Press, 2000), p. 88. The latter is an excellent history of the Italia disaster.

44. Ibid. Roald Amundsen set out in an airplane to find Nobile and was never heard from again.

45. Robert J. Gleason, Icebound in the Siberian Arctic (Anchorage, Alaska: Alaska Northwest Publishing Company, 1977, 1982). All material on the Nanuk is from this book.

46. Merchant Princes, A Company of Adventurers, vol. 3 (New York: Penguin, 1991), p. 244. See also in the Beaver (the Hudson's Bay in-house publication), S.G.L. Hornery, "Post of the Far North Will Have Two-Way Short-Wave Communications," June 1938, pp. 10–13; also, the Beaver, "Atmospheric Defense," December 1945, pp. 40–42.

47. The story of his brush with death is in Richard E. Byrd, Alone (New York: G.P. Putnam's Sons, 1938), p. 165 ff.; Hoyt, The Last Explorer, pp. 310–327.

48. "WFAT-WFBT," QST, March 1929, p. 58; Hoyt, Last Explorer, p. 197. Apparently some planes had hand operated generators for emergencies. After Byrd's little gas engine gave out at his lone outpost (fortunately, perhaps, because it was spewing out carbon monoxide) he used a hand- (and later foot-) operated generator. WFAT were call letters for the Eleanor Boling, WFBT for the SS City of New York.

49. "Expeditions," QST, August 1929, p. 54.

50. Mason, "Arctic Adventure," pp. 9–14. "Corned willie"—corned beef?

51. A.G. Sayre, W2QY, "Ham at 30 Below," QST, January 1939, pp. 9–12, 106.

## Chapter 13

1. Part 97.1, FCC Regulations. The other four principles (paraphrased) are (2) contributing to the advancement of the radio art; (3) encouraging improvement of skills and technical phases of the art; (4) expanding the reservoir of trained operators and technicians; and (5) enhancing international good will.

2. "Part of the Game," QST, August 1926, p. 7. An extreme example of amateur modesty is Harold Bride, the surviving radio operator of the Titanic disaster (see chapter 1, note 22). In contrast, Jack Binns, hero of the Republic disaster, cashed in on his fame and became a successful journalist.

3. United States Weather Bureau, "Monthly Weather Review," volume 41, part 2 (July–December 1913): 357–58, 479–491, 528–529.

4. Allan W. Eckert, A Time of Terror: The Great Dayton Flood (Boston: Little, Brown and Co., 1965), p. 7.

5. Chicago Tribune, March 30, 1913.

6. R.W. Goddard, 5ZJ, "First Aid by Radio," QST, November 1921, pp. 23–24.

7. S. Kruse, technical ed., "A Snowstorm and the ARRL," QST, January 1923, pp. 30–32.

8. Ibid.

9. Ibid.

10. "It Seems to Us," QST, March 1938, p. 9.

11. Ibid.

12. Ibid.

13. David Olson, quoted by Gary Craven Gray, Radio for the Fireline: A History of Electronic Communication in the Forest Service, 1905-1975 (Washington, D.C.: U.S. Department of Agriculture, Forestry Service, 1982), pp. 4–27. This source was published by the forest service as an administrative publication and was distributed primarily within the service; only a limited number of copies were

distributed outside of the bureau. It is not copyrighted. It is a thoroughly researched, well-written monograph, and is my main source of information on forest service radio.

14. David Olson, quoted in ibid., p. 19.

15. Ibid., p. 20.

16. Ibid, pp. 26–27. The author states "No. 2 Burgess Batteries," implying dry cells, but dry cells would not have contained electrolyte; they must have been Edison Storage Cells.

17. Ibid., p. 27.

18. Ibid., p. 110.

19. Ibid., pp. 58–59.

20. Ibid.

21. Ibid., 75, 77–78.

22. Ibid., p. 88.

23. Ibid., p. 89.

24. Ibid., p. 250.

25. For an example of modern technology, see Lynn Ellen Edwards, "The Utilization of Amateur Radio in Disaster Communications" (Boulder: Colorado Natural Hazards and Information Center, Institute of Behavioral Science, Working Paper no. 86, University of Colorado, May 1994), p. 51. Edwards differs between emergencies, which happen at once, and disasters, which are intensified and widespread. I have not differed one from the other. Today cell phones and satellites enter the picture, and two agencies, RACES (the Radio Amateur Civil Emergency Service) and FEMA (the Federal Emergency Management Administration), are deeply involved.

26. K.B. Warner, Secretary-Editor, "The Amateur Scores Again: Dozens of American Amateurs Do Valiant Work When Blizzard Paralyzes Middle West," *QST*, April, 1924, pp. 14–15; *Chicago Tribune*, February 5, 1924, p. 1.

27. Ibid., p. 15.

28. Ibid.

29. Gordon E. Dunn and Banner I. Miller, *Atlantic Hurricanes* (Baton Rouge: Louisiana State University Press, 1960, 1964), p. 271; Fred Doehring, Iver W. Duedall, and John M. Williams, *Florida Hurricanes and Tropical Storms, 1871–1993: An Historical Survey* (Melbourne, Fl.: Florida Institute of Technology, Division of Marine and Environmental Systems, Technical Paper 71, 1994), p. 15; Jay Barnes, *Florida's Hurricane History* (Chapel Hill: University of North Carolina Press, 1998), p. 11.

30. Barnes, p. 111.

31. Ibid., pp. 116–120.

32. "Amateurs Help in Florida Emergency," *QST*, November 1926, pp. 101–102.

33. Ibid.

34. Correspondence, "Navy Day," *QST*, January 1927, p. 59.

35. Ibid.

36. The best history of the Lake Okeechobee disaster is Eric L. Gross, *Somebody Got Drowned, Lord: Florida and the Great Okeechobee Hurricane Disaster of 1928* (PhD dissertation, Florida State University, 1995). See also Robert Mykle, *Killer 'cane: The Deadly Hurricane of 1928* (New York: Cooper Square Press, 2002).

37. Eric L. Gross, *Somebody Got Drowned, Lord: Florida and the Great Okeechobee Hurricane Disaster of 1928*, p. 317. Miller, *Killer 'cane*, (p. 212) estimates that 2,400 died.

38. K.B. Warner, editorial, *QST*, November 1928, pp. 7–8.

39. Ibid.

40. "The West Indies Hurricane Disaster: September 1928. Official Report of Relief Work in Porto Rico, the Virgin Islands, and Florida" (Washington, D.C.: The American National Red Cross, 1929), p. 64.

41. Ibid., p. 64; Gross, p. 514.

42. Gross, p. 507.

43. K.B. Warner, editorial, *QST*, November 1928, pp. 7–8.

44. "The Army-Amateur Radio System Is Revised," *QST*, March 1929, pp. 21–25; F.E. Handy, "When Emergency Strikes," *QST*, April 1938, pp. 35–38, 76, 78, 80, 82, 84–86.

45. Sally Bedell Smith, *In All His Glory: The Life of William S. Paley* (New York: Simon and Schuster, 1990), p. 153.

46. "Columbia Announces Award to Be Given America's Outstanding Radio Operator," *QST*, December 1936, p. 10. Five respected members with some affiliation with the radio profession were to serve as the Board of Award.

47. "W8DPY Wins Paley Award," *QST*, July 1937, pp. 8–9, 76. QST misspells Renovo as Renova.

48. Ibid., p. 76.

49. Ibid., p. 8.

50. Ibid., p. 8.

51. Clinton B. DeSoto, "In the Public Interest, Convenience and Necessity," *QST*, April 1937, p. 11. This is one of the longest non-technical articles to appear in QST, covering pages 11–20 and every other page from p. 74 to p. 110. I have based most of the description of ham activities during the flood upon DeSoto's article. The *St. Louis Post-Dispatch* had excellent coverage of the debacle; see especially the issue of January 25, 1937. Some statistics are from the American Red Cross, *Report of Relief Operations of the American Red Cross: The Ohio-Mississippi Flood Disaster of 1937* (Washington, D.C.: n.d.).

52. American Red Cross, *Report of Relief Operations*; p. 12.

53. American Red Cross, *Report*, p. 90; DeSoto, p. 13.

54. DeSoto, p. 13

55. *QST*, April 1937, p. 15, 94.

56. Ibid., pp. 13–15.

57. Ibid, p. 20, 72, 74, 76.

58. *St. Louis Post-Dispatch*, January 25, 1937.

59. DeSoto, "In the Public Interest, Convenience and Necessity," *QST*, April 1937, pp. 11–20 ff.

60. Ibid., pp. 76, 78.

61. Ibid., 78, 80, 82, 84.

62. *St. Louis Post-Dispatch*, January 28, 1937.

63. DeSoto, 84, 86.

64. *St. Louis Post-Dispatch*, January 26, 1938, p. 7A.

65. DeSoto, pp. 108–110.

66. Malcolm Rohrbough, *The Land Office Business* (New York: Oxford University Press, 1968), p. 61; QST, April 1937 p. 88.

67. DeSoto, pp. 88, 90; C.B.D., "Paley Award Goes to W9MWC," *QST*, August 1938, p. 18. The latter includes the quote from the *Post-Dispatch*. See also *New York Times*, June 5, 1938, p. 162.

68. Everett S. Allen, *A Wind to Shake the World: The Story of the 1938 Hurricane* (Boston: Little, Brown and Co., 1976), p. 37–38. (It has been suggested that the beavers alerted each other by tapping out the message on mud with their flat tails, but I do not believe it!) See also R.A. Scott, *Sudden Sea: The Great Hurricane of 1938* (New York: Little, Brown and Co, 2003).

69. David R. Vallee and Michael R. Dion, "The Great New England Hurricane of 1938," from "Southern New England Tropical Storms and Hurricanes, A Ninety-Eight Page Summary 1909–1997," (Tanton, Massachusetts: National Weather Service, 1998), http://www.erh.noaa.gov/er/box/hurricane1938.htm.

70. This is described throughout Allen's book.

71. Valee and Dion, "The Great New England Hurricane." For accuracy, it should be stated that statistics vary on all aspects of the hurricane.

72. Clinton B. DeSoto, "Amateur Radio Bests Triple Catastrophe," QST, November 1938, pp. 11–18, 76, 78, 80. DeSoto is my principal source for this essay.

73. Ibid., p. 15.

74. Ibid., pp. 16–18, 76, 78, 80,

75. Ibid., p. 16; "1938 Paley Trophy Awarded to W1BDS," QST, July 1939, pp. 23, 74, 76.

76. "1938 Paley Trophy Awarded to W1BDS," QST, July 1939, pp. 23, 74.

77. Ibid.

78. Ibid., p. 11.

79. Clinton B. DeSoto, "QRR, 1932: Amateur Emergency Work During the Past Year," QST, January 1933, pp. 39–42; "Amateurs of Assistance in Emergencies: Notes on Recent Work Accomplished," QST, April 1934, pp. 34–35, 80.

80. Clinton B. DeSoto, "Southern California Amateurs Rise to Earthquake Emergency," QST, May 1933, pp. 9–11; QST, August 1936, p. 31.

81. QST, January 1935, pp. 53–54; Henry Jenkins (W7DIZ), "Amateur Radio Scores Again!" QST, February 1935, pp. 45–46.

## Chapter 14

1. QST, July 1939, p. 86. The advertisement is on page 112 of the May issue.

2. Anita Calcagni Bien and Enid Carter, "The YL's Unite! The Story of the Young Ladies' Radio League," QST, May 1940, pp. 22–27.

3. Ibid. The YLRL column in the June 1999 QST reported on the organization's scheduled summer convention celebrating its 60th anniversary, the present membership being "over 1500."

4. Editorial, "It Seems to Us—," QST, November 1940, p. 7.

5. Ibid.

6. Oakes A. Spalding, W1FTR, "Around the World with the Yankee," QST, October 1941, pp. 9–14, 102, 104. The essay on the Yankee is entirely from this article.

7. Clinton B. DeSoto, "Byrd Antarctic Expedition to Use Amateur Radio," QST, December 1939, pp. 11–15, 25. "Richard E. Byrd: Byrd Antarctic Expedition III, 1939–41," The United States Antarctic Service Expedition, http://www. South-pole.com/p0000109.htm. DeSoto gives the number of personnel at 160, but I am inclined to accept the number 125 as stated in the online article.

8. DeSoto. The other call letters were KC4USB and KC4USC.

9. Clay W. Bailey, "Letter to ARRL," QST, November 1941, pp. 17, 68.

10. Oscar W. Reed, Jr., "Amateurs Provide Red Cross with Communications on Inauguration Day"; R.E. Handy, "Red Cross to Test Amateur Radio's Emergency Communications Facilities—Get Ready Now," QST, March 1941, pp. 36–37; QST, March 1941 pp. 25, 82, 86; QST, "Help in Red Cross—ARRL Test April 4, 5, 6," QST, April 1941, pp. 48–49.

11. A few examples of amateur contributions in disasters are "AEC Members Perform Communications Service in Storm," QST, March 1940, p. 65; "Amateur Radio in Sacramento Valley Flood," QST, May 1940, pp. 67–68; "1940 Flood Activities," QST, June 1940, p. 74; Clinton B. DeSoto, "Atlantic Coast Amateurs Render Emergency Service," QST, October 1940, pp. 28–29, 72; "Advance Planning Pays Dividends," QST, February 1941, QST, p. 68; and Leland H. Smith, "Twister Hits Georgia," QST, April 1940, p. 15, 94; e-mail from Forrest Bartlett, W6OWP, August 18, 2003.

12. Editorial, "It Seems to Us—," QST, April 1942, p. 7.

13. Gordon W. Prange, At Dawn We Slept (New York: Penguin Books, 1984), pp. 499–500; see also Walter Millis, This Is Pearl: The United States and Japan–1941 (New York: William Morrow and Co., 1947), pp. 352–354. The British had been using radar since 1936; it had much to do with their winning the Battle of Britain.

14. Clinton B. De Soto, "V.W.O.A. Honors Amateur Radio," QST, April 1942, p. 27. Strictly speaking, Lockard was not a ham although he was working with radio equipment. The awards also honored two genuine hams who had stuck to their rigs while their ships were under attack. Not until April 25, 1943, did the War and Navy Departments officially announce the existence of radar, although just about every ham and most civilians knew about it. The War Department resented the credit private firms were giving themselves in advertisements about radar, and forthwith placed censorship on the use of the word again. See "Army and Navy Announce Radar!," QST, June 1943, p. 51; "Radar—Now You Read About It and Now You Don't," QST, September 1943, p. 54.

15. Jackie Tilly, KK5NM, "QRV," Opelousa's Area Amateur Radio Club: Wireless Connection," April 17, 1998.

16. April 1942, p. 17.

17. QST, April, 1942, p. 17; October 1942, p. 38.

18. Emil H. Frank, "Soldiers and Sailors and Amateur Radio," QST, January 1942, pp. 32, 60.

19. Clinton B. DeSoto, "U.S.A. Calls and the YLs Answer," QST, May 1942, pp. 9–10.

20. "The War Emergency Radio Service," QST, July 1942, pp. 11–15. Part 15 is presented in total, in fine print, in this article.

21. Editorial, "It Seems to Us—," QST, January 1941, p. 7.

22. Editorial, K.B. Warner, "It Seems to Us—," QST, February 1942, pp. 7–13, 45.

23. Ibid.

24. "The War Emergency Radio Service." The following paragraphs are based upon this article.

25. Editorial, "It Seems to Us—," QST, July 1942, pp. 9–10. Later in 1942 the OCD relaxed its requirement that a single city be involved; it allowed any towns to create WERS but hoped to consolidate them in due time (QST, December 1942, p. 74).

26. "Happenings," QST, August 1942, pp. 15–16; "Happenings," QST, pp. 29–30.

27. John Huntoon, "Planning WERS for Your Community," QST, pp. 22–24, 114, 116.

28. Ibid.

29. "Providence Adopts New Plan for Civilian Defense Radio," QST, March 1942, p. 52.

30. John A. Doremus, "Massachusetts Civilian Defense Radio," QST, September 1942, pp. 11–14.

31. Ibid.

32. Ibid. It is presumed that WEHRS licenses were subsequently issued.

33. "St. Paul Radio Club Code Classes, *QST*, September 1942, p. 80.

34. Rex T. Brown and D.L. Moody, "Akron and the WERS," *QST*, December 1942, pp. 11–15, 116, 118, 120.

35. J. Stautberg and J.E. Weaver, "Queen City Emergency Net, Inc.: A Brief History, *Queen City Emergency Net*, http://qcen.org/history.html.

36. Editorial, K.B. Warner, "We Must Not Fail," *QST*, September 1942, p. 9; John Huntoon, "Training Civilians for Wartime Operating," *QST*, pp. 53–54; John Huntoon, *QST*, October 1942, pp. 45–49; George Grammer, "A Transceiver for WERS," *QST*, pp. 11–15, 84, 88, 90; Don H. Mix, "Building WERS Gear from Salvaged B.C. Sets," *QST*, September 1942, pp. 15–18. *QST* was full of information concerning hams and the war.

37. "Operating News," *QST*, February 1943, pp. 68–69; "It Seems to Us—," *QST*, August 1943, pp. 7–8.

38. "Operating News," *QST*, June 1944, p. 63; Captain Samuel E. Fraim, "Amateur Radio and the Civil Air Patrol," *QST*, January 1943, pp. 50–51; Tech. Sgt. Karl H. Stillo, "Radio in the CAP," *QST*, December 1943, pp. 20–22; Captain J. Wm. Hazelton, "WERS in the Florida State Guard," *QST*, November 1944, pp. 45–47.

39. "Operating News," *QST*, May 1945, p. 56.

40. "It Seems to Us—," *QST*, March 1943, pp. 9–10.

41. Carol A. Keating, "'Ole Mississip'" Rampages Again," *QST*, August 1943, pp. 30–33. Successive paragraphs about the flood are from the same source.

42. WERS in Lake Erie Dike Break," *QST*, September 1943, p. 73.

43. "Extra! Staten Island Shelled; WERS to the Rescue," *QST*, May 1944, p. 61.

44. Carol K. Witte, "Operating News," *QST*, November 1944, pp. 57–59.

45. F.E. Handy, "ARRL Emergency Corps Program," *QST*, December 1945, 49–51.

## Chapter 15

1. Forrest had mastered transcribing Kana code after a fellow ham asked him to listen in on a Japanese station sending in Kana. He heard the station, and was so intrigued that he made a tape of the code and practiced until he was moderately proficient in it (conversation with Forrest Bartlett, January 22, 2004). For information on the Kana Code, see Donald D. Millikin, "The Japanese Morse Telegraph Code," *QST*, September 1942, pp. 23–25, 120, and Charles E. Holden, "The Japanese Morse Radiotelegraph Code," *QST*, October 1943, pp. 30–33. In *QST*, February 1946, p. 26, is this brief note: "Among the nine radio communications employees of Press Wireless now serving in the Philippines who have been awarded by the Philippine Army the Philippine Liberation Ribbon is Forrest B. [*sic*—A.] Bartlett, W6OWP."

2. "Strays," *QST*, August 1945, p. 54. This is quoted from Mary Zurhorst in *Broadcasting*, July 2, 1945. Although some of the statistics of radiomen in the armed forces do not add up to a half million, they are all of questionable accuracy, and I consider 500,000 realistic.

3. Clinton B. DeSoto, "ESMWT at Rutgers," *QST*, September 1943, pp. 43 ff. The program had two predecessors, mentioned in the article. See also Clinton B. DeSoto, "The Signal Corps and the Blue Grass State," *QST* March 1943, pp. 11–15 ff.

4. Clinton B. DeSoto, "U.S.A. Calls and the YLs Answer," *QST*, May 1942, pp. 9–12, 86, 90, 92; Louis B. Dresser, "Women and Radio—Partners in Victory," *QST*, September 1943, pp. 9–16, 100, 102, 103, 104, 106; Anita Bien, "YLRL, QRV!." *QST*, October 1941, pp. 32–37, 78, 80, 82; Jean Hudson, "Teaching Radio in Summer Camps," *QST*, May 1942, pp. 12, 92. In 1933 Miss Hudson won the world's code speed championship; she was just nine years old at the time. These articles describe the role of YLs in considerable depth.

5. "Radio in the Draft Army," *QST*, January 1941, p. 19.

6. *QST*, August 1945, p. 54; "Signal Corps Radio School," *QST*, August 1941, pp. 9–11, 49; Clinton B. DeSoto, "QST Visits Fort Monmouth," *QST*, October 1942, pp. 28–32, 96, 98, 106, 108. The following paragraphs regarding Fort Monmouth are from DeSoto.

7. "QST Visits the Air Forces," *QST*, January 1943, pp. 17–29.

8. Clark C. Rodiman, "QST Visits Gallups Island," *QST*, June 1941, pp. 9–12, 19; Clinton B. De Soto, "QST Returns to Gallups Island," *QST*, May 1943, pp. 14–18, 84, 86, 88, 90.

9. Clinton B. DeSoto, "QST Visits Camp Hood," *QST*, July 1943, pp. 9–16, 82, 84, 86, 88, 90. This article is divided, the first part discussing the role of the TCs (Tank Destroyers), and the second part titled "Radio Training at Camp Hood."

10. Clinton B. DeSoto, "QST Visits the Marine Corps," *QST*, April 1943, pp. 13–17, 96, 98, 100, 102, 104.

11. Ibid.

12. Clinton B. De Soto, "QST Visits the Coast Guard," *QST*, February 1943, pp. 13–18, 102, 104, 106, 108, 110, 112, 114, 116.

13. Ibid.

14. Ibid.

15. *QST*, February 1943, p. 28.

16. Lt.-Cmdr. John L Reinartz, "The Navy and the Amateur," *QST*, September 1940, p. 29; Lt. C.F. Clark, "Voice and Ears of the Fleet," *QST*, October 1940, p. 57.

17. Clinton B. DeSoto, "The Navy Trains Radio Technicians," *QST*, November 1942, pp. 13–18, 116, 118, 120. Some marines also received training at these schools.

18. Clinton B. DeSoto, "QST Visits the Noroton Training Station," *QST*, August 1942, pp. 40, 44, 88, 90, 92, 94, 96, 194. The paragraphs on this subject are all from this article.

19. Lt. Col. Howard R. Haines, ex–W2EIS, "Flying Radiomen of the Ferrying Division," *QST*, June 1944, pp. 16–18, 82, 84.

20. "The Troubles of a Wandering Ham," *QST*, December 1944, pp. 43, 88.

21. Rear Admiral Joseph R. Redman, director of Naval Communications, "Navy Communications and the Amateur," *QST*, October 1945, pp. 14–15 ff. This is the beginning essay for an in depth coverage of the navy, and amateurs in the Second World War. It is titled "Special U.S. Navy Feature Section: The Second World War." It was written by many persons not mentioned, and divided into sections as follows: "The Navy Ashore," pp. 16–51; "The Navy Afloat," pp. 52–57; and "The Navy in Combat," pp. 58–64. Many illustrations, including photographs of radiomen with their amateur call letters, accompany this nearly book-length series of essays. Many examples of amateur courage and capability are presented, including some emphasis on radar. The September 1944 issue of *QST* devoted equal space to the army signal corps, pp. 7–49.

22. "Hams in Combat," *QST*, January 1944, pp. 14–15, followed by a narration "pieced together from letters, from AP dispatches, from prosaic military records" of the ordeal of Staff Sergeant Laurence E. Madison, "Italian Invasion," pp. 15–17; Alvar J. Jujampaa, "Atlantic Convoy," pp. 18–19; J.W. Soehl, "La Fouconnerie by 1600," pp. 20–21. For a narration of radio experience in the European theater, see W.W. Chaplin, *The Fifty-Two Days* (New York: The Macmillan Company, 1944). Three of the four men manning the mobile station were hams. The book is dedicated to "Jig Easy Sugar Queen" (JESQ), the mobile transmitter. A navy man, George Ray Tweed, avoided the Japanese during their occupation of Guam. See A. David Middelton, "'The Ghost of Guam'—KB6GJX: An Interview with W/O George Ray Tweed, USN," in *QST*, March 1945, pp. 38–41; Tweed also wrote of his experiences in *Robinson Crusoe, USN* (New York: Whittlesey House, 1945). See Cyrus T. Read, "A 'Handbook' on Leyte," *QST*, March 1945, pp. 22–23, 102, for a story of ham ingenuity under hostile conditions. This article encapsulated the book by Ira Wolfert, *History Island: The Story of an American Guerrilla on Leyte* (New York: Simon and Schuster, 1945). This was condensed in *Reader's Digest*, March 1945. Another narrative is F. Joseph Visintainer, "Listening Post in the Philippines," *QST*, April 1946, pp. 70–72, 126. And there are many more.

23. "Hams in Combat: Radioman-Gunner in a B-25," *QST*, August 1944, pp. 40–42.

24. "Strays," *QST*, July 1944, p. 62; "Strays," *QST*, August 1944, p. 58.

25. Cpl. Bill Granberg, "This Is Your Armed Forces Radio Station," *QST*, June 1945, pp. 54, 88.

26. Carlton H. March, "On the Burma Road," letter, *QST*, June 1946, p. 71.

27. Major J.W. Hunt, "The Wail of the Kee Bird," *QST*, October 1943, pp. 50–51.

28. Clinton B. DeSoto, "Hamfest in Khaki," *QST*, November 1942, pp. 51–52, 114.

29. "Anglo-American Hamfest," *QST*, December 1944, p. 45; "Second London Hamfest," *QST*, January 1945, p. 49.

30. Lt. Harry Longerich and M. Sgt. Arthur Hansen, "Hamfest in North Africa," *QST*, February 1944, p. 31; "Cairo Convention," *QST*, August 1944, p. 59.

31. "SARO Mid-Pacific Chapter Organization," *QST*, May 1945, p. 58; Corporal Jack Walker, "QSO: Somewhere in the Pacific," in Correspondence from Members, *QST*, July 1944, p. 65.

32. "Ham Hospitality," *QST*, February 1945, p. 23; *QST*, December 1944, p. 23.

33. "Hams in Combat: One Life to Give," *QST*, July 1944, pp. 50–53. While on duty the captain was killed in an airplane crash.

34. "Chicago '"Hamvoree,'" *QST*, July 1944, p. 63.

35. "It Seems to Us: Reopenings," *QST*, October 1945, p. 11; "It Seems to Us: We're Off," *QST*, December 1945, p. 11; "Second Reopening Order," *QST*, p. 31; *QST*, August 1946, p. 36.

36. *QST*, December 1945, p. 11; R.B. Bourne, "The Opening of the Band," March 1946, p. 28.

37. John T. Ralph and H.M. Wood, "Operating the BC-645 on 420 Mc," *QST*, February 1947, pp. 15–21.

38. "The Rag Chewers Club," *QST*, April 1946, p. 81; *QST*, May 1946, p. 81; *QST*, August 1946, p. 37.

39. Henry W. Peterson, "Alaska Earthquakes," *QST*, August 1946, p. 79; "Idaho-Washington Emergency," *QST*, February 1947, p. 77.

40. Albert Hayes, "Winds, Waves and Snakes," *QST*, December 1947, pp. 40–44, 118.

41. Harold M. McKean, "Amateur Radio Operations: Texas-Oklahoma Tornado, Texas City Explosions," *QST*, July 1947, pp. 34–40.

42. Ibid.

43. Terry Lamar, "The Texas City Disaster," *Welcome to the City of Texas City*, http://www.texas-city-tx.org/docs/history/exp.htm; Harold McKean, "Amateur Radio Operations."

44. Two very different books have been written on the Texas City Disaster: Hugh W. Stephens, *The Texas City Disaster, 1947* (Austin: University of Texas Press, 1997) and Bill Minutaglio, *City on Fire* (New York: Harper Collins, 2003). The former mentions amateurs on pp. 66–67 and 94, the latter mentions in passing the Texas-Oklahoma tornado, pp. 110–111 and 188. In neither book is an amateur mentioned by name or details given of his contribution.

45. "When Wires Are Down..." *QST*, June 1948, p. 43.

46. "KLPO, MacMillan Arctic Expedition, 1946," *QST*, September 1946, p. 75.

47. "Ronne Antarctic Research Expedition," *QST*, March 1947, p. 71; "Expeditions," *QST*, February 1948, p. 71.

48. "Expeditions," *QST*, February 1948, p. 71.

49. "Kon-Tiki Communications—Well Done!," *QST*, December 1947, pp. 69, 142, 144, 146, 148.

50. Kon-Tiki Communications—Well Done!," *QST*, December 1947, pp. 69.; letter to Richard Bartlett from Gene Sykes, February 17, 2004. Sykes is also a member of the First Class CW Operator's Club, which is an exclusive British amateur group with stringent requirements for membership; only a few American operators are members.

51. A.L. Budlong, "Moscow: A Report of the Five-Power Conference by ARRL's Representative," *QST*, January 1947, pp. 25–27.

52. Ibid.

53. The conference was well reported in *QST*: "Atlantic City Report," July 1947, pp. 29–31; 100; "Atlantic City Report," August 1947, pp. 28–31; "Atlantic City Report," September 1947, pp. 32–35; "Atlantic City Report," October 1947, pp. 17–21; "Atlantic City Report," November 1947, pp. 32–37.

54. "Inter-American Conference," *QST*, June 1949, p. 16; A.L. Budlong, "The Fourth Inter-American Region 2 Radio Conference," *QST*, September 1949, pp. 35–39.

55. "Happenings of the Months," *QST*, December 1949, p. 27.

56. "It Seems to Us—Unity," *QST*, December 1949, p. 9.

57. Rear Admiral Earl E. Stone, USNR, "The Navy and the Amateur," *QST*, December 1948, pp. 36, 128.

58. "It Seems to Us—A Welcome Hand," *QST*, July 1947, p. 11.

59. "The Military Amateur Radio System" *QST*, February 1949, pp. 34–35; "Military Amateur Radio System" *QST*, November 1949, p. 52. The name was later changed to "Military Affiliate Radio System."

60. Byron Goodman, "The 'Transistor'–an Amplifying Crystal," *QST*, October 1948, p. 48.

61. Obituary of Kenneth Bryant Warner, *QST*, November 1948, pp. 9–13; "Clinton B. DeSoto," obituary, *QST*, June 1949, p. 41.

# Epilogue

1. "Happenings of the Month," QST, June 1949, p. 18; "It Seems to Us," QST, December 1949, p. 9.

2. Information has been furnished by Forrest Bartlett, W6OWP, unless otherwise stated.

3. Based upon ARRL reports on their Web site.

4. E-mail to Forrest Bartlett from Alan Padgett in regard to radio amateurs in the devastated regions, December 28, 2004.

# A Bibliographic Essay

As the reader is well aware by now, this book paints with a wide brush. It covers nearly fifty years of history and embraces the part played by amateurs in many of the headlined stories of the half century 1900–1950. In this book radio operators involved in exploration, disasters, aircraft accomplishments, wartime activities and government policies may bear brief mention while the enterprise they participated in is the recipient of lengthy newspaper and magazine articles, and even of books.

The primary source material for the entire book is *QST*, the official journal of the American Radio Relay League. Every issue was scanned from the first one (December 1915) through volume 34 (December 1950). Another source was *Radio News* from 1914 until 1927, after which it devoted its pages exclusively to AM radio. From *QST*'s articles, announcements, reports of activities, even its editorials concerning everything from congressional hearings to railroad trains marooned in blizzards, I consulted newspapers, magazines, books, transcripts of congressional hearings, and occasionally the World Wide Web to emphasize the significance of ham radio to the happenings. My most consistent newspaper source was the *New York Times*. Another source frequently consulted was Clinton B. DeSoto, *Two Hundred Meters and Down: The Story of Amateur Radio* (Hartford, Conn.: 1936). Finally, Forrest Bartlett, W6OWP, was a constant source of information and clarification for this historian whose technical skills have barely extended to a modern word processor.

A final note: many citations are to articles in *QST* without an author credited. Initials at the end of the article are given, and if they agree with those of a staff member mentioned in the table of contents, I have tried to include their names. Often, however, the initials refer to an unknown author. In such cases I have simply eliminated the author category.

---

## Chapter 1

For Marconi I made use of W.B. Jolly, *Marconi* (New York: 1972), and Gavin Weightman, *Signor Marconi's Magic Box* (New York: 2003). For the early years of radio, I used Orin E. Dunlap, Jr., *Radio's 100 Men of Science: Biographical Narratives of Pathfinders in Electronics and Television* (New York: 1944, 1977); William Maver, Jr., "Wireless Telegraphy–Its Past and Present Status and Prospects," in *Annual Report of the Board of Regents of the Smithsonian Institution for the Year Ending June 30, 1902* (Washington, D.C.: 1903); Paul Schubert, *The Electric Word: The Rise of Radio* (New York: 1971); Joseph B. Lebo, "The Man Before Marconi," *QST*, August 1948; and J.S.V. Allen, "Amateur Radio in 1882," *QST*, De-

cember 1940. My source for Nikola Tesla was Margaret Cheney, *Tesla: Man Out of Time* (New York: 1981). The arguments for Tesla's fabricating the first transmitter and receiver are in *Marconi Wireless Telegraph Co. of America v. United States* (United States Reports: Cases Adjudged in the Supreme Court of the United States, vol. 320, June 21, 1943).

Basic information for the amateur is in the *Radio Amateur's Handbooks* issued by the American Radio Relay League. I consulted the 9th edition (Hartford, Conn.: 1932). Of many early articles on how to construct transmitters and receivers, I made use of G.G. Blake, "How to Make a Wireless Telegraph," in *The Boy's Own Paper* (London, England), November 2, 1904, and A. Frederick Collins, "Design and Construction for a 100 Mile Wireless Telegraph Set" in *Scientific American Supplement*, October 1908. *Scientific American* ran many articles on wireless in the early days.

Of sources for the early uses of radio in ocean transportation, Winthrop Packard, "The World of a Wireless Telegraph Man," *The World's Work*, February 1904, and "The Advance of Wireless," *The World's Work*, February 1905, and Lawrence Perry, "Commercial Wireless Telegraphy," *The World's Work*, March 1903, are useful. Navy interest is discussed in Kenneth Wimmel, *Theodore Roosevelt and the Great White Fleet: American Sea Power Comes of Age* (Washington, D.C.: 1998). Early rescues by wireless are in the *New York Times*, February 13, 1910, and "A Wireless Victory," *Harper's Weekly*, January 30, 1909. As for the *Titanic* disaster, dozens of books have been written about it and most of them are well done. I made use of Robin Gardiner and Dan Van Der Vat, *Titanic Conspiracy* (New York: 1998), the *Times* of London, April 14–28, 1912, and Charles Pellegrino, *Her Name Titanic* (New York: 1990), but I consulted many others.

Principal sources leading up to the Communications Act of 1912 include "Wireless Telegraphy and Wireless Telephony," Hearings under House Resolution 182, 61st Congress, 2nd session, March 29, 1910; Hearings under House Joint Resolution 95, "A Bill to Regulate and Control the Use of Wireless Telegraphy and Wireless Telephony," Committee on Naval Affairs, Subcommittee on Naval Law, 61st Congress, 2nd Session, January 8, 1910. Although the hearings take up many pages, the repetition of arguments pro and con is tiring. Criticism and defense of hams also appeared in journals and newspapers. Robert A. Morton, "The American Wireless Operator," *Outlook*, January 1910, is an example of the criticism. The *New York Times* defended the amateurs (January 31, 1910). *Scientific American* ran many editorials on the subject, especially in 1912 just prior to passage of the act. For the Communications Act of 1912 see U.S. Statutes at Large, Public Law 264, signed by President Taft on August 17, 1912.

## Chapter 2

In a nation with many strong-willed men, but fewer strong-willed men-of-goodwill, Hiram Percy Maxim emerges as a member of the minority. He appears to have lived an exemplary life in every way, this in spite of his background: his father was inventor of the Maxim machine gun. A strong argument could be made that, but for Hiram Percy's leadership, amateur radio could have been so heavily regulated as to have been crushed out of existence. Hiram Percy Maxim deserves considerable space in any history of amateur radio.

To really know him, one needs to know of his background. The biography of his father and brother is well told in Iam McCallum, *Blood Brothers: Hiram and Hudson Maxim—Pioneers of Modern Warfare* (London: 1999) and Hiram Maxim's obituary in the *Times* of London, November 26, 1916. Hiram Percy Maxim related much of his life in *A Genius in the Family: Hiram Stevens Maxim Through His Son's Eyes* (New York: 1936) and *Horseless Carriage Days* (New York: 1936, 1937). A biography of Maxim is Alice Clink Schumacher, *Hiram Percy Maxim* (Cortez, Colorado: 1998). An extensive obituary is in the *New York Times*, February 18, 1936.

The founding of *QST* and the beginnings of the relay concept are noted throughout *QST* prior to the temporary ending of amateur radio with the coming of the First World War. Early competitions, such as the Washington's Birthday relay, are in "Washington's Birthday Amateur Relay Message," *QST*, April 1916, and in *Scientific American*, February 19, 1916; other competitions, including cooperation with the navy and first attempts at a transcontinental relay are reported in *QST* during these years.

The silencing of Texas radio stations during Pershing's pursuit into Mexico is noted in the *New York Times*, March 25, 1916. Pershing's foray into Mexico is well told in Herbert Molloy Maxon, Jr., *The Great Pursuit* (New York: 1970).

## Chapter 3

Several national organizations of hams besides the ARRL were proposed during the years prior to America's entrance into the First World War. Both the *Literary Digest* (January 16, 1915) and the *New York Times* (November 21, 1915; January 14 and March 16, 1916) made assessments of these endeavors and even speculated on the number of hams that existed at the time. They also indicated an understanding of the usefulness of amateurs should war come.

The conflict over communications and neutrality is heavily treated in the *New York Times* (January 29, February 12, July 16, and a series of articles in August 1914). See also Kenneth Macksey, *The Searchers: Radio Intercepts in Two World Wars* (London: 2003). Naval battles in which radio may have played an important part are covered in Keith Yates, *Graf Spee's Raiders: Challenge to the Royal Navy, 1914–1915* (Annapolis: 1995) and Barrie Pitt, *Coronel and Falkland* (London: 1960).

The *Falaba* incident is mentioned in Daniel Allen Butler, *The Lusitania: The Life, Loss, and Legacy of an Ocean Legend* (Mechanicsburg, Pa.: 2000) and in the *New York Times*, April 1, 1915. Differing interpretations of the *Lusitania* tragedy are found in Adolph A. Hoehling and Mary Hoehling, *The Last Voyage of the Lusitania* (New York: 1956), Diana Preston, *Lusitania: An Epic Tragedy* (New York: 2002), and David Ramsay, *Lusitania: Saga and Myth* (New York: 2001). Provocative is Thomas A. Bailey and Paul B. Bryan, *The Lusitania Disaster: An Episode in Modern Warfare and Diplomacy* (New York: 1975).

Controversy over the Tuckerton and Sayville facilities continued with the *Times* running many articles in September 1914, and in April through July 1915. Charles E. Apgar's contribution is well documented in M.G. Abernathy, "A Wireless Detective in Real Life," *Sparks Journal Quarterly*, spring 1977; *Literary Digest*, September 11, 1915; and *QST*, September 1917, p. 18. Suspicion of subversive stations from Maine to Mexico continued to be mentioned in the *New York Times* throughout the rest of 1914 and into 1915. The discovery of the German station in Mexico is narrated by Hayden Pratt in *Sixty Years of Wireless and Radio Experience* (collection of Forrest A. Bartlett).

The closure of all radio activities, both transmitting and receiving, is adequately related in *QST* during its last months of publishing. This includes articles on the surveillance of amateurs, the closure of stations illegally operating, and the recruitment and training of wireless operators during the war. The case of Professor Jonathon Zenneck is related in the *New York Times*, April 10 and 18, 1917; by Hiram Percy Maxim, "The Old Man," *QST*, September 1917; and in the Institute of Radio Engineers' *Proceedings* 16 (August 1928): 1025.

## Chapter 4

This is a strange interlude in the history of amateur radio. Significant strides had been made in technology during the war: the heterodyne was invented by Edwin Howard Armstrong and the vacuum tube went into mass production. Yet the obstinacy of one man, Secretary of the Navy Josephus Daniels, held up the restoration of amateur rights.

The role of radiomen on the *George Washington* is described by Jan Davis Perkins, N6AW, in *Don C. Wallace, W6AM: Amateur Radio's Pioneer* (Vestal, New York: 1991). For a brief biography of Fred Schnell see *QST*, November 1919.

See Josephus Daniels, *Shirt-Sleeve Diplomat* (New York: 1947) and *The Wilson Era* (New York: 1944–1946) for information on the secretary of the navy. The struggle to restore amateurs' rights was preceded by a navy proposal to gain control of all radio. Begin with Hearings Before the Committee on the Merchant Marine and Fisheries, on H.R. 19350, *A Bill to Regulate Radio Communications*, House of Representatives, sixty-fourth Congress, 2nd session, January 11 to 26, 1917. Comments on the bill are in *Scientific American*, February 3 and March 17, 1917. The navy's accomplishments during the First World War are summarized in Hearings Before the Committee

on Merchant Marine and Fisheries on *A Bill to Further Regulate Radio Communications*, House of Representatives, 65th Congress, 3rd session, December 12, 13, 17, 18, and 19, 1918. Anticipation of the lifting of the ban on amateur radio is the main topic of the first, skimpy issues of *QST* issued after the war. They also elaborate upon Mr. Daniels' stubbornness. For the radio activities of Dr. Brinkley, see Gerald Carson, *The Roguish World of Dr. Brinkley* (New York: 1960) and R. Alton Lee, *The Bizarre Careers of John R. Brinkley* (Lexington, Ky.: 2002).

## Chapter 5

The social history of the five decades 1900–1950 is in the History of American Life series (New York: 1940s). Frederick Lewis Allen's *Only Yesterday* (New York: 1931), *Since Yesterday* (New York: 1939), and *The Big Change: America Transforms Itself, 1900–1950* (New York: 1952), and Mark Sullivan's *Our Times* (New York: 1926–1935) hardly touch the quantity of historical literature on those fifty years.

The list of books on the history of radio is lengthy, and because of the rapid changes, many books are outdated. I consulted Sidney W. Head, *Broadcasting in America* (Boston: 1972), Peter M. Lewis and Jerry Booth, *The Invisible Medium* (New York: 1989), Susan Smuylan, *Selling Radio* (Washington, D.C.: 1994) and especially Susan J. Douglas, *Inventing American Broadcasting, 1899–1922* (Baltimore, Md.: 1987) and *Listening In* (New York: 1999).

The beginnings of broadcasting are related in Mitchell V. Charnley, *News by Radio* (New York: 1948). The conflict between BCLs and amateurs receives the best treatment in *QST*. Just a few of many articles are S. Kruse, "The Radiophone and the Code Station: An Argument for Cooperation," in *QST*, March 1922, pp. 21–24. In the same issue is an anonymous article, "With Our Radiophone Listeners: Getting Started Listening," pp. 47–48, 51. An editorial, "The 'Phones and Amateur Radio," March 1922, summarized the problems. Marvin R. Bensman, *The Beginning of Broadcast Regulation in the Twentieth Century* (Jefferson, N.C.: 2000) has written an excellent compilation of that subject.

Amateurs and weather and market reports are discussed in S.R. Winters, "Radiophone Broadcasting of Weather and Market Reports," *Radio News*, February 1922, pp. 717 ff.; and in S.R. Winters, "Government and Amateurs Join," *Radio News*, May 1923, pp. 768 ff. Cooperation with the army is discussed by Herbert H. Foster, "Radio in the National Guard," *Radio News*, January 1923, p. 1259, and in K.B. Warner, "The Army Links Up with the Amateur," *QST*, October 1925, pp. 22–24. Other significant articles on army-amateur relations are Tom C. Rives, captain, signal corps, "Signal Corps Training in Citizen's Military Training Camps," *Radio*

*News,* April 1926, p. 47; and Captain A.C. Stanford, U.S.A. liaison agent, "The Purpose of Army Amateur Affiliation," *Radio News,* April 1927, pp. 33–34. These are a sampling of many articles, editorials, and notices that appeared in *QST* and *Radio News.*

Fred Schnell described his experiences in "The Cruise of NRRL Aboard the USS Seattle," *QST,* January 1926, pp. 9–14; "Schnell Returns," *QST,* November 1925, p. 25; and "Recognition of Our Traffic Manager," *QST,* January 1926, p. 54. Several articles and editorials were devoted to the navy. A significant one is R.G. Mathews, "The Navy and the Naval Reserve," *QST,* August 1929, p. 7. Cooperation with other government units was also noted, for example in J.H. Dillinger, PhD, "The Radio Work of the Department of Commerce," *QST,* June 1921, pp. 18–21; and S. Kruse, "The Bureau of Standards–ARRL Tests of Short Wave Signal Fading," part I, *QST,* November 1920, pp. 5–6, and Part II, *QST,* December 1920, pp. 13–19, 22.

Public relations activities of the ARRL are discussed in "The League's Radio Information Service," *QST,* August 1923, p. 36, and H.F. Mason, "ARRL on the Yukon," *QST,* September 1923, pp. 29–32. Action by communities to stifle amateur radio are discussed in items such as "McWilliams vs. Bergman," *QST,* January 1923, p. 40; "City Ordinances," *QST,* April 1923, pp. 32–33; the editorial "Bugaboo Nr. 1,234,567,890," *QST,* May 1923, p. 52; Paul M. Segal, "Radio Interference Ordinances Cannot Limit Transmitting Station," *QST,* June 1927, pp. 43, 45; and A.R. Budlong, "Municipal Ordinances on Radio Transmission Unlawful," *QST,* January 1928, pp. 28, 70.

The radio conferences were well reported in the radio magazines. Examples are K.B. Warner, "The Washington Radio Conference," *QST,* April 1922, pp. 7–11. Secretary of Commerce Hoover's attitude toward amateurs is spelled out in S.R. Winters, "An Interview with H.C. Hoover," *Radio News,* October 1924, pp. 474–478. All proposed radio legislation was heavily reported in the radio magazines. Examples are K.B. Warner, "The New Radio Bill," *QST,* July 1922, p. 32, 56; and an anonymous article on "The White Bill," *QST,* January 1923, pp. 40–41. See also K.B. Warner, "The Third National Radio Conference," *QST,* December 1924, p. 16; and the anonymous "The Conference in Relation to Amateur Activities," *QST,* p. 8. K.B. Warner also reported on "The Fourth National Radio Conference," *QST,* January 1926, pp. 33–36. The text of the Radio Bill of 1927 is given in full in *QST,* April 1927, pp. 39–44.

# Chapter 6

Renewal of relaying following World War I is in J.O. Smith, "The Operating Department," *QST,*

September 1919, pp. 16–21, and in Fred Schnell, "Ham Traffic in Any Old Shack," *QST,* September 1923, pp. 31–34. Typical of the relay contests was the Washington's Birthday Relay of 1921, as reported by W.K. Kirwan, "Results of the Washington's Birthday Relay," *QST,* June 1921, pp. 22–24. Hiram Percy Maxim's "Dutch Uncle" editorials, by T.O.M. (The Old Man), are sprinkled throughout *QST.* A typical one is his "Rotten Hours," March 1920, pp. 7–8, and "Rotten Air," June 1920, pp. 12–13, 23.

The entrance of women into the amateur ranks is noted in an editorial, "The Ladies Are Coming," *QST,* August 1917, p. 19; in The Old Woman, "Beginning of the End," *QST,* September 1920, pp. 10, 14; and in an editorial, "The Radio Ladies," *QST,* March 1921, pp. 30, 64.

The story of the first official shortwave contacts across the Atlantic is related in Pierre H. Boucheron, "The Amateur Transatlantic Feat," *Radio News,* December 1920; "Godfrey to England to Copy Transatlantics," *QST,* October 1921, pp. 29–32; K.B. Warner, "The Story of the Transcoms," *QST,* March 1921, pp. 5–12, 47; "The Story of the Transatlantics," *QST,* February 1922, pp. 7–11; and in Pierre Boucheron, "Amateurs Span the Atlantic," *Radio News,* February 1922, pp. 697 ff. Paul F. Godley, "Official Report of the Second Transmission Tests," *QST,* February 1922, pp. 14–28, 36–40, 46, gives an excellent summary of the achievement. These are but a few of the many articles on the accomplishment.

Changes in amateur procedures are noted in "International Intermediates Expanded," *QST,* February 1925, p. 22, and in Howard S. Pyle, "The Amateur DX Card: A Study of the International Fad that 8UX Started," *QST,* September 1924, pp. 36–38. The rise and fall of amateur interest in the International Auxiliary Language and Esperanto is discussed by Oscar C. Roos, "A Radio Auxiliary Language for Trans-Oceanic Work," *Radio News,* May 1924, pp. 1565, 1674; Henry W. Wetzel, "The Language of International Radio," *QST,* July 1924, pp. 42–45; and James D. Sayers, "Esperanto Radio World Language," *Radio News,* August 1924, pp. 169, 204–206.

As would be expected, because the ARRL and Maxim led the way in the creation of the International Amateur Radio Union, the steps toward its creation were well-reported in *QST.* Pertinent articles are Maxim, "The International Radio Union," *QST,* May 1924, pp. 16–17, and KB. Warner, "International Amateur Radio Union Formed," *QST,* June 1925, pp. 9–13. The IARU's new constitution appears in full in *QST,* December 1928, pp. 51, 54, 56.

The International Radiotelegraph Conference held in Washington, D.C., in October and November, 1927, is discussed in editorials by K.B. Warner: "Washington," *QST,* December 1927, pp. 9–11; and

QST, January 1928, pp. 9–11; and in his article on the results of the conference for amateurs, "The Amateur and the Radiotelegraph Conference," QST, January 1928, pp. 15–22.

The 1920 St. Louis Convention is discussed by T.O.M. (Maxim) in "Rotten S.O.L.," QST, February 1921, pp. 9–15, 19, 23. The first national convention is chronicled in "The First National Convention," QST, October 1921, pp. 721, and in "The Chicago Radio Convention," Radio News, October 1921, p. 282. The various clubs, such as the "Rag Chewers," "Worked All Continents" and "Worked All States" clubs, are mentioned throughout QST. The founding of the Royal Order of the Wouff Hong is related by F.D. Fallain, "The Story of the Royal Order of the Wouff Hong," in QST, April 1924, pp. 22–26.

## Chapter 7

In the years 1915–1950 QST and other radio magazines devoted probably three-quarters of their space to articles on experiments or new technical developments in radio. Because this is primarily a social history, most of the information contained in these hundreds of articles has been ignored. However, a sampling of experiments, especially those that would be noticed by the public, have been noted. This explains mention of a radio in a car as early as 1910, as described in "Wireless Telegraph Apparatus for Contestants of the Glidden Tour," Scientific American, May 14, 1910, p. 395; K.B. Warner, "The Radio Hound," QST, January 1923, pp. 33–34; Howard F. Mason, "The Radio Lizz," QST, December 1922; and Oliver Wright, "Loops and Fords," QST July 1925, pp. 33–36.

Expeditions to remote parts of the world also profited from radio. QST in some years ran a column titled "Contact with the Expeditions," September 1926, pp. 99–100. Typical articles on exploration are T.S. McCaleb, "Radio with the Rice Amazon Expedition," Radio News, November 1925, pp. 588, 747, 749. Details of an expedition into Borneo are in Harry Wells, "The Story of PMZ," QST, August 1930, pp. 9–14.

The International Pacific Highway Project, which sent a caravan of cars and trucks into Mexico in 1930 and on into Central America in 1931 was well documented by the expedition's radioman, Bertram Sandham, in "With IPH in Mexico," September 1930, pp. 17–20, 72, 84, and in "In the Field with IPH," QST, December 1931, pp. 33–35. The Los Angeles Times ran extensive articles on the expeditions during the months of 1930 and 1931 when it was in the field. Many of the articles were by Harry Carr, a journalist who was with the expedition; examples are his "Links to Two Americas," QST, March 1, 1930, with nearly a dozen dispatches in the following weeks.

Clyde DeVinna, "Hamming with a Portable in Africa," QST, July 1930, pp. 27–32, relates experience with a commercial project; and Georges-Marie Haardt, "The Trans-Asiatic Expedition Starts," National Geographic, June 1931, pp. 776–782, mentions the role of radio in that expedition. The project is also mentioned in "Finding the Expeditions: Motor Car Station," QST, July 1931, p. 50, and QST, August 1931, p. 48. O.W. Hungerford, "VP3THE, The Terry-Holden Expedition," Radio, December 1938, pp. 17–21, 76–78, relates this expedition adequately.

## Chapter 8

The personalities of amateurs are analyzed in E. Lowell Kelly, "Personality over the Air," QST, January 1940, pp. 42–43, 56–57 and in William H. Graham, "Crashing Page One," QST, January 1934, pp. 32, 56, 58. Information on the Olympics of 1932 can be found in "The History of the Olympics: 1932-Los Angeles, U.S.," http://www.history/1900s about.com/library/weekly/aaoook.htm-24k. Amateur's contributions are described in W.A. Lippman, Jr., "W6USA—Amateur Radio at the Olympics," QST, August 1932, pp. 27–28, and "W6USA—The World Was Its Oyster," QST, October 1932, pp. 10, 12, 90.

Statistics of the Chicago World's Fair (Century of Progress Exposition) are on the Web, in the periodicals of the time and in encyclopedias. The amateur contribution is well documented in Wallace F. Wiley, "The World's Fair Radio Amateur Exhibit," QST, December 1933, pp. 29–31; "World's Fair ARRL Convention," QST, July 1933, p. 8; and "Amateur Radio at the Century of Progress," QST, August 1933, p. 28. For Jean Hudson winning the speed title, see I.S. Coggeshall, "When the World's Radio Speed Title Changed Hands," QST, November 1933, pp. 39–40. For the roll of amateurs at the fair in 1934, see "Amateur Radio—A Century of Progress—1934," June 1934, p. 34; "Amateur Radio at the Fair," QST, August 1934, p. 39; and "World's Fair Amateur Exhibit," QST, p. 57.

Information on the San Francisco World's Fair (Golden Gate International Exposition) and the New York World's Fair of 1939 can be found on the Web and in contemporary periodicals. Amateur participation is described in F. Cheyney Beekley, "Amateur Radio at the Fairs," QST, June 1939, pp. 23–24; and "An Animated Radio Diagram," QST, July 1939, p. 26. Sprinkled throughout QST are brief items about amateurs working as aides in ski races, outboard motorboat races, glider competitions, football games, and even rifle matches.

One of the most successful of the New Deal agencies was the Civilian Conservation Corps (CCC). A ham radio net was created for the CCC in some of the army corps areas; in a few places the young men were even offered the opportunity to learn how to be amateurs. Two articles describing these activities are Garland C. Black, "CCC and the Amateur," *QST*, November 1933, pp. 36, 82, and H.O. Bixby, "Third Corps Area Asks Amateur Help," *QST*, October 1934, pp. 16–17.

Every four years through the '20s and '30s amateurs held a Governors'-President's Relay. The preparations, rules and outcome of these are to be found in *QST* a month or two after the relays were held. Amateurs were all the time experimenting with lower and lower meters, down to five. See E.D. Miller, "58-Mc. Rolls Up Its Sleeves: Solid World at 5 Meters at the National Glider Meet," *QST*, September 1932, pp. 29, 88. Descriptions of more experiments with 5 meters are in Jan David Perkins, *Don Wallace* (New York: Vestal, 1991), pp. 111 ff.

The deaths of Will Rogers and Wiley Post made news headlines in 1935. See Bryan B. Sterling and Frances N. Sterling, *Forgotten Eagle: Wiley Post, America's Heroic Aviation Pioneer* (New York, 2001) for the background of this tragedy.

The story of the press opposition to radio (and later television) news has not been dealt with in depth. Mitchell V. Charmley, *News by Radio* (New York: 1948) touches on the conflict. See also T.R. Carskadon, "The Press-Radio War," *The New Republic*, March 1936, pp. 132–135; it summarizes the problem. See also Bice Clemow, "U.P.-I.N.S. Take Lead in Effort to Retain Control of Radio News," *Editor and Publisher*, May 4, 1935, p. 304, and "Transradio Files Suit for Million," *Editor and Publisher*, May 25, 1935, pp. 9, 44. See also Isabelle Keating, "Radio Invades Journalism," *The Nation*, June 12, 1935, pp. 677–678.

The radio conferences, both national and international, were all well covered in *QST*. See A.L. Budlong, "Cairo, Part I," *QST*, January 1938, pp. 113, 66, 70, 72, 74, 76, and 78, and K.B. Warner and Paul M. Segel, "The Battle of Cairo," *QST*, July 1938, pp. 9–12, 45–46, 104, 106. Warner, "The Madrid Conference," *QST*, February 1933, pp. 9–17, chronicles this meeting.

## Chapter 9

Information on sailing ships is in the books by Alan Villiers, of which *By Way of Cape Horn* (London: 1930) and *Grain Race* (New York: 1933) are perhaps the best known. M.B. Anderson's adventures on the *E.R. Sterling* are to be found in the IARU News, *QST*, January 1928, pp. 52, 72, 74. News of expeditions was reported in QST in an "Expeditions" section during the years 1927–1930.

The around-the-world cruises of the *Yankee* are described in Captain and Mrs. Irving Johnson, *Sailing to See: Picture Cruise in the Schooner Yankee* (New York: 1939) *and Westward Bound in the Schooner Yankee* (New York: 1936). Alan Eurich describes his experiences in "CQ WCFT: Further Adventures Aboard the "Yankee," *QST*, March 1938, pp. 11–13, 84, 86, 88. A description of a lonely radio station is Roy E. Abbott, "The World's Loneliest Radio," *QST*, August 1932, p. 59.

For Pitcairn Island's history see Ian A. Dall, *Pitcairn: Children of Mutiny* (Boston: 1973). Alan Eurich, "CQ PITC," *QST*, August 1937, pp. 9–10, 70, 72, describes his experiences on the island. The story of Pitcairn's new radio is by Lew Bellem, "The New PITC: Modern Radio Equipment for Pitcairn Island," *QST*, January 1938, pp. 19–20, 78, 80, 82, 90, 92. Their radio activities on Ducie Island are described in Michael McGiar, M.D., "VP6DI: The Story of the 2002 Ducie Island Expedition," *CQ: Amateur Radio Communications and Technology*, February 2003, pp. 11–13.

The cruise of the Chinese junk *Pang Jin* is by Rex Purcell as told to George W. Polk, "The Cruise of the Pang Jin," *QST*, October 1939, pp. 18–20.

The story of aircraft in the 1920s and '30s is to be found throughout the newspapers and magazines of the time. Human nature insists on "pushing the envelope," and in no endeavor of the time is that more pronounced than in the advance of flying. The tragedies of airplane crashes were reported in the nation's press almost every day. Then there were the disasters of the *Shenandoah*, *Akron*, and *Macon* dirigibles. While it was not responsible for the tragedies, radio played a major part in safety development. *QST* during these years ran many articles on radio advancement that was useful to aircraft improvement. See Joseph Lyman, "Five Meter Tests Overwhelmingly Successful: East Coast Amateurs Stage History-Making Demonstration of 56 mc. Work," *QST*, May 1932, p. 35; and Ross A. Hull, "The 56-mc Eclipse Expedition," *QST*, October 1932, p. 46.

The tragedies involving lighter-than-air craft are well told in Richard K. Smith, *The Airships Akron and Macon* (Annapolis, Md., 1986). Stories involving the *Shenandoah* can be found in the *New York Times*, January 17, 25, September 25, and October 20, 1924. For the *Akron* see the *New York Times*, June 28, July 10, August 8, and September 25, 1931. For the *Macon* see the *New York Times*, April 4, 1933; May 31, August 29, 1935.

Amateurs involved in air races are mentioned in the Communications Department, "The Springfield Air Races," *QST*, August 1930, p. 51 and the *Miami Herald*, January 7–10, 1928. See also "The National Air Races," *Popular Aviation*, November 1928, pp. 13–15. The roll of amateur radio in the sport of gliding is discussed in Ross A. Hull, "Amateur Radio in a New Field," *Popular Aviation*, September 1933, pp. 32, 64.

# Chapter 10

Radio advanced in tandem with aircraft, motorized vehicles, electronics, and medicine during the first half of the twentieth century. The symbiosis of radio and aircraft was especially significant. Both challenged the average citizen's ability to comprehend, to "quite understand" how waves existed in thin air that were capable of carrying language and music to be heard in a receiver scores, possibly hundreds of miles away. And if God had meant for mankind to fly, He would have given us wings! The awe of what was so new and almost incomprehensible resulted in intense interest in the trials, failures, and successes of radio and aircraft. There were hundreds of trials and failures, and some successes.

The "Nancys"—the navy's amphibian aircraft of which just one, the NC-4, completed the journey across the Atlantic, are well described, and their flights documented, in Richard K. Smith, *First Across: The U.S. Navy's Transatlantic Flight of 1919* (Annapolis, Md.: 1973). Controversy over the use of radio in airplanes is discussed in Lawrence A. Hyland, "Radio and Air Commerce," *Aviation*, November 28, 1927, pp. 1282–1285. Several books have been written about Richard E. Byrd. His own book, *Skyward: Man's Master of the Air* (New York: 1928), goes into some detail on the uses of wireless in aircraft.

Radio beacons are discussed in a *New York Times* article about Maitland and Hegelberger, August 17, 1927, p. 1. The Dole Race is adequately taken care of by Robert H. Scheppler, *Pacific Air Race* (Washington, D.C., 1988). Additional material is available from Herman Schaub, secretary of the Society of Air Race Historians. The role of radio in the Dole race is described in J. Walter Frates and A.L. Budlong, "Amateur Radio and the Pacific Flights," *QST*, November 1927, pp. 40–41, 80. The *New York Times* covered the Dole race in issues of August 17 and 20–22, 1927. For the *Times* use of radio, see Emelie and Rexford Matlack, W3CFC, "The Paper, the Station, and the Man—a Brief History of the *New York Times* Radio Stations," *73 Magazine*, February 1980, pp. 54–59.

A brief biography of Kingsford-Smith is in Norman Macmillan, *Great Airmen* (London: 1955), and in the *New York Times*, June 6, 1928. Radio's part in the flight of the *Southern Cross* is in J. Walter Frates, "Following the *Southern Cross* to Brisbane," *QST*, August 1928, pp. 21–22. As for the flight of the *'Untin' Bowler*, the *Chicago Tribune* ran stories throughout July and August 1929. Radio's part is described in F.E. Handy, "KHEJ and the *'Untin' Bowler* Awards," *QST*, October 1929, pp. 21–22, 76. The Army Air Corps' Arctic Experiment is described in F.E. Handy, "ARRL Cooperates with the Arctic Patrol in Mid-Winter Maneuvers," *QST*, May 1930, pp. 29–38. The overall story is in the *New York Times*, January 6, 11, 14, 27, and February 3, 1930.

Anne Morrow Lindbergh's experiences as an amateur appear in several books. *North to the Orient* (New York: 1935), pp. 30–32, 34–35, chronicles her first efforts. Her rig is described in Dorothy Hermann, *Anne Morrow Lindbergh: A Gift of Life* (New York: 1993), p. 77. Other references to radio are in *Hour of Gold, Hour of Lead: Diaries and Letters of Anne Morrow Lindbergh* (New York: 1972), p. 172; and *Locked Rooms and Open Doors: Diaries and Letters of Anne Morrow Lindbergh* (New York: 1974), p. 34, 41. In *Listen! The Wind* (New York: 1938) she describes contacting a ham four thousand miles away.

The crash of the American Air Lines' *Condor* in the Adirondacks, and the part played by hams in the rescue, is all in the *New York Times*, December 29–31, 1934, and January 1, 1935; it is also noted in *QST*, February 1935, pp. 12, 78, 80.

# Chapter 11

Several shelves of books have been written on Arctic and Antarctic exploration. Most are biographies of explorers or narrations of their expeditions. A good overview is Kieran Mulvaney, *At the Ends of the Earth: A History of the Polar Regions* (New York: 2001). See also Charles Officer and Jake Page, *A Fabulous Kingdom* (New York: 2001). For a description of the effects of the long night on the minds and bodies of men, see Frederick A. Cook, introduction by Edward Hoagland, *Through the First Antarctic Night, 1888–1889* (London: 1900).

The question of Shackleton's men and radio is mentioned in Sir Ernest Shackleton, *South: Journals of His Last Expedition to Antarctica* (New York: no date), pp. 315–16, 320, 325, 327–28, 330, 334–35, and 339. See also F.A. Worsley, *Endurance: An Epic of Polar Adventure* (New York: 1931, 1999), and *New York Times*, May 14, 1916, story lifted from the Wireless Press. Additional information is in Leonard Bickel, *Shackleton's Forgotten Men* (New York: 2001), p. 220.

For the life of Sir Douglas Mawson see Philip Ayres, *Mawson: A Life* (Melbourne: 1999). See also Douglas Mawson, foreword by Ranulph Fiennes, *Home of the Blizzard: A True Story of Antarctic Survival* (New York: 1998).

Donald Baxter MacMillan should be better known. He was the outstanding Arctic explorer of the twentieth century. Among his merits was a willingness to try new things, including radio. Information on his life is well presented in Everett S. Allen, *Arctic Odyssey: The Life of Rear Admiral Donald B. MacMillan* (New York: 1963). MacMillan's two books, *Four Years in the White North* (New York: 1918) and *Etah and Beyond* (Boston: 1927) have information on the role played by radio in his expeditions. See also the *New York Times*, June 24, 1923. MacMillan's obituary is in the *New York Times*, September 8, 1970. Amateurs in communication with Arctic explorers are discussed in J.K. Bolles, "Arctic Explorers to Communicate with Amateurs," *QST*, June 1923, pp. 9–10, 69, 73; K.B. Warner, "West Coast Working 'Bowdoin,'" *QST*, November 1923, pp. 15–16; and F.H. Schnell, "White Silence Broken," *QST*, October 1923, pp. 10–12. A good wrap-up of the 1923–24 expedition is "The 'Bowdoin' Returns," *QST*, November 1924, pp. 16–17 and Donald H. Mix, ITS, WNP, "My Radio Experience in the Far North," *QST*, pp. 17–23.

## Chapter 12

Both heavier-than-air craft and lighter-than-air craft participated in the conquest of the Polar regions, and in the 1920s especially, they "pushed the envelope." First to be discussed is the controversial MacMillan expedition of 1925. The story is well told in John H. Bryant and Harold N. Cones, *Dangerous Crossings* (Annapolis, Md., 2000). The story from Byrd's point of view is in Edwin P. Hoyt, *The Last Explorer: The Adventures of Admiral Byrd* (New York: 1968) and Raymond F. Goerler, *To the Pole: The Diary and Notebook of Richard E. Byrd, 1925–1927* (Columbus, Ohio, 1998), and Byrd's own book, *Skyward*. Radio's part in the expedition is documented in K.B. Warner, "Shortwave Communication with WNP," *QST*, July 1925, p. 20. The *New York Times* gave considerable credit to amateurs in articles appearing in July, August, and September 1925. Donald R. MacMillan, "Scientific Aspects of the MacMillan Expedition," *National Geographic Magazine*, September 1925, pp. 349–354; "MacMillan in the Field," *National Geographic Magazine*, October 1925, pp. 457–476, and Richard E. Byrd, "Flying Over the Arctic," *National Geographic Magazine*, November 1925, pp. 519–521, all summarize the accomplishments of the expedition. Radioman Reinartz's failure is presented in *Dangerous Crossings*.

Roald Amundsen's attempt to reach the Pole is described in Clive Holland, editor, *Farthest North* (New York: 1994, 1999), pp. 223–236, and in Lincoln Ellsworth, *Search* (New York: 1932), p. 89. Byrd's flight is chronicled in Hoyt, *Last Explorer*, pp. 104–105; Goerler, *To the Pole*, and K.B. Warner, "Byrd Expedition Sails," *QST*, July 1925, p. 18. Controversial aspects of Byrd's North Pole Flight, as well as his dark side, are in Carroll V. Glines, *Bernt Balchen: Polar Aviator* (Washington, D.C.: 1999). The flight over the Pole by Amundsen's dirigible, the *Norge*, is reported in Holland, *Farthest North*, pp. 251–252, in Ellsworth, *Search*, pp. 135, 142 ff., and in K.B. Warner, "More Arctic Adventures," *QST*, July 1926, p. 17–19.

The adventures of Captain Wilkins are put forth in George H. Wilkins, *Flying the Arctic* (New York: 1928). The radio side of his adventures are detailed in Howard F. Mason, "An Arctic Adventure," *QST*, October 1927, pp. 9–14, in F.H. Schnell, "Amateur Radio to the North Pole Again," *QST*, March 1926, pp. 33–36, and K.B. Warner, "Progress of the Wilkins Expedition," *QST*, May 1926, p. 38. A discussion of the pros and cons of Arctic air flight is in W.L.G. Joerg, editor, *Problems of Polar Research: A Series of Papers by Thirty-one Authors* (New York: 1928), pp. 307–409.

Other polar explorations, and the part played by radio, are mentioned throughout *QST* during the '20s and '30s. Captain Bob Bartlett is mentioned often, and his obituary is in the *New York Times*, April 29, 1946. George Palmer Putnam sent seventeen articles, by way of amateur wireless, from the *Morissey* to the *Times* in July and August 1926.

The tragedy of Umberto Nobile's *Italia* is beautifully told in Wilbur Cross, *Disaster at the Pole* (New York: 2000). The story of the commercial vessel *Nanuk* is told by Robert J. Gleason, *Icebound in the Siberian Arctic* (Anchorage, Alaska: 1977, 1982). The Hudson Bay adoption of shortwave is in Peter C. Newman, *Merchant Princes* (New York: 1991), Byrd's Antarctic expeditions and the part played by radio are mentioned in *Last Explorer*, p. 197, in "WFAT-WFBT," *QST*, March 1929, p. 58, and in "Expeditions," *QST*, July 1929, p. 49, and August 1929, p. 54.

The final comment on polar exploration is A.G. "Gerry" Sayre, "Ham at 30 Below," *QST*, January 1939, pp. 9–12, 106.

## Chapter 13

The Dayton Flood of 1913 is well described in Allan W. Eckert, *The Great Dayton Flood* (Boston: 1965). The contributions of hams to the rescue efforts are often mentioned in passing, but so far I have been unable to find a single article dealing specifically with hams during this disaster. The *Chicago Tribune*, March 30, 1913, for example, makes mention of the wireless help, but that is all. By 1921, however, the contributions of hams were being noticed, as in R.W. Goddard, "First Aid by Radio," *QST*, November 1921, pp. 23–24. An example of amateur help during blizzards is S. Kruse, "A Snowstorm and the ARRL," *QST*, January 1923, pp. 30–32. A second article on the same subject is K.B. Warner, "The Amateur Scores Again: Dozens of American Amateurs Do Valiant Work When Blizzard Strikes Middle West," *QST*, April 1924, pp. 14–15; see also the *Chicago Tribune*, February 5, 1924, p. 1.

Progress toward cooperation with agencies involved during tragedies and disasters is noted in an editorial, "It Seems to Us," *QST*, March 1938, p. 9. The development of radio in the forest service is fully described in Gary Craven, *Radio for the Fireline: A History of Electronic Communication in the Forest Service, 1905-1975* (Washington, D.C.: 1982).

Dedicated amateurs are almost always present during and following floods and hurricanes. New England has had its share. For example, hams did liege service during a flood in 1928 when the Connecticut Valley suffered from days of heavy rain. See David S. Boyden and Robert D. Russell, "Amateur Radio Work in New England Flood," *QST*, January 1928, pp. 99–101.

Several books have been written about hurricanes. I consulted Gordon E. Dunn and Banner I. Miller, *Atlantic Hurricanes* (Baton Rouge: 1960, 1964); Fred Doehring, Iver W. Duedall, and John M. Williams, *Florida Hurricanes and Tropical Storms, 1871-1993: An Historical Survey* (Melbourne, Fl.); and Jay Barnes, *Florida's Hurricane History* (Chapel Hill, N.C.: 1998). For radio's part in the 1926 disaster, see "Amateurs Help in Florida Emergency," *QST*, November 1926, pp. 101–102.

For the Lake Okeechobee disaster there are two good studies. The first is a doctoral dissertation by Eric L. Gross, *Somebody Got Drowned, Lord: Florida and the Great Okeechobee Hurricane Disaster 1928* (Tallahassee, Fl.: 1995), and Robert Mykle, *Killer'cane: The Deadly Hurricane of 1928* (New York: 2002). The radio aspect is in K.B. Warner, editorial, *QST*, November 1928, pp. 7–8. See also *The West Indies Hurricane Disaster: September, 1928* (Washington, D.C.: 1929), p. 64. Further advice on improving services in disasters is in "The Army-Amateur Radio System Revised," *QST*, March 1929, pp. 21–25, and F.E. Handy, "When Emergency Strikes," *QST*, April 1938, pp. 35–38, 76, 78, 80, 82, 84, 86. The Paley Award is mentioned in Sally Bedell Smith, *In All His Glory: The Life of William S. Paley* (New York: 1990), p. 153.

Ohio and Mississippi River floods were well served by amateurs. A lengthy description of ham efforts in the January 1937 Ohio River flood is Clinton B. DeSoto, "In the Public Interest, Convenience, and Necessity," *QST*, April 1937, pp. 11–20 ff. A number of articles covering the flood are in the *St. Louis Post-Dispatch* throughout January 1937. A thumbnail sketch of Shawneetown's history is in Malcolm Rohrbough, *The Land Office Business* (New York: 1968).

The first Paley Award, for a ham helping Shawneetown, is noted in "Paley Award Goes to W9MWC," *QST*, August 1938, p. 18.

The terrible New England hurricane of 1938 is well told in Everett S. Allen, *A Wind to Shake the World: The Story of the 1938 Hurricane* (Boston: 1976), and R.A. Scott, *Sudden Sea: The Great Hurricane of 1938* (New York: 2003). Again, the amateur contribution was recorded by Clinton B. DeSoto, "Amateur Radio Bests Triple Catastrophe," *QST*, November 1938, pp. 11–18 ff.; no author, "1938 Paley Trophy Awarded to W1BDS," *QST*, July 1939, pp. 23, 74, 76.

California disasters are recorded by Clinton B. DeSoto, "QRR, 1932: Amateur Emergency Work During the Past Year," *QST*, January 1933, pp. 39–42, and "Southern California Amateurs Rise to Earthquake Emergency," *QST*, May 1933, pp. 9–11; *QST*, August 1936, p. 31, and "Amateurs of Assistance in Emergencies: Notes on Recent Work Accomplished," *QST*, April 1934, pp. 34–35, 80. The Tillamook Lighthouse emergency is told by Henry Jenkins, "Amateur Radio Scores Again!" *QST*, February 1935, pp. 45–46.

# Chapter 14

The young ladies are mentioned in Anita Calcagni Bien and Enid Carter, "The YL's Unite! The Story of the Young Ladies' Radio League," *QST*, May 1940, pp. 22–27. The loyalty of hams to the nation as war loomed is shown in the editorial "It Seems to Us—," *QST*, November 1940, p. 7. The *Yankee* voyage is described in Oakes A. Spalding, "Around the World with the Yankee," *QST*, October 1941, pp. 9–14, 102, 104.

The Byrd biographies give information on his later expeditions. The radio contribution is in Clinton B. DeSoto, "Byrd Antarctic Expedition to Use Amateur Radio," *QST*, December 1939, pp. 11–15, 25. See also United States Antarctic Service Expedition "Richard E. Byrd, 1888–1957; Byrd Antarctic Expedition III, 1939–41," http://www.southpole.com/p000010.9htm. Appreciation of ham cooperation is displayed in Clay W. Bailey, "Letter to ARRL," *QST*, November 1941, pp. 17, 68.

Cooperation with the Red Cross is described in J.A. Moskey, "The ARRL-Red Cross Preparedness Test," *QST*, October 1941, pp. 57, 59–60, and in other *QST* articles in March and May 1941. Hardly an issue of *QST* was issued without it containing mention of ham aids in disasters, such as Leland W. Smith, "Twister Hits Georgia," *QST*, April 1940, p. 15, 94.

The first mention of "radio locating"—radar— was in an editorial, "It Seems to Us—," April 1942, p. 7. Probably the most thorough analysis of the Pearl Harbor debacle is Gordon W. Prang, *At Dawn We Slept* (New York: 1984), pp. 499–500. The honor bestowed upon Private Joseph L. Lockard is reported by Clinton B. DeSoto, "V.W.O.A. Honors Amateur Radio," *QST*, April 1942, p. 27. The information on Frank Firth and his wife Helen in Honolulu is from Jackie Tilly, "QRV," Opalousa's Area Amateur Radio Club: Wireless Connection," *QST*, April 17, 1998. Young Ladies' (YLs') cooperation is recorded in Clinton B. DeSoto, "U.S.A. Calls and the YLs Answer," *QST*, May 1942, pp. 9–10. Early cooperation with the USOs is described in Emil H. Frank, "Soldiers and Sailors and Amateur Radio," QST, January 1942, pp. 31, 60.

"The Beginnings of the War Emergency Radio Service" (WERS) is reported in QST, July 1942, pp. 11–15. The amateur reaction to government inaction is in an editorial, "It Seems to Us—," QST, January 1941, p. 7, and February 1942, p. 7. The controversy over policy is mentioned in almost every 1942 issue of QST. When policy was finally fixed the journal gave detailed instructions: John Huntoon, "Planning WERS for Your Community," QST, August 1942, pp. 22–24, 114, 116. Progress was reported, as in "Providence Adopts New Plan for Civilian Defense Radio," QST, March 1942, p. 52 and in John A. Doremus, "Massachusetts Civilian Defense Radio," QST, September 1942, pp. 11–14. WERS information was prevalent in subsequent issues of QST.

The training of radio people is emphasized in K.B. Warner, editorial, "We Must Not Fail," QST, September 1942, p. 9; John Huntoon, "Training Civilians for Wartime Operating," QST, pp. 53–54 and QST, October 1942, pp. 45–49. The rigs for WERS work were the subject of several articles, such as Don H. Mix, "Building WERS Gear from Salvaged B.C.L. Sets," QST, September 1942, pp. 15–18.

How peace-time amateur radio worked is discussed in Captain Samuel E. Fraim, "Amateur Radio and the Civil Air Patrol," QST, January 1943, p. 50–51, and Captain J. Wm. Hazelton, "WERS in the Florida State Guard," QST, November 1944, pp. 45–47. "Operating News," QST, May 1945, p. 56, gives statistics. The problems of amateurs and the Mississippi flood of May 1943 are reported by Carol A. Keating, "'Ole Mississip'Rampages Again," QST, August 1943, pp. 30–33. The end of WERS but the continuation of ham activities in disasters is in F.E. Handy, "ARRL Emergency Corps Program," QST, December 1945, pp. 49–51.

## Chapter 15

This chapter concentrates on the training of radiomen during the Second World War, activities during the war, and the resumption of ham activities after the war.

For information on the Japanese Kana code see Donald D. Milliken, "The Japanese Morse Telegraph Code," QST, September 1942, pp. 23–25, 120, and Charles E. Holden, "The Japanese Morse Radiograph Code," QST, October 1943, pp. 30–33. Estimates of the number of radiomen in the armed forces are in "Strays," QST, August 1945, p. 54, which quotes from Mary Zuhorst, Broadcasting, July 2, 1945. Articles on pre-war training are Clinton B. DeSoto, "The Signal Corps in the Blue Grass

State," March 1943, pp. 11–15 ff. and "ESMWT at Rutgers," QST, September 1943, pp. 43 ff.

The role of YL amateurs in the war effort is told in Clinton B. DeSoto, "U.S.A. Calls and the YLs Answer," QST, May 1942, pp. 9–12 ff; Louis B. Dresser, "Women and Radio—Partners in Victory," QST, September 1943, pp. 9 ff, and Anita Bien, "YLRL, QRV!" QST, October 1941, pp. 32–37 ff.

Training at the Signal Corps school at Fort Monmouth is discussed in detail by Clinton B. DeSoto, "Signal Corps Radio School," QST, August 1941, pp. 9–11, and "QST Visits Fort Monmouth," QST, October 1942, pp. 28–32. For radio training in the army air corps, see Clinton B. DeSoto, "QST Visits the Air Forces," QST, January 1943, pp. 17–29. Merchant marine radio training is discussed in Clark C. Rodiman, "QST Visits Gallups Island," QST, June 1941, pp. 9–12 ff, and Clinton B. DeSoto, "QST Returns to Gallups Island," QST, May 1943, pp. 14–18 ff. DeSoto visited other training camps. See "QST Visits Camp Hood," QST, July 1943, pp. 9–16 ff; "QST Visits the Marine Corps," QST, April 1943, pp. 13–17 ff, and "QST Visits the Coast Guard," QST, February 1943, pp. 3–18 ff. Three articles are of significance with regards to navy training: Lt. Comdr. John L. Reinarts, "The Navy and the Amateur," QST, September 1940, p. 29, Clinton B. DeSoto, "The Navy Trains Radio Technicians," QST, November 1942, pp. 3–18 ff, and "QST Visits the Noroton Training Station," QST, August 1942, pp. 44 ff. A wrap-up of amateur contributions to the navy is Rear Admiral Joseph R. Redman, director of Naval Communications, "Navy Communications and the Amateur," QST, October 1945, pp. 14–15, 16–51, 52–57. The signal corps received equal treatment with the initial article by Major General Harry C. Ingles, "U.S. Army Signal Corps," QST, September 1944, pp. 7–49. Lt. Col. Howard R. Haines, "Flying Radiomen of the Ferrying Division," QST, June 1944, pp. 16–18 ff. tells their story.

QST was able to publish a few narrations of wartime ham experiences. Two examples are "Hams in Combat," January 1944, pp. 14–21, and F. Joseph Visintainer, "Listening Post in the Philippines," April 1946, pp. 70–72, 126. There are many memoirs of wartime experiences, a number of which involve radiomen. Two are W.W. Chaplin, The Fifty-two Days (New York: 1944), and Ira Wolfert, History Island—The Story of an American Guerrilla on Leyte (New York: 1945).

Broadcasting to the armed forces is described in Cpl. Bill Greenberg, "This Is Your Armed Forces Radio Station," QST, June 1945, pp. 54–88. Radio as a morale builder is described in Carlton H. March, "On the Burma Road," June 1946, p. 71. Wartime hamfests are discussed in Clinton B. DeSoto, "Hamfest in Khaki," QST, November 1942, pp. 51–52, "Anglo-American Hamfest," QST, December 1944, p. 45, and Harry Longerich and M.

Sgt. Arthur Hansen, "Hamfest in North Africa," *QST*, February 1944, p. 31.

Comments on the end of the war and the renewal of amateur activities are in the editorial "Reopening," *QST*, October 1945, p. 11; "We're Off!" *QST*, December 1945, p. 11; and R.B. Bourne, "The Opening of the Band," *QST*, March 1946, p. 28. Uses of wartime equipment warranted a number of articles, an example of which is John T. Ralph and H.M. Wood, "Operating the BC-645 on 420 Mc," *QST*, February 1947, pp. 15–21.

Natural disasters once again occupied hams. Examples are Henry W. Peterson, "Alaska Earthquakes," *QST*, August 1946, p. 79; "Idaho-Washington Emergency," *QST*, February 1947, p. 77; and Albert Hayes, "Winds, Waves, and Snakes," *QST*, December 1947, pp. 40–44, 88. See Harold M. McKean, "Amateur Radio Operations: Texas-Oklahoma Tornado, Texas City Explosions," *QST*, July 1947, pp. 34–40. Two very different books on the Texas City Disaster are Hugh W. Stephens, *The Texas City Disaster, 1947* (Austin: 1947), and Bill Minutaglio, *City on Fire* (New York: 2003).

"KLPO, MacMillan Arctic Expedition," *QST*, September 1946, p. 75, notes the renewal of Arctic exploration, as does "Ronne Antarctic Research Expedition," *QST*, March 1947, p. 71. The Kon-Tiki expedition was discussed in depth in "Kon-Tiki Communications—Well Done!" *QST*, December 1947, pp. 69 ff.

International communications conferences began shortly after war's end. See A.L. Budlong, "Moscow: A Report of the Five-Power Conference by ARRL's Representative," *QST*, January 1947, pp. 25–27. The Atlantic City conference was reported in successive issues of *QST* from July through November 1947. A.L. Budlong reported "The Fourth Inter-American Region 2 Radio Conference," *QST*, June 1949, p. 16, and September 1949, pp. 35–39.

Changes in army and air force relations with amateurs are in "The Military Amateur Radio System," *QST*, pp. 34–35, and November 1949, p. 52. A portent of things to come was Byron Goodman, "The 'Transistor,'" *QST*, October 1948, p. 48. "Obituary of Kenneth Bryant Warner," *QST*, November 1948, pp. 9–13; "Clinton B. DeSoto," *QST*, June 1949, p. 41.

## Epilogue

This essay is based primarily upon information from Forrest Bartlett, W6OWP. The Fourth Inter-American and Region 2 Radio Conference is reported in *QST*, June 1949. Information on the tsunami is from Alan Padgett to Forrest Bartlett, December 28, 2004.

# Index

Printed in the USA
CPSIA information can be obtained
at www.ICGtesting.com
LVHW082249251123
764912LV00006B/210

9 781476 662756